Giving a Damn

Giving a Damn

Essays in Dialogue with John Haugeland

edited by Zed Adams and Jacob Browning

The MIT Press
Cambridge, Massachusetts
London, England

This book was set in Stone Sans and Stone Serif by Toppan Best-set Premedia Limited. Printed and bound in the United States of America.

Library of Congress Cataloging-in-Publication Data is available.

ISBN: 978-0-262-03524-8

10 9 8 7 6 5 4 3 2 1

Contents

Acknowledgments

First and foremost, we would like to thank the contributors, not just for their contributions but also for their patience with the project and the trust they have put in us as editors. Thanks especially to Joseph Rouse, who consistently encouraged us to make this volume something that would have made John proud.

We also owe thanks to a number of people for commenting on the introduction in various stages of completion: Matthew Dougherty, Jay Elliott, Nat Hansen, Daniel Harris, Joe Lemelin, Eric MacPhail, Kevin Temple, Janna Van Grunsven, and Zach Weinstein.

A final thanks goes to Philip Laughlin at the MIT Press, for all his feedback and support, and to Joan Wellman, for agreeing to allow us to publish John's "Two Dogmas of Rationalism" and his outline of Kant's transcendental deduction.

Note on Abbreviations

Throughout the book, we have used the following abbreviations to refer to John Haugeland's works:

TU: "Truth and Understanding" (dissertation, UC Berkeley, 1976)

AI: *Artificial Intelligence: The Very Idea* (MIT Press, 1985)

HT: *Having Thought* (Harvard University Press, 1998)

PMR: *Philosophy of Mental Representation* (edited by H. Clapin; Clarendon Press, 2002)

DD: *Dasein Disclosed* (edited by J. Rouse; Harvard University Press, 2013)

Introduction

Zed Adams and Jacob Browning

1 Introduction

I suggest that we can't really have understood the human mind—and, in particular, with regard to its very distinctive capacity (perhaps only a few thousand years old) to seek objective truth—until we have understood its capacity for faithful commitment. But this is quite different from anything like a belief or a desire, let alone competent negotiation of the immediate environment. This essential topic for cognitive science has, I think, not yet even made it to the scientific horizon.
—John Haugeland, "Andy Clark on Cognition and Representation" (*PMR*, 35–36)

John Haugeland thought that a fundamental problem with contemporary philosophy is that we are working with an impoverished set of tools. The problem is that if we don't have the right tools, it's going to be awfully hard to get the job done. As a solution, Haugeland proposed to massively expand the array of concepts that we work with in philosophy, to include a much wider range of resources for understanding the mind, the world, and how they relate.

Haugeland's willingness to radically expand our conceptual tool kit is bound to strike many as an ontological extravagance. To this sort of worry, Haugeland had a ready response:

I have occasionally wondered what makes ontology so expensive, or what limits our budget, or, indeed, what exactly we are spending. Having dwelt more summers in the southwest than the northeast, I feel no great longing for desert landscapes; and frankly, Ockham's razor cuts little ice out there anyway. Perhaps, however, I can abide a prudent California compromise: We shall assume no entity before its time. (*HT*, 113)

Over the course of his career, Haugeland proposed a number of new "entities" that he thought helped us to better understand ourselves and our place in the world.[1] He also offered a diagnosis of why these entities have been neglected. In "Two Dogmas of Rationalism" (published for the first time in this volume), he discusses the two philosophical presuppositions that he thinks are most to blame for this neglect: *cognitivism* and *positivism*. Cognitivism is the view that all intelligent thought can be expressed in propositions.[2] Positivism is the view that the world is exhausted by the facts.[3] The problem with these presuppositions is that they narrowly focus our attention on a specific aspect of human life—making claims about facts—and treat it as the only way we engage with the world. As a corrective to this narrow focus, Haugeland draws our attention to some of the existential dimensions of human life. Specifically, he draws our attention to the centrality of *understanding*, our basic ability to make sense of things. For Haugeland, there is not one type of understanding but as many varieties as there are varieties of human social practices. As he puts it, "Understanding hammers is knowing how to hammer with them, understanding a language is knowing how to converse in it, understanding people is knowing how to interact and get along with them, and so on" (*DD*, 195). Understanding is about being able to skillfully cope with the world in all its complexity.

Haugeland's determination to expand our conceptual tool kit led him to write about a variety of topics and figures that are often seen as lying far apart: from topics in cognitive science and the philosophy of mind to figures such as Martin Heidegger and Thomas Kuhn. The primary goal of this introduction is to articulate some of the connections between these topics and figures so as to provide an account of how Haugeland integrated them into a unified philosophical view. Our strategy is to revisit a number of Haugeland's most influential essays and organize them into a series of complementary topics: holism, sociality, embodiment, truth, and commitment. In each section, we set out the background context for his writings on that topic, as well as what we take to be the big ideas that come out of his engagement with it. Our aim is to show how Haugeland's ideas have provided a lasting legacy for thinking about the nature of human mindedness. In some cases, such as his criticism of the computational model of the mind, or his defense of embodied cognition, Haugeland's once controversial ideas are now far more widely accepted. But in other cases, Haugeland's claims are so bold that they are difficult (even for us) to accept at face value. Regardless of whether his ideas are now more accepted or outlandish, our goal is to present them in a way that makes them philosophically enticing. We aim to motivate others to read or reread Haugeland not just as a significant figure in cognitive science and philosophy of mind at the turn of the millennium but as a philosopher whose work continues to speak to our central concerns about mindedness today.

Before we start, one caveat is worth mentioning: in our reading, Haugeland's own view underwent a significant revision over the course of his career. Accordingly, we will speak of his "early" view and his "late" views. We also spell out some of the continuities and discontinuities between these views. Taken as a whole, Haugeland's philosophical career is a vivid illustration of his own late view that an existential commitment to getting things right is a risky, demanding, and ultimately necessary prerequisite to genuine understanding.

2   Holism

The trouble with artificial intelligence is that computers don't give a damn.
—John Haugeland, "Understanding Natural Language" (*HT*, 47)

Haugeland's writing career began with a series of critical discussions of the prospects for the success of the computational theory of the mind, or, as he called it, GOFAI ("Good Old-Fashioned Artificial Intelligence"). Haugeland's writings on this topic are remarkably evenhanded: he makes every effort to spell out the attractions of GOFAI and never attempts to argue that it is an incoherent research program.[4] His book on the topic, *Artificial Intelligence: The Very Idea* (1985), is notable in this regard. Although Haugeland ultimately remains skeptical about the prospects for GOFAI's success, much of the book is dedicated to giving a detailed account of what makes GOFAI philosophically rich and exciting. Perhaps most importantly, Haugeland's book uses GOFAI as a test case for a set of philosophical commitments whose influence extends far beyond debates in artificial intelligence.

Haugeland's critical discussion of the philosophical underpinnings of GOFAI takes place in the context of his inheritance of the work of one of his mentors, Hubert Dreyfus. Dreyfus originally published his own pivotal critique of GOFAI, *What Computers Can't Do*, in 1972, at the height of the boom years of GOFAI. When it was first published, Dreyfus's book received a dismissive, almost hostile, reaction within the artificial

intelligence community as being the work of an armchair theorist attempting to disprove an empirical hypothesis through a priori arguments.[5]
Haugeland's own work on artificial intelligence took place almost a generation later, when GOFAI remained philosophically influential (as evidenced,
for instance, by Jerry Fodor's work), but also when there was a growing
realization that GOFAI had failed to deliver on many of the promises made
during the boom years. Haugeland's hands-on experience as a programmer,
as well as his time at the Center for Advanced Study in the Behavioral Sciences, informed his own account of what makes GOFAI so exciting, as well
as his diagnosis of what might be insurmountable barriers to its success. It
was during this period that Haugeland published *Artificial Intelligence*, as
well as the two editions of his edited volume of essays on cognitive science,
Mind Design and *Mind Design II* (1981 and 1997). These volumes include
essays on GOFAI, as well as on connectionist and dynamic models of
cognition.

One persistent problem for GOFAI that was much discussed during this
period was the so-called *frame problem*.[6] The frame problem concerns how
an AI system can "home in on the relevant factors without spending an
enormous effort ruling out alternatives" (*AI*, 204). The challenge the frame
problem poses is that our knowledge of the world is holistic: changes in one
piece of knowledge also affect other pieces of knowledge, often in far-off
fields. For example, learning that the dog ate the pie requires that you
update your grocery list for the party tomorrow, your plans for seeing a
movie rather than baking this evening, where you will store future pies that
you make, and so on. Humans find it easy to make these connections, but
the standard GOFAI model requires going through every piece of stored
information to test whether the hungry dog's action affects it (or somehow
structuring all knowledge such that these connections are evident). These
sorts of solutions proved to be extremely computationally demanding to
realize in practice, leading some to wonder whether that might be a sign of
an underlying problem for GOFAI in general.

As a way of drawing a larger philosophical moral from the frame problem, Haugeland generalizes the notion of holism at play in it. As is typical
of Haugeland, he begins with a focused engagement on a technical problem: in this case, the problem of what it would take for a computer program
to be able to engage in ordinary conversation. He argues that the holism
involved in this sort of engagement extends to many aspects of the

world—such as social practices, the distinction between different points of view, and embodiment—that are not part of standard GOFAI accounts. Haugeland then uses this as a jumping-off point for drawing some larger morals about the nature of understanding and its dependence on being engaged with the world. Specifically, he focuses on the role that giving a damn plays in connecting what might otherwise seem to be completely disparate bits of information. This does not mean Haugeland's criticisms are devastating for any and all possible attempts to build something with a mind (Haugeland himself thinks that minds are a kind of artifact); rather, it is an attempt to articulate an account of understanding that can act as a standard for what qualifies as having a mind in the first place.

In this section, we summarize one of Haugeland's specific critiques of GOFAI, the charge that caring about one's self and the world is necessary for understanding natural language. Unlike other prominent critiques of GOFAI, Haugeland's critique begins by conceding that being able to pass the Turing Test—that is, being able to engage in an ongoing conversation about everyday topics—represents evidence of genuinely understanding a natural language.[7] As he puts this concession, "The most ordinary conversations are fraught with life and all its meanings" (*HT*, 59). He then goes on to argue that since GOFAI systems do not care about anything—nothing matters to them—it is unlikely that any of them will ever pass the Turing Test. The bulk of Haugeland's critique involves articulating the varieties of care that he thinks are required for genuine understanding. Haugeland's most direct presentation of this critique is his article "Understanding Natural Language" (1979).

Understanding Natural Language

No system lacking a sense of *itself* as "somebody" with a complete life of its own (and about which it particularly cares) can possibly be adequate as a model of human understanding.

—John Haugeland, "Toward a New Existentialism" (*HT*, 3)

In his early essay "Understanding Natural Language," Haugeland argues that understanding a natural language essentially depends on caring about one's self and the world. He argues for this by distinguishing between three distinct ways in which this sort of care is integral to understanding. Haugeland refers to all three of these as forms of holism, the underlying idea being that making sense of individual segments of natural language requires seeing them as parts of larger wholes, wholes that themselves involve one's relationship to one's self and the world.

Haugeland's essay proceeds by first introducing a sort of holism that does *not* depend on giving a damn, before discussing three sorts that do. He refers to the first sort as *intentional* holism and the subsequent three sorts as *commonsense*, *situation*, and *existential* holisms. We will discuss each in turn.

Intentional Holism

Intentional holism concerns the way in which understanding the meaning of a particular passage of text depends on fitting the passage into a much larger pattern of linguistic activity, so as to see the passage as saying something sensible in context.[8] To illustrate this sort of holism, Haugeland introduces an analogy between linguistic activity and playing chess.[9] He then raises the question of "how one might *empirically* defend the claim that a given (strange) object plays chess" (*HT*, 47) and gives the following three-step answer:

1. Give systematic criteria for physically identifying the object's inputs and outputs
2. Provide a systematic way of interpreting them as various moves (such as a manual for translating them into standard notation); and then
3. Let some skeptics play chess with it. (*HT*, 47–48)

Haugeland's point is that there is a crucial difference between understanding this strange object's inputs and outputs taken individually versus taken together. Taken individually, they seem meaningless; but if we take them together, it might become possible to interpret them as sensible moves in a chess game. And if it is possible to do this in a way that allows you to actually play chess with this strange object, then it is reasonable to conclude that they *are* chess moves. Analogously, particular fragments of text might seem meaningless when taken on their own; but if we take them together, as part of a much larger pattern of linguistic activity, it might become possible to interpret the text as saying something. And if it is possible to do this in a way that allows you to have an ongoing conversation with them (e.g., in a Turing Test–like scenario), then it is reasonable to conclude that these fragments of text *are* meaningful.

This is a clear example of a general point about holism that Haugeland returns to repeatedly throughout his career, that what an object *is* can be constituted by the way it fits into a larger *pattern*.[10] For example, consider how the inputs and outputs of the strange chess-playing object might take place in a variety of media, such as plastic, metal, or stone tokens. Haugeland's fundamental point is that what it is for these tokens to be pieces in a game of chess is independent of their material constitution.[11] Their constitution as chess pieces depends not on whether they are plastic or metal or stone but on their being interpretable as making "sensible (legal and plausible)" moves in a chess game (*HT*, 48). For example, someone might start playing a particular game of chess by using a plastic bottle cap as her king but later continue playing the same game by using a metal bottle cap as the same king. In such a case, what makes these two caps one and the same chess piece has nothing to do with the physical constitution of the caps, but everything to do with the shared role they play in this particular game of chess.[12] Here is how Haugeland later puts the point:

Chess games are a kind of *pattern*, and chess phenomena can only occur within this pattern, as *subpatterns* of it. The point about different media … is that these subpatterns would not be what they are, and hence could not be recognized, except as subpatterns of a superordinate pattern with the specific structure of chess—not a pattern of shape or color, therefore, but a pattern at what we might call "*the chess level*." To put it more metaphysically, the game of chess is *constitutive* for chess phenomena, and *therefore* some grasp of it is prerequisite to their perceivability as such. (*HT*, 248)

In short, for Haugeland, what it is to be a chess piece is fundamentally holistic in the sense that it is constituted by the emergence of a particular subpattern in a larger pattern of chess-playing activity. To recognize something as being this or that sort of chess piece, one must have a prior grasp of the game of chess. Haugeland thinks that this sort of intentional holism is a widespread phenomenon (which we discuss further in subsequent sections), but in "Understanding Natural Language," he introduces it only to distinguish it from the three other sorts of holism that are his main concern.

Commonsense Holism

Commonsense holism concerns the way in which understanding the meaning of a particular passage of natural language often depends on a background familiarity with the world and what matters in it. This is the first sort of holism that Haugeland discusses that shows how giving a damn is essential for genuine understanding. Importantly, it is a form of *real-time* holism, as opposed to the *prior* holism of intentional interpretation. Here is how Haugeland explains this contrast:

> The holism of intentional interpretation is *prior* holism, in the sense that it's already accommodated *before* the interpretation of ongoing discourse. An interpreter *first* finds an over-all scheme that works, and *then* can interpret each new utterance separately as it comes. ... By contrast, common-sense holism is *real-time* holism—it is freshly relevant to each new sentence, and it can never be ignored. (*HT*, 49)

The point of this contrast is to bring out how understanding what a particular word or passage is used to mean in a particular context often requires attention to what matters in that particular context. Haugeland thinks that this sort of process pervades our use of language but is most evident in situations in which we are required to do things like resolve ambiguities, tell the difference between whether a word is being used literally or metaphorically, figure out whether the speaker is being serious or joking, and so on. Haugeland thinks that underlying these abilities is a kind of practical knowledge that people have just in virtue of their life experience.

To illustrate this sort of holism, Haugeland introduces the following sentence: "I left my raincoat in the bathtub, because it was still wet" (*HT*, 49). He then points out that the "it" in this sentence could equally well refer to the raincoat or the bathtub, since these are both things that are often wet. Yet it is simply obvious to most English speakers that this "it" refers to the raincoat and not the bathtub. Haugeland thinks that the best explanation for the obviousness of the one interpretation over the other is a background familiarity with both wet raincoats and wet bathtubs, and a tendency to give a damn about the first being wet but not the second. That is, we all already know the point of putting a wet raincoat in a dry bathtub, whereas it is unclear what the point would be of putting a dry raincoat in a wet bathtub. This is a vivid illustration of how coping with commonsense holism involves caring about what matters in the world.[13]

Situation Holism

Situation holism concerns the way in which understanding passages of natural language—especially the plots of stories—often depends on an ability to keep track of the different points of view conveyed by the passage, as well as to appreciate why these different points of view matter to the story.

To illustrate this sort of holism, Haugeland tells the following Middle Eastern folktale:

One evening, the Khoja looked down into a well, and was startled to find the moon shining up at him. It won't help anyone down there, he thought, and quickly he fetched a hook on a rope. But when he threw it in, the hook snagged on a hidden rock. The Khoja pulled and pulled and pulled. Then suddenly it broke loose, and he went right on his back with a thump. From where he lay, however, he could see the moon, finally back in the sky where it belonged—and he was proud of the good job he had done. (*HT*, 53)

Making sense of this story requires keeping track of two distinct points of view: what really happened, and what Khoja imagines happened. Although the story reports that the Khoja "was startled to find the moon shining up at him" from the bottom of the well, most readers recognize that he is only seeing a reflection of the moon in the well. The same applies to the Khoja's pulling on the rope and then seeing the moon in the sky. Understanding what really happened—that the Khoja's pulling on the rope did not actu- ally move the moon—depends on commonsense holism because it depends on knowing that the moon would not fit in a well and cannot be moved by pulling on a rope. Crucially, however, a full understanding of the story requires more than commonsense holism because it also requires keeping track of what the Khoja imagines happened. Understanding the Khoja's pride, for instance, requires appreciating that the Khoja himself thought that the moon was in the well, and that his pulling on the rope did move it. This is a vivid illustration of how coping with situation holism involves caring about the differences between multiple points of view on the same situation.[14]

Existential Holism

Existential holism concerns the way in which understanding passages of natural language—especially the morals of stories—often depends on appreciating how what is said involves someone's "self-image or sense of identity [being] at stake" (*HT*, 56). To illustrate this sort of holism, Hauge- land introduces the following fable from Aesop:

One day, a farmer's son accidentally stepped on a snake, and was fatally bitten. Enraged, the father chased the snake with an axe, and managed to cut off its tail; and thereupon, the snake nearly ruined the farm by biting all the animals. Well, the farmer thought it over, and finally took the snake some sweetmeats, saying: "I can understand your anger, and surely you can understand mine. But now that we are even, let's forgive, forget, and be friends again." "No, no," said the snake, "take away your gifts. You can *never* forget your dead son, nor I my missing tail." (*HT*, 57–58)

If one tried to make the moral of this story explicit, one might summarize it as the idea that "both children and limbs are irreplaceable." The problem with this summary is that it not only "misrepresents the moral, but … lacks it altogether—it is utterly flat and lifeless" (*HT*, 58). Perhaps most significantly, this summary overlooks what is so mistaken or misguided about the farmer's gift of sweetmeats. To understand the story, we must be able to identify with the farmer, to appreciate his rage and revenge, as well as why he might eventually try to make amends with the snake. Only on the basis of identifying with the farmer and appreciating why his losses might lead him to attempt to make amends does the snake's rebuke become meaningful: it is a criticism of both the farmer's and our own failure to appreciate the full weight of the loss of a child. As Haugeland puts it, "Children can well mean more to who one is and the meaning of one's life than even one's own limbs. So who are *you*, and what is *your* life? The folly—what the fable is really 'about'—is not knowing" (*HT*, 58). The story is ultimately about how embarrassing it is for the farmer to have put a price on his son's life in the first place, and how that embarrassment reveals a failure to recognize the true value of something irreplaceable.

Existential holism is a form of holism because it requires appreciating a story in terms that extend far beyond merely understanding what happened. Understanding the moral of a story does not only depend on our background familiarity with the world (as commonsense holism does) or our ability to keep track of different points of view (as situation holism does). It also depends on our existential condition, as mortal beings struggling to make coherent lives in the face of a hostile world. Haugeland writes:

A single act cannot be embarrassing, shameful, irresponsible, or foolish in isolation, but only as an event in the biography of a whole, historical, individual—a person whose personality it reflects and whose self-understanding it threatens. (*HT*, 58)

Taken together, commonsense, situation, and existential holism bring out the real challenge represented by the Turing Test. The challenge is that stories *matter* to us, and this mattering is essential for understanding the stories. Haugeland concludes that "only a being that cares about who it is, as some sort of enduring whole, *can* care about guilt or folly, self-respect or achievement, life or death. And only such a being can read" (*HT*, 58–59). In sum, genuinely understanding natural language requires giving a damn.

The larger conclusion that Haugeland draws from these considerations is the significance of understanding for mindedness more generally. Here is how he puts this point:

No current approach to artificial intelligence takes *understanding* seriously—where understanding itself is understood as distinct from knowledge (in whole or in part) and prerequisite thereto. It seems to me that, taken in this sense, *only people* ever understand anything—no animals and no artifacts (yet). It follows that, in a strict and proper sense, no animal or machine genuinely believes or desires anything either—How could it believe something it doesn't understand?—though, obviously, in some other, weaker sense, animals (at least) have plenty of beliefs and desires. This conviction, I should add, is not based upon any in-principle barrier: it's just an empirical observation about what happens to be the case at the moment, so far as we can tell. (Haugeland 1997, 27)

In the next section, we begin to spell out Haugeland's own account of mindedness.

3   Sociality

Humanity is not a zoological classification, but a more recent social and historical phenomenon.
—John Haugeland, "Toward a New Existentialism" (*HT*, 1)

For Haugeland, human minds are social artifacts.[15] A mind is not something one is born with; it is an accomplishment, one that results from abiding by the norms of a socially instituted practice. Becoming a person capable of thinking is, in this sense, a status that one achieves or a social role that one comes to occupy. In his early work, Haugeland holds that the constitution of minds is *entirely* the result of conforming to socially instituted norms; this follows from his early commitment to thinking that all constitution is socially instituted (cf. *DD*, 8). In his later work, Haugeland revises his early view, holding that there is more to the constitution of minds than social institution, and existential commitment also plays a fundamental role. (We discuss his revisions to his early view in sec. 5.)

Haugeland's account of the sociality of minds brings together elements from both the Continental and analytic traditions. From the Continental side, Haugeland draws on Heidegger's notion that humans are fundamentally conformist beings who habitually fall into societal ways of behaving.[16] From the analytic side, Haugeland draws on Wilfrid Sellars's notion that concepts are not inner entities but public rules that govern an interactive game of giving and asking for reasons.[17] In Haugeland's combined account, acquiring a mind is a matter not just of conforming to social norms but also of enforcing those norms in one's interactions with others and the world.

In this section, we summarize Haugeland's early views on the sociality of minds by looking at two influential essays: "Heidegger on Being a Person" (1982) and "The Intentionality All-Stars" (1990). This summary explains why Haugeland thinks that having a mind is essentially social, and sets up Haugeland's later revisions to his early view.

Heidegger on Being a Person

According to [my] analysis, a person is not fundamentally a talking animal or a thinking thing but a ... crucial sort of subpattern in an overall pattern instituted by conformism and handed down from generation to generation.

—John Haugeland, "Heidegger on Being a Person" (*DD*, 16)

In this essay, Haugeland offers "a nonstandard and rather freewheeling interpretation" (*DD*, 3) of Heidegger's account of personhood. According to his reading of Heidegger, being a person is less like being a specimen of a species and more like being an instance of a game played according to socially instituted rules. Individual people are "cases" (*DD*, 10) of personhood or mindedness; people "have" minds in a sense that is similar to which they might "have" tuberculosis. Unlike having tuberculosis, however, having thought is not something that one can simply "catch." Rather, it is the result of actively participating in a social practice.

As a way of explaining how participation in a social practice can result in the emergence of mindedness, without presupposing this capacity as a prerequisite for participation in the practice, Haugeland observes that humans are innately conformists.[18] For Haugeland, being a conformist "means not just imitativeness (monkey see, monkey do) but also censoriousness—that is, a positive tendency to see that one's neighbors do likewise and to suppress variation" (*DD*, 4). This tendency leads groups of conformists to develop patterns of stable behavioral dispositions. When these patterns become entrenched in a community to the extent that they are not merely exhibited by community members, but also enforced by members and handed down from generation to generation, they become norms. Norms are, in this sense, "a kind of 'emergent' entity, with an identity and life of their own" (*DD*, 4). Importantly, to have such a life, these socially instituted norms need not be explicitly articulated. Haugeland uses an imaginary variant of chess to illustrate this point:

> Imagine, for instance, that the rules of chess were not explicitly codified but were observed only as a body of conformists' norms—"how one acts" when in chess-playing circumstances. ... We call a sort that is involved in many interrelated norms a *role* (e.g., the role of the king in chess). (*DD*, 6)

In this imaginary variant of chess, the rules of the game function as norms even though they are not explicitly articulated but merely exhibited and enforced by a community of players. In such a community, it becomes possible for something to become a king just in virtue of being manipulated in accordance with the socially instituted norms for king pieces. By extension, Haugeland holds that what it is for someone to become a person—to have a mind—just is for him or her to act in accordance with a set of socially instituted norms, none of which need be explicitly articulated.

There is, however, a fundamental difference between being a king piece in a chess game and being a person, a difference that this example makes manifest: whereas chess pieces are *manipulated* according to norms, persons *act* according to norms. Persons, unlike chess pieces, can themselves be held responsible for acting in accordance with the norms of their community. Moreover, this difference is crucial to their status as persons, because it is only through being held responsible for their actions that they are individuated as individual people (i.e., as individual "cases" of having thought, or individual subpatterns in communal patterns of conformity). For example, when a child steals a piece of candy with his or her fingers, the child is held responsible, not the fingers or the community. The child is one "'unit of accountability' and therefore the 'subject' of the behavior because it is what takes the heat and learns from 'its' mistakes" (*DD*, 10–11). It is through being held responsible as a unit of accountability that a child becomes intelligible as an individual case of personhood.

Another way to put this last point is that, for Haugeland, being a person results from being held responsible for occupying a social role. However, one of the distinctive things about humans is our capacity to occupy multiple roles, either at different times or at one and the same time. This opens up the possibility of two different ways in which one can be a person: either as "disowned" or as "self-owned" (*DD*, 14–15; "self-owned" is Haugeland's early translation of *eigentlich*). A disowned person is someone who acts in accordance with whatever socially instituted norms are most pressing at that time and place. A self-owned person, by contrast, is someone who undertakes the activity of "seeking out and positively adjudicating the conflicting requirements of one's various roles" (*DD*, 14). A person who undertakes this self-critical activity becomes a doubly unified unit of accountability: first, as an individual unit of accountability who is held responsible for occupying a social role; and second, as an individual who strives to unify the various social roles that she or he might occupy at different times and places. Whereas a disowned person might vacillate between a variety of potentially inconsistent roles, a self-owned person is capable of overriding norms imposed by society—no matter how pressing—that do not cohere with an envisioned unified whole. This striving for coherence involves modifying or rejecting the normative demands of social roles so as to become a unity that persists across changes in outside pressure to conform to this or that social norm. A person who is able to bring about such a persistent unity lives a "resolute" or "authentic" life (*DD*, 15; "authentic" is the standard translation of *eigentlich*).

The crucial thing to note about Haugeland's early account of personhood is that the self-owned life is merely a coherent way of living within preexisting social institutions. There is no account of persons who inaugurate new ways of life that move beyond established social institutions. As we shall see, Haugeland's later view focuses on making room for this revolutionary kind of personhood.

In sum, in Haugeland's early view of personhood, having thought results from abiding by the norms of a socially instituted practice. In "The Intentionality All-Stars," he further develops this account by spelling out the sort of social practice that is involved.

The Intentionality All-Stars

If vapid holism is to be the key to intentionality, the nonaccidental "larger-pattern" must be specified more fully; and this fuller specification, moreover, should account for the peculiarity of intentionality as a relation, for its normative force, and for what differentiates original intentionality from derivative. Even with all of these stipulations, however, there remain a number of distinct candidates—alternative specifications of the relevant "whole." And differences among them, it seems to me, are characteristic of differing basic approaches to intentionality.

—John Haugeland, "The Intentionality All-Stars" (*HT*, 130)

In this essay, Haugeland develops his early account of the sociality of minds by comparing and contrasting it with two other accounts of having thought—or, as he puts it here, of having intentionality. "Intentionality" refers to aboutness, "that character of some things (items, events, states, ...) that they are 'of,' are 'about,' or 'represent' others" (*HT*, 127).[19] The fundamental challenge Haugeland discusses is that of understanding how something can have intentionality in a nonderivative sense, that is, how something can be a *source* of intentionality. It is, after all, easy to understand how something can have intentionality in a derivative sense: for example, we can randomly pick two lumps of Play-Doh and simply stipulate that the first refers to the Empire State Building and the second refers to the Chrysler Building. But the challenge is to explain where the intentionality invoked by our stipulations derives from in the first place. Haugeland refers to this nonderivative sort of intentionality as "original" intentionality (*HT*, 129).

Haugeland's names for the three approaches he discusses in this essay are "neo-Cartesian," "neo-behaviorist," and "neo-pragmatist" (the last being his own early view).[20] All three approaches share the goal of trying to offer a worked-out theory of original intentionality. All three also accept a kind of "vapid materialism" (*HT*, 128) as their starting point, according to which original intentionality is possessed by patches of matter (and not, for instance, by some sort of immaterial substance that intrinsically but inexplicably possesses original intentionality). And all three share the strategy of identifying a larger whole or pattern that these patches of matter are part of, as well as using that larger whole to explain the emergence of original intentionality. Haugeland refers to this strategy as "vapid holism," the underlying idea being that "the intentionality of any individual state or occurrence always depends on some larger pattern into which it fits" (*HT*,

130). In short, the three approaches that Haugeland discusses are all worked-out proposals for showing how vapid holism renders original intentionality compatible with vapid materialism; these approaches are distinguished in terms of the sort of larger pattern that each invokes.

In this essay, Haugeland does not argue against neo-Cartesianism[21] or neo-behaviorism[22] but mainly introduces them to set up what is distinctive about his own early view, neo-pragmatism.[23] In the next three subsections, we briefly summarize the three larger patterns that these different approaches invoke as ways of attempting to explain original intentionality.

Neo-Cartesianism

Prominent neo-Cartesians include Jerry Fodor, Hartry Field, and Zenon Pylyshyn. The larger pattern these neo-Cartesians focus on to explain original intentionality is the interaction of discrete inner states of formal systems. ("Neo-Cartesianism" is the name for the view underlying GOFAI, introduced in sec. 2.) The neo-Cartesians propose that if the inner states of a system interact with one another in the right sort of way, then those states have intentionality. As a way of spelling out what "the right sort of way" means for them, neo-Cartesians propose a *principle of interpretation* (*HT*, 134), according to which we can discover that the inner states of a system have intentionality through an activity akin to code breaking. In short, neo-Cartesians hold that if we can interpret the inner states of a system as, by and large, saying something reasonable, then that *is* what they are saying (or, in other words, what they are about).

A preliminary worry about the neo-Cartesian principle of interpretation is that it might seem too permissive, permitting us to interpret any inner state as meaning anything we like. The worry is that "interpretation" here does not involve "discovering" the intentionality of the inner states of systems so much as "imposing" it on them. As a way of avoiding this worry, neo-Cartesians propose that successful interpretations must be simultaneously beholden to two independent constraints: consistency and truth.[24] That is, across the larger pattern of inner states in question, one and the same type of inner state must consistently be interpreted as saying the same thing; and these meanings must combine to say something true. These two constraints make it difficult to achieve a successful interpretation of large patterns of inner states, thereby motivating the idea that interpreters cannot simply impose whatever meaningfulness they would like onto these patterns but must genuinely seek to discover a meaningfulness that is already there.

A second worry about the neo-Cartesian principle is that it might seem that the intentionality it discovers remains derivative, not from the interpreter but from whoever or whatever is responsible for putting together this pattern of inner states in the first place. For example, with sufficient industriousness, a reader who does not understand English might arrive at a successful interpretation of the marks printed on the pages of this book, according to which they make a series of reasonable claims about Haugeland's philosophical views.[25] Nevertheless, the worry is that the intentionality of these marks remains derivative from the authors of the essays collected here.

As a way of avoiding this second worry, neo-Cartesians propose that the inner states of a system have original, and not merely derivative, intentionality when these inner states are themselves "semantically active" (*HT*, 137) in the sense that they are "constantly interacting in regular ways that are directly and systematically appropriate to their [semantic] contents—in the course of reasoning, problem solving, decision making, and so on" (*HT*, 136). The marks on the pages of this book are not at all active in this sense. With this additional constraint on interpretation in place, neo-Cartesians can happily concede that the printed marks in this book only possess derivative intentionality, and that it is only when the inner states of a system actively interact in ways that are appropriate to their semantic contents that these states possess original intentionality.

Neo-behaviorism

Prominent neo-behaviorists include Donald Davidson, Daniel Dennett, and Robert Stalnaker. The larger pattern that these neo-behaviorists focus on to explain original intentionality is the interaction of a system with its environment. They propose that if a system interacts with its environment in the right sort of way, then that system has intentionality. As a way of spelling out what "the right sort of way" means for them, neo-behaviorists propose a *principle of ascription* (*HT*, 142), according to which thoughts are ascribed to systems as part of making sense of the system's ability to skillfully cope with its environment. In short, if the attribution of thoughts to a system helps us to make sense of the system's overall behavioral competence, then we are justified in taking the system to have these thoughts.

As with neo-Cartesianism, a worry about neo-behaviorism is that it might seem too permissive, allowing us to ascribe thoughts to systems that we would otherwise have no reason to take to have minds. For example, it seems to justify taking an ordinary spring mousetrap to have the "desire" to kill mice and the "belief" that springing closed on mice is a means of satisfying this desire. As a way of avoiding this worry, neo-behaviorists propose that successful ascriptions of thoughts must be simultaneously beholden to two independent constraints: persistent goals and behavioral success. That is, across the larger pattern of a system's behavior, ascribing thoughts to the system should involve attributing goals that range across a variety of its interactions with its environment, and these interactions should aim to achieve these goals (and success or failure in achieving these goals through various means should inform future interactions). For example, to be justified in ascribing to a mousetrap the desire to kill mice as a persistent goal, this goal should be manifest in the trap's behavior in a wider variety of circumstances than just when mice put pressure on the trap's trigger (e.g., the trap should do more than simply sit still when mice are nearby, but should actively set out to kill them). Furthermore, if we are to be justified in taking behavioral success to inform the mousetrap's interactions with its environment, the trap should snap closed only when a mouse steps on it, rather than when other things happen to put pressure on the trigger. Of course, it is possible for a system with the goal of killing mice to make mistakes, but such mistakes "are identifiable as such only against the backdrop of a pattern that is largely clear and reliable" (*HT*, 145).

In sum, for neo-behaviorists, to be justified in ascribing original intentionality to a system, the system's behavior must exhibit a kind of "semantic intrigue" (*HT*, 144), according to which ascribing a persistent set of goals to the system gives us an otherwise unavailable insight into the system's interactions with its environment. The presence of semantic intrigue "marks the difference between gratuitous ascriptions and those that are genuinely required to maximize manifest competence" (*HT*, 144–145).

Once again, as with neo-Cartesianism, Haugeland does not argue against neo-behaviorism in this essay. He introduces it mainly as a way of setting up his own early view.

Neo-pragmatism

Prominent neo-pragmatists include Wilfrid Sellars, Robert Brandom, and Haugeland himself (in his early incarnation). The larger pattern that neo-pragmatists focus on to explain original intentionality is social conformism. This is a fundamentally different sort of pattern from those invoked by the preceding two approaches, insofar as the principles of interpretation and ascription both assume that "the only standard of what to count as intentional is comparison with ourselves" (*HT*, 158). Neo-pragmatism, by contrast, aims to identify a pattern that is identified not by comparison with ourselves but by having the "distinctive ontological foundation" (*HT*, 158) of consisting in stable behavioral dispositions that are enforced by members of a society and handed down from generation to generation. For neo-pragmatists,

> The fundamental pattern becomes a culture or way of life, with all its institutions, artifacts, and mores. The idea is that contentful tokens, like ritual objects, customary performances, and tools, occupy determinate niches within the social fabric—and these niches "define" them as what they are. Only in virtue of such culturally instituted roles can tokens have contents at all. (*HT*, 147)

Unlike neo-Cartesians, who locate original intentionality "inside the head" in the interaction of inner states (or inner tokens), neo-pragmatists locate original intentionality in the manipulation of outer tokens, such as linguistic tools (i.e., words and sentences). For neo-pragmatists, the intentionality of inner thoughts derives from the content of these outer tokens, which itself derives from the roles they play in socially instituted practices.

Neo-pragmatists propose to understand intentionality on the model of tool use, according to which what a tool is "for" is determined not simply by what it does but by the role assigned to it in a socially instituted practice. Just as Phillips-head screwdrivers are "for" screwing in Phillips-head screws, words and sentences are "for" communication. What makes linguistic tools distinctive, however, is that they are "*double-use* tools" (*HT*, 153) in the sense that they are designed not just to be used by members of the community but also to be responded to by other members. Consider, for example, how a neo-pragmatist might account for the intentionality of alarm calls in terms of the social expectations surrounding them. A boy who cries "Wolf!" does so with the expectation that the other members of his society will respond appropriately, just as the other members of his society expect that he will only make this call under the appropriate conditions. Abiding by or violating the norms of the use of this sort of linguistic tool determines someone's "status" (*HT*, 154) as a language user: for example, a boy who cries "Wolf!" under inappropriate conditions comes to have the status of being unreliable. This status is propriety determining even when it is not manifest, in that once the boy comes to have the status of being unreliable, it is appropriate to disregard his alarm calls, even if there is, in fact, a wolf present. Neo-pragmatists propose to generalize this account of propriety-determining nonmanifest statuses to account for the variety of linguistic practices that we engage in, including that of factual assertion.

The most important thing to note about the neo-pragmatist account of original intentionality is that it makes intentionality *entirely* socially instituted. If the use of a linguistic tool completely accords with the norm for its use in a society, then it is correct. There is no room for the possibility of an utterance that is socially correct but incorrect about the world. As we shall see, Haugeland's later view centers on making room for this possibility.

4   Embodiment

I want to suggest that the human mind may be *more* intimately intermingled with its body and its world than is any other, and that this is one of its distinctive advantages.
—John Haugeland, "Mind Embodied and Embedded" (*HT*, 223)

For Haugeland, the human mind depends not just on social practices but also on the human body and the human environment, especially the human built environment. In spelling out the nature of this dependence, Haugeland does not merely argue that minds depend on the material world in the minimal sense that "without any matter, there wouldn't be anything else contingent either" (*HT*, 128). Rather, he aims to spell out a more surprising sense in which "the significant complexity of intelligent behavior depends intimately on the concrete details of the agent's embodiment and worldly situation" (*HT*, 211). Haugeland's ultimate goal is to show how the mind extends far beyond the head, such that not just other parts of the body but also aspects of the environment are constituents of it.

This aspect of Haugeland's view is, once again, usefully contrasted with GOFAI accounts of the mind. Haugeland's account challenges three aspects of GOFAI accounts: the idea that the mind is a (i) self-contained, (ii) medium-independent (iii) symbol manipulator. GOFAI accounts hold that the mind is self-contained in the sense that what goes on inside it is, in principle, independent of what goes on outside it. They hold that it is medium independent in the sense that the particular sort of material in which the mind is instantiated is irrelevant to its proper functioning; minds are, in this sense, multiply realizable in a wide variety of material media. Finally, GOFAI accounts hold that the mind is a symbol manipulator in the sense that they think mental activity primarily consists in the manipulation of inner representations according to rules. Haugeland's own work played a pivotal role in developing an alternative view of the mind, one that has come to be called "embodied cognition."[26]

In this section, we summarize Haugeland's most developed statement of his views on the dependence of the mind on the body and the environment, his essay "Mind Embodied and Embedded" (1995).

Mind Embodied and Embedded

Well, *are there transducers* between our minds and our bodies? From a certain all-too-easy perspective, the question can seem obtuse: *of course* there are. Almost by definition, it seems, there *has to be* a conversion between the symbolic or conceptual contents of our minds and the physical processes in our bodies; and that conversion just is transduction. But [I am], in effect, denying this—not by denying that there are minds or that there are bodies, but that there needs to be any interface or conversion between them.

—John Haugeland, "Mind Embodied and Embedded" (*HT*, 224)

In this essay, Haugeland aims to expose a Cartesian assumption about the mind that continues to constrain the ways that many philosophers and cognitive scientists think about the mind's relationship to the world, including many who take themselves to reject a Cartesian view of the mind. The Cartesian assumption is that intelligent behavior essentially involves an "interface" (*HT*, 213) between the mind and world. As a way of exposing this Cartesian assumption, Haugeland introduces a contrast between two different models of perception, which we will call *the interface model* and *the integration model*. Haugeland thinks that the interface model so dominates theorizing about perception that it seems impossible to imagine an alternative. Accordingly, in offering his own argument for the integration model, "The challenge is as much to spell out what [it] could mean as to make a case for it" (*HT*, 208).

The interface model conceives of perception as essentially involving the conversion of nonsymbolic input from the world into symbolic representations in the mind. This model remains fundamentally Cartesian even if one holds that the conversion is an entirely material process. It is Cartesian in that, first, it rests on a principled distinction between nonsymbolic worldly input that is "outside" the mind and symbolic activity that is "inside" the mind, and, second, it holds that the "intelligent" part of intelligent behavior essentially involves the manipulation of symbolic representations in the mind. Accordingly, one significant challenge that the interface model assumes is the task of explaining how nonsymbolic worldly inputs are converted into a symbolic form that can be manipulated by the mind.

The integration model, by contrast, conceives of perception as a "tightly coupled, high-bandwidth" (*HT*, 223) interaction between the body and its environment. According to this model, making sense of behavior essentially involves both bodily and environmental elements. This model rejects the assumption that intelligent behavior must involve symbolically representing the world. It proposes instead that intelligent behavior can consist in a reliable and resilient ability to skillfully cope with an environment, and that specific features of bodies and environments are often constituent parts of this ability. Haugeland gives a number of examples to illustrate what he means by this sort of tightly coupled, high-bandwidth interaction, but perhaps his most provocative and influential example is the following:

Now let me tell you how I get to San Jose: I pick the right road (Interstate 880 south), stay on it, and get off at the end. Can we say that the *road* knows the way to San Jose, or perhaps that the road and I *collaborate*? I don't think this is as crazy as it may first sound. The complexity of the road (its shape) is comparable to that of the task and highly specific thereto; moreover, staying on the road requires constant high-bandwidth interaction with this very complexity. In other words, the internal guidance systems and the road itself must be closely coupled, in part because much of the "information" upon which the ability depends is "encoded" *in the road*. Thus, much as an internal map or program, learned and stored in memory, would ... have to be deemed *part of* an intelligent system that used it to get to San Jose, so I suggest the *road* should be considered *integral* to my ability. (*HT*, 234)

In this example, Haugeland's ability to get to San Jose essentially depends on being able to follow Interstate 880 south to its end. He is able to figure out how to get there without consulting any inner representation of the spatial relationship between his current location and his destination.

Rather, all he has to do is stay on this particular road and avoid whatever obstacles present themselves along the way. Crucially, in this example, the road is functionally equivalent to an inner representation of the spatial relationship between his current location and his destination, one that, if relied on to get there, we would have no hesitation in considering to be a constituent part of the mental process of figuring out the route. Accordingly, Haugeland concludes that we should equally well consider the road to be a constituent part of his mental process of figuring out the route.

Implicit in this example is a general argument for the conditions under which Haugeland thinks that we should consider parts of the body and environment to be parts of a mind. Here is this argument, in outline:

P1 A principle of functional equivalence, according to which something that is functionally equivalent to something that we would have no hesitation in considering to be part of a mind should equally well be considered to be part of a mind.

P2 There are some things outside the head that satisfy this principle of functional equivalence.

C The mind sometimes extends outside the head.

Since the publication of "Mind Embodied and Embedded" in 1995, this sort of argument for the "extended mind" has been developed and defended by a number of figures, most prominently by Andy Clark and David Chalmers (Clark and Chalmers 1998).

To appreciate how radical Haugeland's argument is, however, it is worth noting that the principle of functional equivalence that it rests on does *not* require that mental things outside the head must be symbolic representations (such as sentences or maps). Rather, aspects of the world can be functionally equivalent to symbolic representations without themselves being symbolic. In other words, in opposition to the interface model of perception, Haugeland holds that the world itself can be integrated into mental processes without needing to be converted into symbolic form. Perhaps the clearest examples of this are worldly paraphernalia that we have designed and built. As a way of bringing out how it is possible for these aspects of our environment to be part of the mind without being symbolic, Haugeland offers a general characterization of meaningfulness (or intentionality, broadly construed), one that includes both symbolic representations and worldly paraphernalia. As he puts it, "The meaningful in general is that which is significant in terms of something beyond itself, and subject to normative evaluation according to that significance" (*HT*, 232). In this characterization, Interstate 880 is meaningful because it was designed and built with the goal of getting people to San Jose (among other places), such that it is normatively assessable in terms of its success or failure at achieving this goal. It is in virtue of having this sort of goal directness built into them that Haugeland thinks worldly paraphernalia can be meaningful without being symbolic.

A further difference between the interface and integration models concerns how they each conceive of the relationship between perception and action. For the interface model, perception is in principle separate from action, because, in itself, perception is concerned solely with the creation of inner symbols that represent the outside world. What an agent chooses to do with these representations is a separate issue. For the integration model, by contrast, perception is not separate from action, because it treats perception as essentially involving inhabiting and exploring an environment. In Rodney Brooks's memorable phrase, for the integration model, "the world is its own best model" (Brooks 1990, 5; cited in *HT*, 232). According to it, perception is an embodied and embedded skill whereby a being with a specific sort of body is attuned to the relevant features of a specific sort of environment.[27] This attunement is fundamentally practical insofar as it concerns the aspects of an environment that are useful to the animal in question. For example, for different animals, "nooks can afford shelter and seclusion, green leaves or smaller neighbors can afford lunch, larger neighbors can afford attack, and so on—all depending on who's looking and with what interests" (*HT*, 222). Humans are similarly attuned, the only difference being the extent to which our environment is not merely useful but also meaningful, in the sense that we have literally built our goals into it. Here is how Haugeland illustrates this point:

Consider, for example, agriculture—without question, a basic manifestation of human intelligence, and dependent on a vast wealth of information accumulated through the centuries. Well, *where* has this information accumulated? Crucial elements of that heritage, I want to claim, are embodied in the shapes and strengths of the plow, the yoke, and the harness, as well as the practices for building and using them. (*HT*, 235)

For Haugeland, human intelligence is built into the social patterns, practices, and worldly paraphernalia of our environment. Underlying this claim is a broader notion of intelligence, one that echoes Haugeland's broader notion of meaningfulness, according to which "intelligence is the ability to deal reliably with more than the present and the manifest" (*HT*, 230). Whereas the interface model holds that dealing with "more than the present and the manifest" requires inner symbolic representations, the integration model holds that this is equally well something that nonsymbolic worldly paraphernalia make possible. In the case of the road, for instance, although only a small bit of it is present and manifest to a driver at a time, following it directs the driver toward a destination that is not at all present or manifest.

In sum, Haugeland thinks that we need a broader notion of mindedness that includes more than just the manipulation of inner symbolic representations. In this broader notion, mindedness can constitutively involve both bodily skills and worldly paraphernalia.

5   Truth

The problem is: How can truth and understanding be objective, given that they are deeply dependent on humanity?
—John Haugeland, "Truth and Understanding" (*TU*, i)

The previous two sections spelled out some of the ways in which Haugeland thinks that minds are integrated into the world. Given this integration, it is difficult to see how his view has room for minds to become decoupled from the world in such a way that they might misrepresent how things are.[28] The underlying problem is that it seems impossible for either the mind or the world to possess any kind of independent integrity; and if this is impossible, then it is hard to see how there is room for the contents of thoughts to be independent from how the world is, or for the world to be an independent criterion for assessing the accuracy or inaccuracy of thoughts. As Richard Rorty once put the predicament for views that assimilate mental activity to the use of worldly tools, "There is no way in which tools can take one out of touch with reality" (Rorty 1999, xxiii).[29]

Unlike Rorty, however, Haugeland is not satisfied with an account of mindedness that does not allow for the possibility of recognizing that an entire social practice is out of step with the world. This means that

Haugeland's own early view needs to be revised, or at least supplemented, in a way that allows for this possibility. His solution is to provide an account of a distinctive type of social practice, one that he calls "genuine science." As with many of Haugeland's terms and phrases, "genuine science" is intended as a generic label for any and all social practices that allow for the possibility of discovering that they are, in fact, out of step with the world. As a generic label for whatever it is that these practices have in common, Haugeland uses "objectivity" to refer to the way in which they purport to represent how the world really is, not merely how they take the world to be.

Haugeland's engagement with the problem of objectivity inherits Immanuel Kant's philosophical project of explaining how it is possible for thought to be about objects in an independent world. As Haugeland puts it, his work "is a direct descendant of Kant's inquiry into the forms of sensibility and understanding as conditions of the possibility of objects as objects" (*DD*, 188). Importantly, like Kant, Haugeland takes the problem to be not whether objectivity is possible but how it is possible:

> The problem is not *whether* there is objectivity, truth, a world, understanding *of* nature, and so on, but what these phenomena amount to, and what they presuppose. The differences between factuality and prejudice, honesty and deceit, reality and delusions, science and fantasy, etc., are too pervasive, too patent, and too important to deny. They may not be exactly what they have seemed, or presuppose what has been supposed, but they are surely "there" and surely indispensable. (*TU*, 7)

In short, Haugeland thinks that there is a substantial philosophical problem in showing how it is possible for truth and understanding to be dependent on human social practices while at the same time about an objective world that is independent of how we take it to be.

In his later writings, Haugeland engages with this problem of objectivity directly. He does so by offering a strikingly original account of truth, one that he refers to as a "beholdenness theory" (*HT*, 348). In this section, we summarize Haugeland's beholdenness theory of truth by drawing on the final four essays of *Having Thought*, with a special emphasis on what we take to be Haugeland's magnum opus, "Truth and Rule-Following" (1998).

The Problem of Objective Perception

It is crucial, however, not to suppose at the outset that *objects* are an unproblematic fixed point, and the only issue is how perception gets to be *of them*.

—John Haugeland, "Objective Perception" (*HT*, 246)

To focus attention on the specific problem that he thinks needs to be addressed, Haugeland introduces the phrase "objective perception" to refer to a distinctly human form of perception that is "*of objects as objects*" (*HT*, 241). "Objects" here refers not merely to corporeal things but more broadly to whatever it is that humans can perceive, including facts, properties, events, and so on. The problem of objective perception is the problem of understanding how it is possible for perceptual acts to have *determinate* objects that function as *normative* criteria for evaluating those acts. These two aspects of the problem are related: the determinacy aspect concerns the way in which perceptual acts aim to represent specific, distinct objects; the normative aspect concerns the way in which perceptual acts can be evaluated in terms of whether they accurately or inaccurately represent their distinct objects. We will discuss these two aspects of the problem in turn.

The Determinacy of Perception

The "objecthood" of perceptual objects and the "of-ness" of perception go hand in hand, and are intelligible only in terms of one another, something like the interdependence of target and aim, or puzzle and solution. So, the deeper question is: *How* and *why* is such a structure—what we might call *the structure of objectivity*—imposed on the physics and physiology of sensation?

—John Haugeland, "Objective Perception" (*HT*, 246)

The determinacy aspect concerns the question of what makes a particular perceptual act about a specific, distinct object, as opposed to any of the other things that it might be about. One way that Haugeland brings out the difficulty of this problem is by returning, once again, to the example of chess, and raising the question of what is involved in perceiving a knight fork.[30] This example illustrates the problem of determinacy in at least four ways.

First, any particular act of perceiving a knight fork is the culmination of a long and involved causal process. The problem is determining the particular part of this process that the act represents.[31] For example, if the chess game is played in daylight with a normal chess set, seeing the knight fork is the culmination of a causal process that includes the sun's emission of visible-wavelength light, the selective reflection of that light by the chess pieces and board, the selective absorption of that reflected light by one's retinas, and so on. The problem is figuring out what it is that makes this perceptual act about *the knight fork*, as opposed to some other part of this causal process.

Second, even if we figure out how this perceptual act is focused on the chess pieces and board, the act of looking at them could just as easily lead one to perceive any number of their aspects, such as the choke-ability of the pieces, the inflammability of the board, the movability of the pieces on the board, and so on.[32] Once again, the problem is figuring out how this perceptual act is specifically about the knight fork *as* a knight fork, as opposed to some other aspect of the chess pieces and board.

Third, the ability to perceive this arrangement of pieces on this board as a knight fork presupposes an ability to see other arrangements of other pieces on other boards as knight forks as well (i.e., as a distinctive, repeatable kind of pattern). These other pieces need not be made of the same material, nor need they be in the same positions, for them to make up a knight fork. This brings out how part of the problem of objective perception lies in figuring out what *unifies* all these different pieces and positions as being the same sort of perceptual object.[33]

Fourth and finally, since a knight fork is not an individual piece or a particular square on the board but is instead made up of a relationship between spatially detached pieces in a peculiar configuration, this example shows why it would be a mistake to take the objecthood of objects of human perception as something simply given to us by the world,[34] by, say, "tacitly presuppos[ing] that being an object is tantamount to being a temporally and spatially cohesive corporeal lump" (*HT*, 246). The problem with this presupposition is not just that the objects of human perception extend beyond lump-like things. The problem is that an adequate account of objective perception must explain the *integrity* of objects of perception (i.e., what holds them together as determinate objects) regardless of what sort of objects they are.

Taken together, these four considerations help to illustrate why Haugeland thinks that an adequate account of perceptual representation must not take the objecthood of objects of human perception for granted but rather needs to explain how it is possible for perceptual acts to be about determinate objects at all.

The Normativity of Perception

If the possibility of error is systematically eliminated, then it's vacuous to speak of correctness or recognition at all.

—John Haugeland, "Pattern and Being" (*HT*, 272–273)

The normativity aspect of the problem of objective perception concerns the question of what makes it possible for objects of perception to function as criteria for evaluating acts of perception. It might seem odd to refer to the *objects* of perception as *criteria*, but it reveals how we evaluate acts of perception according to how accurately or inaccurately they represent their objects; when acts of perception are so evaluated, the objects of perception are themselves the standards (or criteria) of evaluation.

One way to appreciate this aspect of the problem is to consider the case of a system that is not normatively beholden to determinate objects and to see why Haugeland thinks that such a system is thereby incapable of objective perception. Here is an example he uses to this end:

Consider an automatic door: it responds to an approaching pedestrian by opening, and to the pedestrian's passing through by closing again. Of course, it might respond identically to a wayward shopping cart, or a large enough piece of plaster falling from the ceiling; and we can even imagine it being triggered by the magnetic fields of a floor polisher passing near its control box, across the hall. Are such incidents *misrecognitions*? Has the door *mistaken* plaster or a floor polisher for a pedestrian, say? Obviously not, for pedestrians, plaster, and polishers are nothing to a door. Therefore, even in the usual case, we cannot say that it has recognized a pedestrian. (*HT*, 272)

Automatic doors reliably open in response to both moving pedestrians and wayward shopping carts. But it would be nonsense to reward them for opening in response to the first or punish them for opening in response to the second. This is the most obvious sense in which neither pedestrians nor shopping carts matter to automatic doors. It is for this reason that Haugeland thinks we have no grounds for identifying either people or carts as *the* sort of object that automatic doors aim to perceive. And it is because they do not have determinate objects of perception that they are incapable of objective perception.

The example of automatic doors is doubly useful, because what they lack is precisely what Haugeland thinks is required for objective perception. Haugeland proposes that the objecthood of objects of perception depends on our ability as conformists not only to abide by and to enforce norms but also to institute norms. Haugeland refers to the human ability to institute norms as adopting a *stance* toward the world, where a stance "is a kind of posture or attitude that somebody can take toward something, a specific way of regarding and dealing with it" (*HT*, 283). Crucially, Haugeland thinks that among the norms that stances are capable of instituting are perceptual norms, norms that determine what it is to be a certain sort of object of perception. For example, Haugeland thinks that being able to have a knight fork as an object of perception requires adopting a chess-playing stance toward the world, one that involves an understanding of the game of chess. If one cannot adopt this stance, then knight forks are simply not available as possible objects of perception.

What makes Haugeland's proposed solution to the problem of objective perception truly radical is his further claim that all the objects of perception are like knight forks. Whether one is seeing a black plastic chess piece in the shape of a horse *as* black, *as* plastic, *as* horse shaped, or *as* a knight depends on taking a stance toward it. Haugeland thinks that stance taking is, in this sense, required for all forms of objective perception.

Crucially, stance taking explains not just the determinacy but also the normativity of objective perception, because taking a stance involves not just instituting but also enforcing the norms that are constitutive of what it is to be a particular sort of perceptual object. Consider, for example, how a chess-playing stance involves an intolerance of illegal moves, in the sense of not allowing oneself or other players to make such moves in a chess game. Here is how Haugeland puts the point:

Any player of chess … has an investment in the legality of all the moves in the game, regardless of whose moves they are. … In other words, if you are to keep playing—you who are *involved* in this game as a player—then you must *insist* that both your own and your opponent's moves be legal. This insistence on legality is a kind of first-person involvement that is even more fundamental to game playing than is trying to win. (*HT*, 340)

To spell this out, consider how adopting a chess-playing stance involves a reliable and resilient ability to tell the difference between legal and illegal moves and to insist on legality, on pain of giving up playing the game. This ability is reliable in the sense that players are consistently able to do it; it is resilient in the sense that players persevere in the face of difficulties or challenges to doing it. And it involves insisting on legality because one is simply failing to play chess if one permits illegal moves. Being able to adopt the chess-playing stance is, for these reasons, an achievement, one that Haugeland thinks is different in kind from the sorts of skillful coping exhibited by nonhuman animals.

In sum, Haugeland thinks that the ability to take a stance toward the world is what explains how we are able to perceive knight forks *as* knight forks, as well as all the other possible objects of human perception.

The Problem with Neo-pragmatism

Convinced that genuine science is more than a social institution, anti-pragmatists (such as Heidegger and I) [hold] that there must be two fundamentally distinct sorts of normative constraints: social propriety and objective correctness (truth).

—John Haugeland, "Truth and Rule-Following" (*HT*, 317)

Haugeland's early and late views on objective perception both emphasize the significance of stances for making sense of the determinacy and normativity of perception. They differ, however, with regard to the extent to which they take stances to constitute what it is to be an object of perception. Whereas his early view holds that what it is to be an object of perception is *entirely* constituted by the instituted norms that a perceiver adopts toward the world, his late view aims to show how the constitution of objects of perception in (what he calls) "genuine science" involves "more than mere institution" (*HT*, 318). In this subsection, we explain Haugeland's reasons for revising his early view. In the next subsection, we explain his late view.

To understand Haugeland's reasons for revising his early view, it is crucial to appreciate just how seriously he takes the determinacy aspect of the problem of objective perception. For Haugeland, an adequate solution to the problem of objective perception cannot begin by supposing that the objecthood of the objects of perception is simply given to us by the world. His early view centers on avoiding this supposition at all costs. The examples that he invokes (such as perceiving the state of play in a chess game) are all designed to emphasize the role that we play in determining what it is to be an object of perception. For example, consider what he says about the case of perceiving a bodily gesture as a greeting:

What it is to greet someone, and what it is to be a circumstance in which a greeting is appropriate, are nothing other than what the community members accept and deem as such—which is to say, they are themselves instituted along with the normative practices in which they occur, and by the same socializing process. (HT, 311–312)

What it is for something to be a greeting is entirely constituted by social norms. If someone's gesture accords completely with the norms for being a greeting, then that gesture is a greeting; there is no room for the possibility of gestures that accord completely with the norms for being a greeting but are not greetings.[35] Haugeland's early view generalizes this sort of account and holds that the same sort of constitutive dependence is true for all forms of objective perception.

Haugeland's reasons for revising his early view can now be straightforwardly stated. In his early view, "the instituted [norms] themselves have no *independent* criterial status at all" (HT, 314). The problem is that the instituted norms for what it is to be a greeting (for instance) are not *themselves* beholden to the world, in the sense that there is no room for the possibility of discovering that these norms are themselves wrong about how the world is. As Haugeland puts this point, "There is no room for a greeting … to have been performed correctly, according to the instituted norms for greetings, and yet to have got those [norms] wrong" (HT, 314). The problem here is not just that Haugeland's early view reduces truth to consensus; the problem is that it makes no room for the possibility of social practices that aim to arrive at instituted norms that are themselves beholden to how the world is. Haugeland refers to such practices as "genuine science." His early view does not have any resources for explaining what makes genuine science distinct from manifestly consensus-based social practices (such as the practice of greeting or the practice of playing chess).

In sum, Haugeland's early view treats the norms for the constitution of all objects as being on par with the norms for being a greeting or being a knight fork. It does not discuss scientific practices that essentially involve raising the question of whether there might be some better way of understanding what we take ourselves to be perceiving within a certain domain; there is no room, for instance, within the chess-playing stance for raising the question of whether there might be some better way of understanding what we take to be knights or knight forks. In short, his early view is silent on the question of how the world might *resist* the instituted norms we adopt to make sense of it.[36] The question of whether there might be a better set of instituted norms for perceiving what we take ourselves to be perceiving is a question that simply does not come up in the context of the chess-playing stance, but is a question that Haugeland thinks stands at the center of genuinely scientific practices. His late view centers on making room for this question, without thereby supposing that the objecthood of objects is simply given to us by the world.

A Beholdenness Theory of Truth

What *objectivity* demands ... is that the "objects" of objective tellings should have a determinacy and normative standing *independent* of the performance norms for those tellings, such that the tellings could be performed properly, and still get their objects wrong.

—John Haugeland, "Truth and Rule-Following" (*HT*, 314)

Haugeland's beholdenness theory of truth and objectivity is an original contribution to philosophy, one that we can offer only a cursory summary of here. Our hope is that this short summary will encourage others to read Haugeland's much longer exposition of his theory in "Truth and Rule-Following," as well as provide some guidance for doing so.

One way to begin to appreciate Haugeland's theory is to note the ingenious way in which he returns to the example of playing chess. He begins by cautioning that "the use of games as examples is fraught with philosophical peril" (*HT*, 320).[37] He then exploits this peril by turning the chess example on its head and using it to illustrate some of the ways in which genuine science is fundamentally different from playing chess. He does this by asking us to imagine a bizarre variant of chess, one that he calls *automatic, esoteric, empirical* chess (hereafter AEE chess).

Haugeland's variant is automatic in the sense that the pieces move on their own. It is esoteric in the sense that the self-moving pieces need not be instantiated in anything like the sorts of things that chess pieces are ordinarily instantiated in. Rather, part of the challenge of being able to play this bizarre variant of chess is coming to acquire the ability to perceive these pieces as pieces, and moves as moves, in whatever media they might occur. Importantly, there need not be any independent check on whether someone has acquired these perceptual skills; these skills "will be 'compelling' just in case, via them, the players can and do actually play [this variant of] chess with one another" (*HT*, 328). Finally, it is empirical in the sense that the rules of the game are nothing at all like the normal rules of chess. Rather, the very point of playing this game is to work with others to figure out what the rules are, that is, what rules best make sense of the automatic movements of the pieces in their esoteric media. Accordingly, with AEE chess, "the interesting challenge is not so much winning the game as figuring out what it is—that is, how to play it. Or—more to the point—it can be part of the challenge to ascertain whether there's a game at all or not, and, if so, what belongs to it and what doesn't" (*HT*, 330). Crucially, not just any rules that one might come up with will work; since the pieces are not moved by the players, the real difficulty is identifying what the pieces are, as well as how they are moving. For this reason, being able to play AEE chess at all is an accomplishment; as Haugeland puts the point, "Finding a game that is playable—is ... a kind of *achievement*, one that includes an element of 'discovery' about the world" (*HT*, 331). The discovery is that the pieces are moving themselves in accordance with rules at all (i.e., are parts of an intelligible game), and that these are the same rules that the players come up with in playing the game.

Haugeland's preliminary point in introducing AEE chess should now be evident: it is an unusual way of describing in extremely general terms the scientific practice of working with others to arrive at empirical explanations of the behavior of observable phenomena (in terms of lawlike regularities that predict what will and will not happen), in a way that does not take the objecthood of the objects of scientific explanation to be simply given to us by the world. What makes AEE chess truly philosophically provocative, however, is Haugeland's further suggestion that, just as with normal chess, players of AEE chess need to be intolerant of illegal moves. That is, players of AEE chess need to have a reliable and resilient ability to tell the difference between legal and illegal moves, and to insist on legality, on pain of giving up playing the game. This suggestion is provocative because illegal moves in AEE chess are not due to any of the players: since the pieces are self-moving, if a piece makes an illegal move, then that can only be because *the world* is breaking the rules. That is, since the whole point of playing AEE chess is to work with others to figure out what the rules are, the discovery of an illegal move in AEE chess is the discovery of a way in which the world is resisting the instituted norms we adopt to make sense of it. It is, in other words, an example of precisely the sort of beholdenness to the world that Haugeland's early view lacks.

The ability to recognize illegal moves in AEE chess is probably the short-est possible introduction to Haugeland's beholdenness theory of truth and objectivity, because it vividly illustrates the sorts of skills that Haugeland thinks are required to engage in genuine science. In particular, it illustrates why Haugeland thinks that these skills must go far beyond differential responsiveness. He elaborates on this point by reflecting on an apparent paradox that is involved in perceiving illegal moves (in either normal or AEE chess):

> The very idea of an illegal move flirts with paradox. That there should be chess phe-nomena at all *presupposes* that they accord with the rules (standards) that constitute chess as such. Nothing is so much as intelligible, let alone recognizable, as a chess phenomenon outside of the domain constituted according to those standards. ... Hence, strictly speaking, illegal moves are impossible. Yet they must be recogniz-able—and hence (in at least that sense) something that conceivably "*could*" occur—if the ruling-out is to be nonvacuous. That's the paradox. (*HT*, 332)

Haugeland's solution to this apparent paradox is to introduce a distinction between two types of skills: *mundane* and *constitutive* skills. Whereas "*mundane* skills ... are the resilient abilities to recognize, manipulate, and otherwise cope with phenomena within the game ... in effect, the ability to engage in play" (*HT*, 323), "a *constitutive* skill is a resilient ability to tell whether the phenomena governed by some constitutive standard are, in fact, in accord with that standard" (*HT*, 323). The distinction between these two types of skills makes the perception of illegal moves possible, because it explains how the ability to recognize pieces and moves extends beyond the space of legal moves. Whereas mundane skills make it possible to recog-nize a piece as, say, a rook, and its movement as diagonal, constitutive skills make it possible to recognize this combination as illegal. Haugeland refers to the space of recognizable but illegal moves as "the excluded zone" (*HT*, 337).

Combining the ideas just introduced brings out how the excluded zone in AEE chess identifies the precise way in which a genuine scientific prac-tice "can stick its neck out empirically, by giving constituted phenomena [the] power to resist or refute [instituted norms]" (*HT*, 338). This is because the ability to recognize illegal moves in AEE chess is the ability to recognize that the instituted norms of playing AEE chess are themselves wrong about the world. As Haugeland puts this conclusion, the excluded zone in AEE chess explains how

objective phenomena [can be] both accessible as normative criteria and literally *out of control*. The constitution of the domain determines what it is for them to be or to behave in this way or that; but whether they then *do* or not is "up to them"—and skillful practitioners can *tell*. (*HT*, 347)

In sum, the ability to recognize illegal moves in AEE chess shows how it is possible for truth and understanding to be dependent on human social practices while at the same time beholden to how the world is. It is in virtue of explaining how understanding can be beholden to the world in this manner that Haugeland's late view fundamentally moves beyond neo-pragmatism.

6   Commitment

Promising and many other valuable institutions ... are made possible by socially imposed norms. ... What sets Nietzsche [and his account of promising] apart, however, is that [his] account is only preparatory for another and quite different understanding of what promising can be and mean: the self-responsible exercise of an autonomous protracted will—the act of a sovereign individual with the *right* to make promises. Social sanctions, while prerequisite for the genesis of this capacity, are left entirely behind in its maturity.

—John Haugeland, "Two Dogmas of Rationalism" (this vol., 302)

Haugeland's beholdenness theory of truth has two sides. On the one hand, the theory offers an account of the nature of the objective world. Haugeland proposes that the objective world is not simply given to us and does not preexist human social practices of making sense. Rather, the objective world is constituted by a set of instituted norms that are criterial for what is to be an object of perception. But the constitution of the objective world is only half the story, because these norms also require a subject to institute them. As he puts it, "The constituted objective world and the free constituting subject are intelligible only as two sides of one coin" (*HT*, 6). So, on the other hand, the theory also offers an account of the nature of subjects that are capable of objective perception. Haugeland's account requires that these subjects be able "to take responsibility for the norms and skills in terms of which one copes with things" (*HT*, 2). His name for undertaking this responsibility is "existential commitment" (*HT*, 2).

Haugeland's discussion of existential commitment takes place in the context of debates about how to properly understand the significance of Heidegger's work for cognitive science and the philosophy of mind. Haugeland's primary opponents in these debates are those who think that nothing is lost, and perhaps much is gained, if the existentialist tendencies in Heidegger's work are overlooked. For example, Dreyfus (with Jane Rubin) argues in an appendix to *Being-in-the-World* (1991) that Heidegger's existentialist tendencies are confused, and concludes by suggesting that "later Heidegger seems to have recognized this," and for that reason, "he gives up his existential account" (336). Haugeland rejects this way of inheriting Heidegger, on the grounds that existential commitment is "the fulcrum of [Heidegger's] entire ontology" (*DD*, 188).

In this section, we spell out why Haugeland thinks existential commitment is essential for objective perception. We begin by looking at "Authentic Intentionality" (2002), in which Haugeland argues that existential commitment is a necessary part of genuine science. We end by looking at Haugeland's longest completed work on Heidegger, "Truth and Finitude" (2000), in which he argues that an awareness of death is a condition for the possibility of truth existing at all.

Authentic Intentionality

Cognitive science and artificial intelligence cannot succeed in their own essential aims unless and until they can understand and/or implement genuine freedom and the capacity to love.

—John Haugeland, "Authentic Intentionality" (*DD*, 274)

In this essay, Haugeland summarizes his own "positive account of what original intentionality requires and what makes it possible" (*DD*, 262). The summary is structured around spelling out the role of commitment in his account, with the ultimate goal of highlighting why original intentionality requires existential commitment. In the previous section, we saw how Haugeland's late view holds that original intentionality imposes demands on the object to abide by the norms of genuine science. But his late view makes equally strong demands on the subject to lay down, preserve, and revise these norms. Haugeland spells out these demands on the subject by distinguishing three levels of self-criticism that thinkers are capable of: first, enforcing the norms of a social practice; second, modifying the norms of a practice to improve the likelihood of its success; and third, being willing to give up entirely on a practice when recognizing that "success" within it would not involve arriving at the truth.

The first level of self-criticism involves the sort of censoriousness discussed in section 3. It is primarily manifested through the ability to criticize oneself or others for failing to abide by the norms of a social practice. Haugeland thinks that Thomas Kuhn's "normal science" (*DD*, 265) is a vivid illustration of how demanding this sort of self-criticism can be, insofar as it requires practitioners to "carefully scrutinize actual procedures to ensure that they are in accord with its norms of proper performance" (*DD*, 265).[38] Normal science illustrates this sort of scrutiny in a variety of ways, from expecting experimental results to be replicable to the use of blind peer review in academic publishing.

The second level of self-criticism involves "critically scrutiniz[ing] not only individual performances but also the very practice itself" (*DD*, 266). The idea is that being a scientist requires more than merely ensuring that the norms of a scientific practice are followed; it also requires critically examining the norms themselves. As Haugeland puts it, "Scientists turn their scrutiny not on individual performances but on the very norms that determine procedural propriety for all performances of that sort" (*DD*, 266). The choice of norms is, by and large, determined by the contribution that they make to the success of a scientific practice. For example, if it turns out that blind peer review tends to have a stifling effect on the production of knowledge, then this norm might be revised or abandoned altogether.

The third and final level of self-criticism involves the willingness to put an entire scientific practice into question. That is, scientists must be prepared to recognize that despite their best efforts to modify the norms of a practice, it might be impossible to refine them in a way that works to get at the truth. In such a case, it becomes necessary to give up the entire practice. This sort of self-criticism involves a manifestly existential commitment, in that

it is an *honest commitment*—in the sense of resolve or dedication—to making something work, on pain of having to give the whole thing up. Such honest commitment is "double-edged"—it cuts both ways—in that, first, it requires honest and dedicated effort to making it work, and yet, second, it also requires the honest courage, eventually, to admit that it cannot be made to work—if it cannot—and then to quit. (*DD*, 274)

This final level of commitment is distinct from the first and second levels, because it threatens the *very being* of the scientist *as* scientist: "Who, after all, are you, professionally, if your professional specialty dies?" (*DD*, 271). The understanding of objects and the self-understanding of the scientist are thoroughly intertwined in a scientific practice, and both stand and fall together. It is in this sense that authentic intentionality requires undertaking an existential commitment. This existential commitment involves the risk that one's scientific way of life—and therefore one's identity as a scientist—may have to be abandoned.

The first and second levels of self-criticism concern what Heidegger calls the *ontic* level of reality. They involve getting the facts right about the entities within a practice. But the third level of self-criticism concerns the *ontological* level of asking whether the practice should exist at all. This involves asking what, if anything, the practice really tracks. Haugeland thinks that questions at the ontological level require an extreme form of existential commitment, one that is freely undertaken in the sense that the choice to inaugurate, participate in, or abandon a practice is not forced on one by the world. As a way of bringing out the type of freedom involved in this extreme form of existential commitment, Haugeland connects it with a more general human ability: the ability to love something or someone. It is in this sense that Haugeland thinks that, at bottom, "*Love* is the mark of the human" (*HT*, 2).

Truth and Finitude

I have already mentioned alchemy as a dead science, but I did not ask how it died. No doubt the world-spirit was somehow moving on, but down here on the ground, alchemy went out of business because it could not keep its promises. Does physics make promises? Of course it does—more precise and more demanding than any other empirical enterprise in human history. That is why it is the king of the sciences; yet, as we know, kingdoms do not always last.

—John Haugeland, "Death and Dasein" (*DD*, 185)

For Haugeland, entities exist only for beings who comport themselves toward them. In the case of chess, this is trivially true: rooks exist only within the practice of playing chess. But Haugeland takes this idea to include *all* the entities that humans are capable of thinking about, including scientific entities; it is existential commitment that brings entities into being and makes them intelligible. This leads to a final point Haugeland takes from Heidegger: the way in which an awareness of the possibility of death is a condition for the possibility of truth existing at all.

Haugeland begins his most extended discussion of this topic, "Truth and Finitude," by noting that many other Heideggerians read the existential dimensions of Heidegger's thought as something that should be overlooked or ignored. Haugeland argues against this reading, on the grounds that all of Heidegger's existential themes—responsibility, ownedness, anxiety, and death—are ultimately necessary for understanding truth.

This is not to say that they are necessary for understanding every aspect of human life. In much of our day-to-day existence, we need only care about whatever our society tells us to care about, impersonally abiding by social norms. Impersonally caring about these norms need not be uncritical (as the first level of self-criticism discussed earlier illustrates). But this sort of criticism does not involve ourselves as *individuals*—as the specific person each of us is. Rather, it only involves us as a particular kind of person—a student, a teacher, an administrator, and so on. We do not need to take responsibility for these practices as distinctively our own; they are just the social realities we inhabit as conformists within a society already teeming with well-established institutional roles.

However, certain social practices do make demands on us personally. Specifically, practices that are beholden to truth require an existential commitment on our part. Who we are and what we do are intertwined with these entities. When we encounter a breakdown in these practices, it strikes us as a threat to us as a person. As Heidegger puts it, it makes us *anxious*. Haugeland writes, "In anxiety, a person's individuality is 'brought home' to the person in an utterly unmistakable and undeniable way" (*DD*, 207–208). In a breakdown, the demand to account for—and possibly correct—our understanding of the world must be faced head-on.

To understand the existential aspects of Haugeland's thought, it will help to move beyond the case of genuine science and consider instead the case of a lover and beloved. The lover is committed to a unique entity in the beloved. The lover is committed not merely in an impersonal sense, toward a generic instance of a thinking being, but in the existential sense of caring about the beloved as the singular being who makes her happy, to whom she has a distinctive commitment, who makes her life meaningful, and who feels the same about her (or so she imagines).

Anxiety arises when the lover's understanding of the beloved is challenged. For example, if the lover sees the beloved kissing someone else on the lips, she might experience this act as incompatible with her understanding of their relationship. Some incompatibilities can be brushed away as simple mistakes (she mistook a stranger for the beloved) or as acceptable (the beloved was joking around, or acting in a play). These correspond to the sort of double-checking involved in the first level of self-criticism discussed earlier. Other incompatibilities may require revising some of the norms governing their relationship: for instance, they may find that a commitment to monogamy has a stifling effect on the happiness of their relationship, and therefore cease to abide by this norm. This corresponds to the sort of revisions involved in the second level of self-criticism. But not every incompatibility or every revision can be accepted. Some put the whole relationship into question and can lead the lover to ask herself whether the beloved really loves her, or whether she really loves the beloved. What distinguishes a genuine relationship from a fraud is the same thing that distinguishes genuine science from a counterfeit: beholdenness to the truth and, with it, the possibility of failure. In the fraudulent case, the lover can always convince herself that the incompatibility is not a big deal, that it is excusable, that it never happened, and so on. The relationship keeps going because she is irresponsible or, as Haugeland puts it, "bullheadedly refus[es] even to *see*" (*DD*, 216).

In the case of genuine love, the lover must accept that there is a real possibility that she cannot be beholden to the truth *and* persevere in the relationship. The commitments made in a relationship—some explicit, some implicit—are the promises that establish the standards for the relationship to be genuine. Accepting that the beloved may not keep these promises—that the relationship might fail—is what Haugeland calls *responsibility*. Responsibility requires that the lover treat the beloved as genuinely free from how she represents the beloved as being. Haugeland writes, "Taking responsibility resolutely means living in a way that explicitly has everything at stake" (*DD*, 216). Being responsible means being open to the possibility that the incompatibilities may be impossibilities: that either the lover or the beloved cannot—or will not—remain committed to the relationship. If either consistently fails to keep these commitments, then the relationship should be abandoned. The recognition that the relationship may fail is accepting its *finitude*. Haugeland argues that responsibility requires accepting that our practices exist only insofar as we persevere in them, and we should only persevere in them so far as they are beholden to the truth. For Haugeland, taking any project seriously—"owning it" as one's own—requires accepting the possibility of its demise ("ownedness" is Haugeland's late translation of *eigentlich*).

But the lover's acceptance of the possibility of the relationship's demise, however, also entails accepting *her own* demise, insofar as the identity of anyone who is in a serious relationship is inextricably bound up with that relationship. The same is true in science: the death of a scientific practice means the death of a way of life. A scientist's identity—who she is, what matters, what really exists—is all tied up with her scientific project. The Freudian who gives up psychoanalysis loses more than her job; she loses her grasp on what it is to be human. But this possibility of giving up on something we care about deeply is requisite for being beholden to it in the way that is required for truth. Haugeland writes:

Ontological truth is beholden to *entities*—the very same entities that ontical truth is beholden to and via the very same means of discovery. The difference lies in the character of the potential failure and the required response. A failure of ontical truth is a misdiscovery of an entity, such as a factual mistake. With more or less work, it can be identified and corrected, and life goes on. A failure of ontological truth is a systematic breakdown that undermines everything—which just means a breakdown that *cannot* be "fixed up with a bit of work." So the only responsible response (eventually) is to take it all back, which means that life, *that* life, does *not* "go on." (*DD*, 218)

Although we are often oblivious to these concerns, their life-and-death dimensions become obvious in a breakdown (or breakup). The relationship ends, the scientific project falls apart, the department dissolves. In each case, *that* life is over; it cannot be continued. Haugeland writes, "Resolute being toward death is the condition of the possibility of ontological *truth*" (*DD*, 219).

This is Haugeland's ultimate response to his own early view: while a kind of derivative intentionality can be found in sociality, original, authentic intentionality requires that humans be willing to stake their own individual lives on the success of the social practices and institutions that they think matter. Discussing this point at the end of "Two Dogmas of Rationalism," Haugeland writes, "Taking—nay, not just taking but actively pursuing—this fundamental risk, a risk even to career and profession, demands something like personal responsibility and integrity—if not of all scientists always, at least of some scientists sometimes" (this vol., 308). Haugeland's late view introduces a revolutionary dimension to the core of human life: it is because we are willing to risk all, and genuinely open to the possibility of failure, that truth is *possible at all*. "The essential move must be to understand truth itself as a distinctively human achievement" (this vol., 306). It is only because we *give a damn* that truth, objectivity, philosophy, love, morality, art, and other institutions of human value exist at all. And it is only because we are existentially committed to these institutions—that we are willing to risk everything for them—that they persevere.

Notes

1. Haugeland uses many terms in his own, idiosyncratic, manner, which it helps to keep in mind when reading him. He uses "entity" in the broadest possible sense, to refer to anything that can be thought about.

2. Haugeland's most sustained discussion of this presupposition is "The Nature and Plausibility of Cognitivism" (in *HT*).

3. Haugeland's most sustained discussion of this presupposition is "Representational Genera" (in *HT*).

4. See, e.g., Haugeland's (2004) review of Vincent Descombes's *The Mind's Provisions*, in which Haugeland distinguishes his own critique of GOFAI from Descombes's on precisely these grounds.

5. Chapter 9 of McCorduck 2004 gives a useful summary of this aspect of the reception of Dreyfus's book.

6. Haugeland discusses the frame problem in detail in "An Overview of the Frame Problem" (1987).

7. In this respect, Haugeland's critique of GOFAI is notably different from that offered by John Searle's (1980) "Chinese room" thought experiment. For Haugeland's criticisms of Searle's critique of GOFAI, see Haugeland 1980, 2002.

8. In this essay, Haugeland discusses linguistic meaning at a fairly "big-picture" level. He does not purport to be offering a worked-out theory of linguistic meaning; when he speaks of "the meaning" of a passage, we take him to be referring to the content of a passage when it is uttered or written.

9. As we will see, Haugeland's fondness for using chess, and imaginary variants of chess, as philosophical objects of study was a consistent theme throughout his career. This is only the first of many chess examples that we will discuss in this introduction.

10. Haugeland's essay "Pattern and Being" (collected in *HT*) is his most in-depth discussion of this point.

11. As he later puts this point, "The fact that [chess pieces] coincide for a while with—are 'implemented in' or 'carried by'—plastic figurines (or whatever) is a practical convenience and an ontological distraction" (*HT*, 329).

12. This is a simple example of the sort of case that Haugeland discusses in much more detail in "Weak Supervenience" and "Ontological Supervenience" (both collected in *HT*) as a way of arguing against token-identity theory.

13. Since the publication of "Understanding Natural Language," commonsense holism has been much more extensively explored in the work of contextualists in the philosophy of language. For a useful overview, see Borg 2012.

14. Since the publication of "Understanding Natural Language," situation holism has been much more extensively explored in the work of relevance theorists in the philosophy of language. For a useful overview, see Scott-Phillips 2014.

15. Haugeland uses phrases like "having a human mind," "having a mind," "having thought," "having intentionality," "being a person," and "being human" interchangeably, and we will follow him in this usage.

16. See Heidegger 1927.

17. See Sellars 1963.

18. Here is how he puts this observation in a later work: "People—and only people—can be members of a huge variety of different institutions and organizations, including quite a few at the same time. Now, why is that? The answer—incontestable, I think, once you attend to it—is that human beings can learn to abide by norms, mostly from their elders and peers. Indeed, to a large extent, they can't help it, not, mainly, because compliance is imposed on them (though there's some of that), but because they are … norm-hungry. It is precisely by virtue of being 'normalized'— which is to say, standardized in highly specific ways—that people can reliably function as … institutional 'cogs.' … Such normalizing is a pervasive and basic feature of humankind." Haugeland, "Andy Clark on Cognition and Representation" (*PMR*, 30–31).

19. With Daniel Dennett, Haugeland coauthored a useful introduction to the topic of intentionality. See Dennett and Haugeland 1987.

20. These names derive from the earlier, less-worked-out views that these approaches inherit and develop.

21. Haugeland's most extensive critical engagement with neo-Cartesianism is his *Artificial Intelligence: The Very Idea (AI)*.

22. Haugeland's most extensive critical engagement with neo-behaviorism is his long book review of Dennett's *Brainstorms*. See Haugeland 1982.

23. Moreover, at the end of the essay, Haugeland suggests the possibility that the three approaches might not conflict and might merely describe different kinds of original intentionality. According to this suggestion, his own early view would be privileged only in that it describes a "higher" form of intentionality (*HT*, 161).

24. Haugeland uses this proposal to make the general point that "what makes any holism nontrivial is the simultaneous imposition of two independent constraints" (*HT*, 134).

25. For the sake of argument, let us assume that the marks in this volume can be so interpreted.

26. Shapiro 2011 is a useful introduction to this alternative research program.

27. Haugeland draws on J. J. Gibson's (1979) notion of "affordances" as a way of spelling out this idea of attunement.

28. Since Haugeland holds that representation cannot exist without the possibility of misrepresentation, this would make representation impossible as well.

29. Rorty himself did not take this predicament to be problematic. Haugeland's later rejection of neo-pragmatism is based, in part, on the recognition that this predicament *is* problematic for any adequate account of truth and objectivity.

30. A knight fork is a situation in chess in which a knight simultaneously threatens two opposing pieces.

31. Dretske (1981) gives an influential introduction to this aspect of the problem of objective perception. Haugeland critically engages with Dretske's discussion of the problem in "Objective Perception" (in *HT*).

32. Davidson (1980) gives an influential introduction to this aspect of the problem of objective perception.

33. Smith (1998) gives a useful introduction to this aspect of the problem of objective perception.

34. This starting point is shared by many of Haugeland's other colleagues at the University of Pittsburgh, including Wilfrid Sellars, Robert Brandom, and John McDowell. For a useful introduction to this aspect of the Pittsburgh School of Philosophy, see Maher 2012.

35. In Haugeland's view, it is even slightly misleading to say that perceiving a greeting involves perceiving a bodily gesture as a greeting, since he holds that it is

possible to perceive the greetingness of greetings directly—one need not first see something as a bodily gesture and then see it as a greeting. Haugeland discusses this point at length in "Pattern and Being" (in *HT*).

36. Strictly speaking, Haugeland's early view does allow for the world to resist the institution of norms in one sense: norms might make impossible demands on participants in the social practice that institutes these norms or the equipment it involves. For example, the world would probably resist playing chess with a million rows of pawns on each side or with planet-sized pieces. This aspect of Haugeland's view is usefully discussed by Joseph Rouse (2002, 244–246). However, we do not think that this qualification undermines the contrast Haugeland draws between manifestly consensus-based social practices and genuine science; insofar as social practices such as playing chess do not *aim* to arrive at norms that are beholden to how the world is, they simply presuppose such norms in order for game play to be possible at all.

37. This cautionary note is, at least in part, a criticism of his own early view, which uncritically uses the example of perceiving chess pieces and moves as a model for all objective perception.

38. Alongside Heidegger 1927, Kuhn 1962 is the work that most influenced Haugeland's views.

References

Borg, E. 2012. *Pursuing Meaning*. Oxford: Oxford University Press.

Brooks, R. 1990. Elephants don't play chess. *Robotics and Autonomous Systems* 6 (1): 3–15.

Clapin, H., ed. 2002. *Philosophy of Mental Representation*. Oxford: Oxford University Press. (Abbreviated as *PMR*.)

Clark, A., and D. Chalmers. 1998. The extended mind. *Analysis* 58 (1): 7–19.

Davidson, D. 1980. *Essays on Actions and Events*. New York: Clarendon Press.

Dennett, D., and J. Haugeland. 1987. Intentionality. In *The Oxford Companion to the Mind*, ed. R. L. Gregory. Oxford: Oxford University Press.

Dretske, F. 1981. *Knowledge and the Flow of Information*. Cambridge, MA: MIT Press.

Dreyfus, H. L. 1972. *What Computers Can't Do*. New York: Harper & Row.

Dreyfus, H. L., and J. Rubin. 1991. Appendix. In *Being-in-the-World: A Commentary on Heidegger's Being and Time, Division I*. Cambridge, MA: MIT Press.

Gibson, J. J. 1979. *The Ecological Approach to Visual Perception*. Boston: Houghton Mifflin.

Haugeland, J. 1976. Truth and understanding. PhD dissertation. University of California, Berkeley. (Abbreviated as *TU*.)

Haugeland, J. 1980. Programs, causal powers, and intentionality. *Behavioral and Brain Sciences* 3 (3): 432–433.

Haugeland, J., ed. 1981. *Mind Design: Philosophy, Psychology, Artificial Intelligence*. Cambridge, MA: MIT Press.

Haugeland, J. 1982. The mother of intention: Review of Dennett's *Brainstorms*. *Noûs* 16 (4): 613–619.

Haugeland, J. 1985. *Artificial Intelligence: The Very Idea*. Cambridge, MA: MIT Press. (Abbreviated as *AI*.)

Haugeland, J. 1987. An overview of the frame problem. In *The Robot's Dilemma: The Frame Problem in Artificial Intelligence*, ed. Z. W. Pylyshyn, 77–93. Norwood, NJ: Ablex.

Haugeland, J., ed. 1997. *Mind Design II: Philosophy, Psychology, Artificial Intelligence*. Cambridge, MA: MIT Press.

Haugeland, J. 1998. *Having Thought: Essays in the Metaphysics of Mind*. Cambridge, MA: Harvard University Press. (Abbreviated as *HT*.)

Haugeland, J. 2002. Syntax, semantics, physics. In *Views into the Chinese Room: New Essays on Searle and Artificial Intelligence*, ed. J. M. Preston and M. A. Bishop, 379–392. Oxford: Oxford University Press.

Haugeland, J. 2004. Closing the last loophole: Joining forces with Vincent Descombes. *Inquiry* 47:254–266.

Haugeland, J. 2013. *Dasein Disclosed*. Cambridge, MA: Harvard University Press. (Abbreviated as *DD*.)

Heidegger, M. 1927. *Sein und Zeit*. Tubingen: Max Niemeyer.

Kuhn, T. 1962. *The Structure of Scientific Revolutions*. Chicago: University of Chicago Press.

Maher, C. 2012. *The Pittsburgh School of Philosophy: Sellars, McDowell, Brandom*. New York: Routledge.

McCorduck, P. 2004. *Machines Who Think*, 2nd ed. Natick, MA: A. K. Peters.

Rorty, R. 1999. *Philosophy and Social Hope*. New York: Penguin.

Rouse, J. 2002. *How Scientific Practices Matter: Reclaiming Philosophical Naturalism*. Chicago: University of Chicago Press.

Scott-Phillips, T. 2014. *Speaking Our Minds*. New York: Palgrave Macmillan.

Searle, J. R. 1980. Minds, brains, and programs. *Behavioral and Brain Sciences* 3: 417–424.

Sellars, W. 1963. Empiricism and the philosophy of mind. In *Science, Perception, and Reality*, ed. W. Sellars, 253–329. London: Routledge & Kegan Paul.

Shapiro, L. A. 2011. *Embodied Cognition*. New York: Routledge.

Smith, B. 1998. *On the Origin of Objects*. Cambridge, MA: MIT Press.

I Heideggerian Themes

1   Anonymity, Mineness, and Agent Specificity: Pragmatic Normativity and the Authentic Situation in Heidegger's *Being and Time*

William Blattner

To cast oneself into such a role is to take on the relevant norms—both in the sense of undertaking to abide by them and in the sense of accepting responsibility for failings.

—John Haugeland, "Dasein's Disclosedness" (*DD*, 39)

One of the defining characteristics of the "California" approach to *Being and Time* (*SZ*),[1] pioneered by Hubert Dreyfus and his first generation of doctoral students, John Haugeland, Charles Guignon, and John Richardson,[2] is the central role accorded to Heidegger's concept of "the Anyone" (*das Man*).[3] California Heideggerians understand the Anyone as a pattern of social normativity that establishes the contours of our self-understanding and world. Heidegger writes: "Distantiality, averageness, leveling-down as ways of being of the Anyone constitute what we know as 'the public.' It first regulates all interpretation of Dasein and of the world and maintains its correctness in all matters" (*SZ*, 127).

Here Heidegger clearly announces that the Anyone "regulates" Dasein's daily existence. Although the paragraph from which these two sentences are drawn denies that the Anyone has an "exceptional and primary being-relationship to 'the facts,'" this does not mean that the Anyone is not both essential to and constitutive of Dasein. Thus "the Anyone is an existentiale and belongs as an originary phenomenon to the positive constitution of Dasein" (*SZ*, 129). Haugeland provided a characteristically pithy formulation of the idea in his 1982 "Heidegger on Being a Person":

The total assemblage of norms for a conforming community largely determines the behavioral dispositions of each non-deviant member; in effect, it defines what it is to be a "normal" member of the community. Heidegger calls this assemblage the anyone. (Perhaps Wittgenstein meant something similar by "forms of life.") I regard it as the pivotal notion for understanding *Being and Time*. (*DD*, 5)

One wants to know how individual persons fit into such a picture, given that human life is always carried out in the context of social regulation by the Anyone. One might be tempted to express this question in Heidegger's idiom thus: "Who is in-the-world?" However, this formulation is not right, given Heidegger's statement that "the Who is not this one and not that one, not one oneself and not several and not the sum of them all. The Who is the neuter, the Anyone" (*SZ*, 126). For this reason, I will not ask who is in-the-world but rather ask who or what the self is in *SZ*. Heidegger introduces the language of the self in §27 in the midst of his discussion of the Anyone:

The self of everyday Dasein is the Anyone-self, which we distinguish from the authentic self, that is, the self that has been properly seized upon. As Anyone-self, the current Dasein is scattered into the Anyone and must first find itself. (*SZ*, 129)

Heidegger both deploys a concept of the self alongside that of the Anyone and distinguishes between the Anyone-self and the authentic self.

The interpretation of the relationship between the individual self and the Anyone in California Heideggerianism has been guided by an important assumption about the way in which the Anyone regulates Dasein's everyday possibilities. Dreyfus, Haugeland, and Guignon all agree that the Anyone is, among other things, a repository of "generic" or "anonymous" possibilities of human life. When Heidegger writes that the Anyone "first regulates all interpretation of Dasein," California Heideggerians interpret this to mean that the background norms of our culture establish a set of "social roles" that make up the diverse possibilities of living that are available to an individual person or self in his or her culture. Heidegger's technical term for such a social role is "for-the-sake-of-which," the Californians argue. The roles of father, professor, coach, friend, and neighbor are possibilities of living in terms of which a person can understand himself or herself. They are also "anonymous" or "generic" in that they are entirely general. They are not "tailored" to individual persons; they can be occupied by anyone who can acquire and deploy the skills that constitute enacting the role. Thus Dreyfus writes:

As we have already seen, the for-the-sake-of-whichs available to Dasein are not first created by you or me, but rather are public possibilities provided by society. They make no essential reference to you or me. (Dreyfus 1991, 158)

To be a carpenter, one must acquire the cutting, measuring, sanding, and fitting skills that carpenters deploy in the course of their business. Deploying those skills for the purposes to which they are put in the social role of a carpenter makes one a carpenter. If (as California Heideggerians agree) understanding is know-how or competence, then, Haugeland argues, we may construe self-understanding as "knowing how (in each case) to be 'me.'"

And what know-how is that? According to Heidegger, any and all know-how that I may have is ipso facto some portion of my knowing how to be me. If I understood race cars in the way that mechanics do, then I would know how to be a race-car mechanic—which, in part, is what I would be. (*DD*, 13)

All know-how, all skill or ability, inheres in a social role. Mastering those skills and abilities is mastering the role. And mastering the role is being a carpenter, a father, or what have you.

Now, notice that if all the abilities-to-be (*Seinkönnen*)[4] that Dasein can master adhere to social roles, there are no special or distinctive abilities left over to constitute me, the individual self. What then am I, in contrast with you? Guignon adopts an interpretation of the self as simply the "crossing point" of multiple social roles, the co-instantiation of multiple roles in a single agent.

In Dilthey's language, the self is a representative of social systems and a crossing point of the structural types built into the cultural world. ... Since the structuring of roles and the criteria for operating in the world are applicable to anyone whatsoever, I am not in any sense unique in my ordinary ways of Being. ... It follows, then, that my Being in everydayness is "representable" or "delegatable" (vertretbar) (126); because the self is nothing other than an exemplification of forms of life that are essentially public, anyone can fill in for me or take my place. (Guignon 1983, 108–109)

Since each of the roles that I instantiate is anonymous or generic, I can be "represented" by others in everything I do. I am a professor, and part of being a professor who has been assigned section 1 of Intro to Philosophy is that I must lead a class session on Wednesday, September 4. There is nothing special about the role that requires me to lead the session. I could ask my colleague Kate to lead the class for me that day. She can represent me in the role.

Surely the proposal to generalize this sort of representability to all an individual's roles and relationships is an overgeneralization, however. It is just implausible to say that I can be represented in everything I do. Can I really delegate my role in love, friendship, and fatherhood to others? I cannot say to my friend Steve, "Look, I can't make the anniversary dinner with my wife this evening. Would you go in my place?" There must be something more to my individuality than simply being the coexemplification of multiple social roles. We must say something about what makes any case of coexemplifying multiple roles the living of a human life by a single person, me. That is, we must say something about what Heidegger calls "in-each-case-mineness" (*Jemeinigkeit*, or just "mineness" for short).

1   Guilt, Conscience, and Responsiveness to Norms

Haugeland's first attempt to address this question turned on the notion of a "unit of accountability," but as we shall see, he developed this notion initially in too behavioristic a fashion.

Each unit of accountability, as a pattern of normal dispositions and social roles, is a subpattern of Dasein—an institution. But it is a distinctive institution, in that it can have behavior as "my" behavior, and can be censured if that behavior is improper; it is a case of Dasein. (*DD*, 11)

What makes a piece of behavior "mine" is that "I" can be censured for it. What does it mean to say that "I can be censured" for a piece of behavior? Haugeland uses disciplining children as an example: when a child puts his hand in the cookie jar in contravention of parental rules, the parent might spank the child's bottom. (Haugeland was reared in the 1940s and '50s!) Why spank the bottom when it was the hand that committed the crime? Because the bottom and the hand "belong to the same person." But what does that mean? It means that inflicting a deprivation on the bottom ("censuring it," in Haugeland's language) tends to generate conformity in the hand. In this initial statement of the idea, then, the unity or integrity of the individual person consists in its responding to deprivations inflicted on any part of it. This way of framing the idea, however, is reductively behavioristic, and in particular, it does not distinguish responding to punishment from taking responsibility for conforming to the norms whose violation leads to punishment.

By 1990 Haugeland had begun to emphasize this constitutive investment of the individual self in the roles that he or she plays: "To cast oneself into such a role is to *take on* the relevant norms—both in the sense of undertaking to abide by them and in the sense of accepting responsibility for failings" (*DD*, 38). Let us view Haugeland's statement here as a challenge: we must say something about the individual self's responsiveness to the norms that constitute the Anyone. Unsurprisingly, it was Haugeland himself who made the signal contribution to addressing this challenge, though it took him another ten years.

The first iterations of California Heideggerianism regarded Division II of *Being and Time* as an existentialist addition to Division I, something that one could ignore if one's interest were primarily in the traditional philosophical questions of the nature of truth, the constitution of the self, and the intelligibility of the world. Division II was treated as pursuing vaguely Kierkegaardian or Nietzschean reflections on existential authenticity. Non-California Heideggerians had long held to the indispensability of Division II, but typically the connection was taken to be something rather different. For example, if everything Dasein does in its everyday life is fleeing from anxiety, hiding from a deep truth, then the "hermeneutics of suspicion" (Ricoeur) in Division II unmasks the charade of everyday human life. Haugeland's breakthrough insight was to see that what Heidegger calls "guilt" and "conscience" are conditions of the possibility of the phenomena analyzed in Division I. They are not (or perhaps not only) modes of access to some dark underbelly of everyday life but phenomena in virtue of which individual agents are able to live everyday lives in the first place, irrespective of any deep truth from which they may also be hiding.

Dasein does not just robotically execute normative rules. Yes, primarily and usually we do unthinkingly conform to the normative expectations of our social environment. I do not have to think to avoid stripping down to my underwear during an Intro to Philosophy lecture, nor do I have to think before I enter a building via the door, rather than the window. One of the principal phenomenological insights of Division I is that we mostly conform to social norms unreflectively. Nonetheless there is a difference between conforming to a norm and executing programming.[5] To conform to a norm is to be responsive to normative valences (stripping down while lecturing would be shocking) or statuses (the instructor is permitted to call on students during the lecture). Unlike natural properties and powers, normative valences and statuses can be defied. Normative valences and statuses call upon agents to comport themselves in specific ways, and these calls can be defied or flouted or simply ignored. There must be some facet of Dasein's disclosedness that unveils normative valences and statuses to us. Haugeland identified disposedness (*Befindlichkeit*, which he renders as "findingness" in "Truth and Finitude") as the medium of such disclosedness. A concrete determination of disposedness is an attunement. Attunements "tune Dasein in" to what matters in a situation; that is, they reveal normative valences and statuses in the broadest sense of the term.

To be tuned in to what matters is not just to "detect" some feature of the situation. It is to be responsive to it, to be guided by it. We are not guided by normative valences in the way in which iron filings can be guided by the movement of a magnet. We can defy normative valences. To find oneself standing under a commitment is to disclose the possibility of flouting it as unacceptable. "But a responsiveness that finds what is ruled out in the responding entity's own actions to be unacceptable *to that entity itself* is *responsibility*" (*DD*, 204). Haugeland does not interpret any particular piece of Heideggerian terminology as designating this responsibility, but I have argued elsewhere that this is what Heidegger means by "guilt."[6] Haugeland interprets "guilt" as what he calls "existential responsibility," which is "responsibility for its own self as a whole, for who it is" (*DD*, 208). Heidegger, however, does not put any reference to the "whole self" into his official definition of "guilt": "Hence, we determine the formal existential idea of the 'guilty' thus: being-the-ground for a being that has been determined by a not—that is, *being-the-ground of a nullity*" (*SZ*, 283), which he also subsequently glosses as the "essential nullity of projection" (*SZ*, 285). To understand what this means, we must understand in what way Dasein is the "ground of a nullity."

The key to understanding the word "null" is that the German word (*nichtig*) is just the adjectival form of "not" (*nicht*). To be null means to be "not-ish,"[7] or "characterized by a 'not'" (*SZ*, 283), as Heidegger glosses it. "Projection" (*Entwurf*) refers to our pressing forward into (including choosing) our possibilities. So to say that our "projection is essentially null" is to say that our forward motion in life is characterized by a "not." What "not" is this? One "stands in each case in one or another possibility, [one] constantly is *not* another, and [one] has relinquished it in [one's] existentiell projection" (*SZ*, 285). When Heidegger subsequently describes Dasein's freedom as residing "in the choice of the one [possibility], that is, in bearing not having chosen and not being able to choose the other [possibilities]" (*SZ*, 285), he is harking back to a passage where he sent a clear signal about his Kantian interpretation of freedom:

Possibility as an existentiale does not mean the free-floating ability-to-be in the sense of the "indifference of the will" (*libertas indifferentiae*). As essentially disposed, Dasein has in each case already stumbled into determinate possibilities; as the ability-to-be that it *is*, it has already let such [possibilities] pass by; it constantly relinquishes possibilities of its being, takes hold of them, and errs. That means, however: Dasein is being-possible that is delivered over to itself, through and through *thrown possibility*. (*SZ*, 144)

One is not "able" or free to choose the other possibilities in the sense that one has already "relinquished" or forsworn them. So the "not" that characterizes Dasein's projection is the impossibility of escaping its subjection to the pull of the normative valences in terms of which the possibilities open to it are disclosed.

We are always subject to normative imports that call upon us to live or act in various ways. Although this "being called upon" is not literally a sort of speech, it can be modeled on the speech act of summoning. The attunement of shame discloses something about or associated with me as shameful; shamefulness calls for hiding or covering up. Shamefulness addresses us and summons us to a determinate response. This structure of address and summons is essential to our very being: to be an agent is to be thrown into life in such a way that the imports disclosed by one's attunements summon one to action. Heidegger calls this summoning "conscience," for as conscience in the ordinary sense summons us to respond to our moral guilt, Heidegger's "transcendental conscience"[8] summons us to respond to our "essential" guilt (as Heidegger puts it in *SZ*, 286), our being bound by norms.

This is one of Haugeland's deepest insights about *Being and Time*: Heidegger analyzes subjectivity or agency in terms of his concepts of guilt and conscience. The very concepts of subjectivity and the first person are fraught within the tradition of California Heideggerianism. Indeed, their absence has been one of the principal focal points of criticism of the movement. Those who have seized on this point, however, such as Frederick Olafson,[9] have tended to insist on reintroducing a fairly traditional conception of experience or subjectivity into the reading of *Being and Time*. The goal ought not to be to smuggle a Cartesian notion of subjectivity back into

Heidegger but to figure out the place of the first person and a notion of subjectivity based on it within the framework of California Heideggerianism. Some have tried to address this lacuna in California Heideggerianism by arguing that what Heidegger calls "authenticity" is the first-person point of view on one's own life and actions.[10] I have argued against this strategy elsewhere and will not repeat those criticisms here.[11] Instead I want to turn to and exploit a strategy of pragmatic analysis developed by other students of Haugeland's.

2 Agent Specificity and the Logic of Resoluteness

In their account of pragmatic normativity in 'Yo!' and 'Lo!', Rebecca Kukla and Mark Lance distinguish between "agent-specific" and "kind-relative" normative statuses.[12] For reasons that will soon become clear, I will alter their terminology and use "agent-particular" where they use "agent-specific." But first, a normative status is a status, that is, a state or condition of a participant in a practice, that is normative, that is, which has to do with how the participant is called upon to act, think, feel, or be.[13] Core examples of normative statuses are commitment and entitlement. If I promise to pay you back the ten dollars I am borrowing from you, I have undertaken a commitment to do so, and you have acquired an entitlement to my doing so (or perhaps to the money). Now, as Kukla and Lance explain, kind-relative normative statuses "apply to people in virtue of their membership in some general kind," and the authors provide as an example a prohibition on felons voting. This prohibition applies to all actual and possible felons who are citizens of the community that adopts the prohibition. Because Omar Little is a felon, he may not vote in Baltimore, not because he's Omar Little but because he belongs to the class of felons. Agent-particular normative statuses "apply to a particular agent (or some particular agents) in light of concrete, particularized facts about her normative position, and have no implications even in the ideal for others."[14] Kukla and Lance give as an example the commitments engendered by promises: only Reese Bobby himself can promise to attend his son Ricky's race; Lucy Bobby cannot make the promise on Reese's behalf. Further, once made, the promise is a commitment to Ricky and no one else. Consider the instructive contrast case of a contract that specifies that the rights of one party will transfer to its "heirs, successors, and assigns": in such a case, the promise is to an individual or particular institution, but only insofar as it exemplifies certain general (kind or class) characteristics. It is not a promise to the individual or particular institution per se. When Bill and Hillary stood at the altar and promised to be faithful to each other until "death do us part," they undertook a commitment to each other. Bill cannot subsequently transfer that commitment to his brother Roger.

With this machinery up and running, I want now to return to the California Heideggerian notion of an "anonymous" or "generic" social role. We can make it more precise in the form of what I will call a "social position."[15] A social position is a set of kind-relative normative statuses. To be a professor is to be obligated to teach one's courses and grade one's students' work; to be entitled to demand certain sorts of formal address by students; and so on. These are all kind-relative statuses because they apply to Professor Davis in virtue of his being a member of the faculty at his institution. He shares these statuses with other members of the faculty at his institution. Some social positions are defined in such a way that only a single individual can occupy them at one time, but this restriction does not change the logic of kind relativity.[16] In this language, the insistence among early California Heideggerians that the possibilities of living and conduct available to a person are all "generic" or "anonymous" amounts to the thesis that existential possibilities are what I am calling "social positions."

Are the possibilities in terms of which we understand ourselves always social positions, or is there space for something more individual in our self-understanding? I want to argue that there is room for something more individual. To do so, I need to distinguish what I will call "instantiation" from "specification."[17] The most obvious way for there to be an agent-particular normative status is for an individual or individuals to instantiate a (set of) kind-relative status(es). Consider the spousal obligation to be faithful to one's partner. It is a kind-relative obligation one incurs upon getting married. So, in virtue of having gotten married, Bill and Hillary have agent-particular obligations to be faithful to each other. That is, the kind-relative obligation is particularized or instantiated for Bill and Hillary in relation to each other, but not in relation to anyone else, actual or possible, in virtue of occupying the mutual roles of spouse.

Now think about Professor Jones, who has long offered to read drafts of his students' research papers, so that it has come to be expected of him that he will do this. The normative status, "Jones is expected to read Smith's paper," is connected with Jones's social position as professor but is an extension or development of that position and not really kind relative. It has to do, as we like to say, with the "kind of professor" Jones is—except that this language is misleading: there is no well-defined kind like that. It is not the same sort of specification that is in play, say, when a member of the faculty is designated as "primary graduate faculty," so that she is both entitled and generally expected to sit on dissertation committees. It has to do, rather, with who Jones is as a professor, the way he has specified the role in his own case. Let us call this "specification."

Let me offer an example of the sort of dynamic I have in mind by way of specification.[18] At Georgetown, where I teach, every student must take two general-education courses in philosophy. Most of the students are academic high achievers with well-developed skills. It makes sense, then, for introductory philosophy courses to dive right into classical sources, such as Plato and Aristotle, Aquinas, Descartes, and so on. Most of our introductory courses do. Georgetown is also an athletic school with highly competitive Division I teams in many sports, and so the university recruits exceptional athletes. Not all these athletes are as well prepared to wade into classical literature as the general student body (though some are). It is entirely consistent with the role of the professor to say to oneself, "Being a student-athlete is hard, but they have to do all the same academic work as non-athletes, and in this case that means slogging through Aristotle and Hume." (That is, after all, the official view.) One might, alternatively, try to reach this group of students by designing an introductory course that could draw them into philosophical reflection while working with their distinctive strengths and weaknesses. Six years ago, I found myself called upon to do just this, and I responded by developing an introductory course in philosophy of sport that, I hope, fits the bill.

Let me highlight a few aspects of this example. First, what I did is in no way required of being a college professor. One can be not merely a satisfactory but even a brilliant professor while not experiencing the specific normative demand I felt. Second, what I did is nonetheless an intelligible extension or development of the generic role of the college professor. I neither transcended the role nor revolutionized it; I specified it. Third, the demand I felt was no doubt the product of my own personal history, my experience as a youth baseball coach and as the father of a high school athlete, as well as the "positive coaching training" I begrudgingly agreed to take while a youth coach.[19] This is all to say that my experience of being called upon to do this had as much to do with playing the role of a professor as with who I found myself to be at that time six years ago.

Let us turn to the text and look at a well-known passage with this notion of specification in mind. To make the subsequent discussion of the passage easier, I will break it up into parts, marked here by bracketed numbers:

[1] Resoluteness is, according to its ontological essence, in each case the resoluteness of a current factical Dasein.[20] The essence of this entity is its existence. Resoluteness "exists" only as understanding-self-projecting resolution. [2] However, in terms of what does Dasein disclose itself in resoluteness? Upon what should it resolve? [3] The answer can *only* be given in the resolution itself. It would be a complete misunderstanding of the phenomenon of resoluteness if one thought that resoluteness were merely grasping possibilities by taking them up, possibilities that have been laid before it and recommended. [4] *Resolution is precisely and in the first place the disclosive projecting and determining of the current factical possibility.* [5] To resoluteness *belongs* necessarily the *indeterminateness* that characterizes every factical-thrown ability-to-be of Dasein. Resoluteness is sure of itself only as resolution. However, the *existentiell indeterminateness* that in each case determines itself for first time in the resolution has nonetheless its *existential determinateness*. (SZ, 298)

Part [2] of the passage states the question to be addressed: what does Dasein resolve on, into what does it project itself, in resoluteness? That is, which possibilities of living does it throw itself into? Part [3] excludes one answer to this question, namely, that resolute Dasein merely takes up possibilities that are on offer in its cultural context, possibilities that have been recommended to it. We can understand this exclusion as rejecting instantiation as the model of the pragmatic normativity of the authentic situation. Part [4] states Heidegger's positive view: resolving is determining how to project oneself into possibilities. He calls it the "disclosive projecting and determining of the current factical possibility." Notice that "possibility" is in the singular. He does not say that resoluteness projects and determines the situation, or the self, or any other more "mass-like" or aggregate phenomenon. Resoluteness determines the current factical possibility, that is, the particular possibility that one enacts in the situation. In resoluteness Dasein makes its current factical possibility determinate. The logic here is that of the determinable and determining. Resoluteness determines a determinable possibility; that is, Dasein specifies the possibility in terms of which it understands itself. This specification of the current factical possibility is tied, moreover, to the individuality of the individual. This is the upshot of the tone-setting part [1] of the paragraph: "Resoluteness is ... in each case the resoluteness of a current factical Dasein." In resoluteness, I determine or specify some possibility of conduct or living, and I do so in light of who I am.

If the determination that constitutes resoluteness is tied to the individuality of the individual Dasein, then why does Heidegger insist on the indeterminateness of resoluteness in part [5] of the passage? One might read this as implying that the resolute self is purely formal or "empty," indeterminate in the sense of having no determinate content. I suggest an alternative reading: resoluteness, as a concept or as a general phenomenon, is indeterminate, has no content. In other words, the mere classification "resolute" or "authentic" does not on its own imply any determinate content in an individual's living. Contrast this with our usual understandings of morality: a moral life is a specific sort of life, a life guided by a definite set of norms and ideals. Resoluteness is not like morality in this way. Resoluteness is a formal ideal in that what it requires is simply that one specify the current possibility into which one projects oneself and that one do so in a way that lives up to the concrete normative demands that issue from who one individually is. So resoluteness is "existentially" (i.e., formally) determinate, but indeterminate in an "existentiell" way. Resoluteness is a formal ideal, rather than a substantive one. But it does not follow that, as a resolute or authentic self, I am myself formal and empty.

Note that two pages after the passage just quoted at length, Heidegger writes, "*In contrast, the situation is essentially closed off to the Anyone. It knows only the 'general position'*" (SZ, 300). The Anyone is indeed a repository of anonymous or generic possibilities of living, and hence it "knows" only the "general position," or social statuses. It does not know the specified situation in which resolute Dasein acts, for that requires that the individual self respond to the normative claims that arise for it, demanding that it specify its social roles and fit them to the concrete, factical, particularized, and specific situation in which it acts. To refine this further, let us now consider two objections to my notion of specification.[21]

3   Objections

3.1   Specification and Idiosyncrasy

One might suspect that what I am calling "specification" is just a necessary by-product of an individual person enacting a role. That is, given that I am not a robot, how can I fail to enact the role of professor of philosophy in my own distinctive way? I have my own style of lecturing, my own ways of talking with students in office hours, leading classroom discussions, commenting on students' written work, and so on. Is "specification" not simply the fact that an individual person is enacting a role, rather than a machine executing a rigid algorithm? To address this worry, we need to distinguish three different sorts of phenomena here: (1) the accidental detail of all human action; (2) aspects of one's "personal style," if one has one, that are not merely accidental but also not a response to any normative call; and (3) normative specifications of one's self-understanding.

Because social positions are enacted or lived out by concrete individuals, their enactment necessarily takes on much particular and rich detail. When I do anything, I do so in a highly particular way. I walk to the lectern in my Philosophy of Sport classroom with my briefcase in my right hand, sunglasses up on top of my head or hanging from my shirt, my hips moving in just this way, and so on.[22] Some of this detail is entirely ephemeral, just the way I happen to do something today. Some of it is idiosyncratic, by which I mean that it is characteristic of me but doesn't carry much if any meaning. I do typically hang my sunglasses from the top button on my shirt, until I can put them away safely in their case, and I suppose (though I cannot confirm) that I typically walk in a certain way. Sometimes such idiosyncratic detail rises to the level of what we might call a "personal style." By this I mean aspects of my personal comportment that reflect a certain amount of work or refinement (in the sense of having been refined, honed). I have a style of lecturing that I have worked on over the years, just to find my "comfort zone" and make myself clear to my students. None of this idiosyncrasy or personal style rises to the level of what I want to call a "normative specification of my social position." (Or at least there is nothing in the way I've just described these phenomena that suggests that they have so risen.)

In the case of the call to develop a course in the philosophy of sport, I was specifically challenged to stretch the limits of what I understood myself to be as a philosophy professor.[23] To feel called upon to move outside the bounds of one's current enactment of a social position means that one can also fail to do so; one can "flee" from the call, as Heidegger puts it. Another way to look at this dynamic is to see that when one is challenged and called upon to specify a social position, there is something at stake or at issue in how one responds. There is nothing at stake in how I hang my sunglasses, nor anything at issue in how I move my hips as I walk. But whether and how I respond to the call to reach out to the student-athletes at Georgetown says a lot, at least to me, about who I am as a professor. This is what makes the call existentially normative, rather than merely practical.[24] For these reasons as well, when I do not live up to the call I have heard, I feel as if I have failed, not just that some practical requirement has gone unfulfilled. This goes a long way to explaining why, according to Heidegger, resoluteness involves anxiety, or at least "readiness for anxiety" (SZ, 295–296). In such a call, the uncanniness or unsettledness of the self is exposed: who I am to be as a professor is at issue, and thus my self-understanding is at stake.

3.2 Resoluteness and *Phronesis*

Now, one might wonder how hard-and-fast the distinction I rely on in the previous section is, between existential normativity and practicality. Consider the example of my lecturing style. I have developed this style over many years as a response to the ways in which my students seem to learn best. Thus my lecturing style is in part a response to the practical solicitations of the situation. In a particular case, I find that explaining a point this way, rather than that, clarifies it better for my students. The situation thus solicits or calls upon me to explain the point this way. More than twenty years of such solicitations, and I have developed a style of lecturing that is an adaptation of my ways of thinking and understanding to the sorts of students I teach. If I were to move to another university with different sorts of students, I would likely find that my current style is not a great fit, and the situation would then solicit me to change and develop in response to new needs and opportunities. These situational solicitations are normative calls, in an important sense of the term. It is possible to fall short of responding to them well, and if I don't respond at all, I will experience myself as falling short in my work. To address these considerations, let us turn to one of Hubert Dreyfus's interpretations of resoluteness.

Dreyfus has proposed that the resolute individual is an Aristotelian *phronimos*.[25] As Dreyfus explains it, the resolute agent is an "expert" or "virtuoso,"[26] someone who responds immediately and without deliberation to the current practical situation in its particularity. To make his case, Dreyfus cites the very passage from *Being and Time* I quoted earlier (*SZ*, 298). Rather than relying on the general rules that beginners and even competent practitioners use, the expert is able to respond deftly and precisely to the details of the particular situation in which he or she acts.

> Like the *phronimos*, the resolute individual presumably does what is retroactively recognized by others as appropriate, but what he does is not the taken-for-granted, average right thing—not what one does—but what his past experience leads him to do in that particular Situation. Moreover, as we have seen, since the Situation is specific and the *phronimos*'s past experience unique, what he does cannot be the appropriate thing. It can only be an appropriate thing. (Dreyfus 2000, 163)

Dreyfus might argue that the sort of dynamic on display in my example fits the model of the *phronimos*. Once I let my colleagues know what I am up to, what I plan to do is "retroactively recognized by others as appropriate," even if no one else came up with the idea or even would have. So maybe Dreyfus is right that resoluteness is *phronesis*, or at least that my actions to develop my Philosophy of Sport course were examples of *phronesis*.

What I am calling "specification" is not the same thing as the sort of Aristotelian *phronesis* Dreyfus describes,[27] for Dreyfus's exposition of *phronesis* is too vague to capture the specification of a social position as I have described it. The central problem is Dreyfus's word "appropriate." I am entitled to teach this new course, but I am not obligated to do so. This is true of all my colleagues as well. But I alone felt called upon to develop and offer the course. "Commitment" and "entitlement" are too rigid for this sense of being called upon, but finding it "appropriate" is too loose. It is easy to miss this distinction if one is in the grip of what Kukla and Lance call "the imperatival fallacy."[28] If we think of the social position of being a professor as made up entirely of deontic statuses, that is, commitments, entitlements, permissions, obligations, and the like, then it is difficult to understand in what way I have in this case specified the role of being a professor. I am committed to teaching the courses assigned to me and leading the class sessions of those courses. If I ask to teach Philosophy of Sport and my chair assigns the course to me, I am thereby committed to teaching Philosophy of Sport and leading its sessions. This is in principle no different than if I ask to teach Heidegger's *Being and Time* and my chair assigns that course to me. I am entitled to ask to teach any plausible philosophy course, I am permitted to assign two papers or three at my discretion, and so on. It is hard to know what more to say in conventional deontic vocabulary.

Let us spread our wings a bit here and use a richer normative vocabulary. There is reason to teach the course: there is an identifiable need for the course, and the course performs a describable and valuable function. These are reasons that, once I point them out, many of my colleagues might well recognize. Some of my colleagues, I am fairly certain, think that there is no reason to teach the course. They think that student-athletes should have to take the normal philosophy courses and that to cater to these students specifically is to compromise the integrity of the department's educational mission. We can have this debate because the question of whether there is a reason to offer this course is a question about a public normative status. It is also agent neutral in an important sense: if my colleague Mark is convinced that there is a reason to offer the course, he might say, "Someone ought to teach this course," or "We, the department, ought to offer it."

Note that it is perfectly possible for all of us in the philosophy department to acknowledge such a reason, but for none of us to feel called upon to offer the course in question. Say, for example, we have no one teaching a course on Nietzsche. Most of us (though probably not all of us) would believe that we should offer such a course. Only a few of us—the most die-hard Europeanists among us—would argue that we are obligated to offer a course on Nietzsche and thus are derelict in our duty as a philosophy department in not doing so. Most of us would think that we have good educational reason to offer the course and are therefore falling short of an important ideal as a department in not offering it. But suppose none of us feels individually called upon to offer the course.

I want now to try to make the most difficult point about all of this. When I felt called upon to develop and offer Philosophy of Sport, I experienced the call as issuing from what I understood myself to be as a philosophy professor at GU. It was not just an option, an intriguing possibility, as, say, I once experienced the idea of teaching a course on Spinoza and Leibniz. The language of "being called upon" to do something suggests the pull, and not just the attraction, of the task. One cannot capture the content of the call I heard in the form of an imperative or in other deontic language. To be called upon to do something that is not obligatory is more than just to have reason to do it. I hear this call and others do not in virtue of who I understand myself to be as a philosophy professor, a self-understanding clearly not shared by all my colleagues. If we restrict ourselves to the deontic vocabulary of commitments and entitlements, or alternatively fall back on the vague language of what is appropriate, we cannot see the specification of my self-understanding that takes place here.

For all these reasons, Dreyfus's language of the *phronimos* as an "expert" or even "virtuoso," while apt for many sorts of cases, does not capture the specific dynamics of the kind of case in which I am interested. Dreyfus's Heideggerian *phronimos* responds to the practical solicitations of a situation in a manner that her colleagues, fellow members of her community, or even her contemporaries, recognize to be apt or deft. This is because the solicitations of the general position in which she acts are solicitations addressed to everyone and no one, or, more precisely, are kind-relative, in Kukla and Lance's vocabulary. We all (or most of us) acknowledge the situation's call to offer a Nietzsche course, but none of us find ourselves specifically drawn into a transformation of who we understand ourselves to be as professors.

4   Conclusion and Next Steps

In conclusion, let me connect the analysis I have offered to Heidegger's language of "mineness" and "ownership" (*Eigentlichkeit*, "authenticity"). For me to be in a position to be called upon to specify the role of philosophy professor in the way in which I did, my conduct, my life, must be mine. The idea that my life must be mine might mean two things. It might mean simply that I must be receptive to being addressed by normative imports. For me to be a philosophy professor, the generic or kind-relative imports that inhere in the social position must be able to be instantiated in my own case, which means in turn that I must be able to "hear" the summons to attend this class. There is something necessarily "first-personal" about enacting social positions. We are not robots who mechanically carry out algorithms that specify our roles. We are persons who are addressed by normative statuses. Heidegger insists, however, that the traditional philosophical focus on the first personality of experience, while not false, is superficial and obscures an important existential feature of human life.

That Dasein is "in each case mine" means, further, that I can always take ownership of a social position I enact. We take ownership of a social position when we respond to concrete normative calls to specify it in a distinctive manner that deviates from the established kind-relative social norms. Taking ownership of a social position certainly requires first-personal responsiveness to norms, but it demands much more. In my Philosophy of Sport example, I responded to being called upon to develop a course that was certainly not required by my role. Once I have heard this call and responded to it, the way in which I enact the social position of professor has been transformed. I transform what it is for me to be a philosophy professor without (necessarily) transforming what it is for one to be a philosophy professor. I specify the role for my own case; I do not (necessarily) change the generic role as it subsists in the culture.

Now, I am not saying that this is all there is to Heideggerian authenticity or even resoluteness. There is a lot more going on in those thicker concepts. Until we have gotten clear on the pragmatic normativity of the authentic situation in *Being and Time*, however, we stand essentially no chance of understanding the more far-reaching concepts of authenticity and resoluteness. It is surely important that resolute Dasein must be prepared to take back its resolution, that is, that there is an essential form of risk involved in resoluteness. In the terms developed above, when I respond to a normative call to specify my self-understanding, I run the risk of failing to live up to who I am called upon *to be*. This is not just an everyday sort of risk, the risk of failing at some discrete task (giving this lecture) or project (writing this book). It is the risk of failing to be me, failing to become who I already am (called upon to be). Resoluteness requires running forth into death,[29] as Heidegger calls it, and that running forth into death has something to do with being prepared to risk everything on an understanding of what it is to be me, or even to be human, or finally even to be, *simpliciter*. These are themes that Haugeland explored with great penetration in "Truth and Finitude" and "Dasein and Death." The next task in working out the approach to the authentic situation in terms of specification is to see what impact it should have on our understanding of running forth into death and resoluteness that runs forth, that is, authenticity.

Acknowledgments

I would like thank the three audiences who heard earlier versions of this essay and discussed its ideas with me: the Philosophy Department of the University of Chicago on the occasion of the Inaugural John Haugeland Memorial Lecture (May 2011); the participants in the Thirteenth Annual Meeting of the International Society for Phenomenological Studies (July 2011); and the participants in Anton Friedrich Koch's Oberseminar of the Philosophisches Seminar der Universität Heidelberg (June 2012). In particular, special thanks are due to Andy Blitzer, Toni Koch, Rebecca Kukla, Mark Lance, Oren Magid, Simone Neuber, and Kate Withy. Finally, the editors of this volume provided insightful feedback on an earlier draft of this essay; it is better for their intervention.

Notes

1. Martin Heidegger, *Sein und Zeit*, hereafter cited as *SZ*. I provide my own translations of the text. Readers may correlate passages with the standard English translation by Macquarrie and Robinson (which I have consulted in producing my own translations).

2. Dreyfus 1991. Although Dreyfus's commentary was not published until 1991, it circulated in draft form for many years and expressed Dreyfus's thinking, which guided and was influenced by his students. Haugeland's, Guignon's, and Richardson's early approaches to Heidegger may be found here: Haugeland, "Heidegger on Being a Person" (in DD); Guignon 1983; and Richardson 1986.

3. Macquarrie and Robinson: "the 'They.'"

4. Macquarrie and Robinson: "potentialities-for-Being."

5. I am not taking a stand on whether a sufficiently powerful computer can be programmed so that, at another level of description or from another stance, it is conforming to norms.

6. Blattner 2015.

7. An insight I got from Haugeland in his seminars but cannot find in print.

8. The term "transcendental conscience" is due to Rebecca Kukla (2002).

9. See Olafson 1994 and the responses to it, Carman 1994 and Dreyfus 1996.

10. Crowell 2001; Carman 2005.

11. Blattner 2015.

12. Kukla and Lance 2009.

13. In oral presentations of this paper, I have found that this use of "normative" is puzzling to some. Especially in German (as I found out during a discussion of the paper in Anton Friedrich Koch's Oberseminar at the University of Heidelberg in June 2012), the word "norm" suggests or implies generality. So in German, for example, the verb *normen* means "to bring into alignment with generalized standards," such as the well-known DIN standards promulgated by the Deutsches Institut für Normung. The way I am using "norm" here—and in doing so following a strategy that may loosely be identified with the Pittsburgh school—it refers to any claim on an individual's actions, feelings, thoughts, or being that she stands under or takes herself to stand under. Simone Neuber suggested *Anspruch* (which roughly means "claim") as an appropriate German translation of the idea, and Koch added that I am arguing for *ungenormte Ansprüche*, claims that are not normed or standardized.

14. Kukla and Lance 2009, 23–24.

15. I introduced this notion first in Blattner 2000, though in a less precise form.

16. The president of the United States is defined so as to be a unique individual (at most; there are situations in which the office would be vacant). Nonetheless, the presidential social statuses that Barack Obama has, such as the entitlement to nominate a person to fill a vacancy on the U.S. Supreme Court, are statuses he has in virtue of occupying the office, not in virtue of being Barack Obama. If he were to resign and be succeeded in office by Joe Biden, Biden would then have those statuses. Indeed, legal commitments undertaken by one president are inherited by his or her successor.

17. Mark Lance has pointed out to me that Henry S. Richardson (1990) uses the same language to draw the same distinction. Richardson's concerns are different from mine, however: he asks whether we can understand the specification of an ethical norm as rational, whereas ethical norms are secondary to the issue I am exploring, and I am not interested in whether specification can be seen as rational. Despite these differences, it is roughly the same idea.

18. I use a personal example, at the risk of navel-gazing, since I find it far easier to describe the dynamics in the first person.

19. Katherine Withy (2011) argues that Heidegger does not provide a proper analysis of the notion of a personal (as opposed to cultural or generational) history and as a consequence cannot really develop an adequate analysis of character and character change. She proposes adding some Aristotelian machinery to *Being and Time* to fill this gap.

20. "Factical" means determinate in a specifically Daseinish way, whereas "factual" means determinate in the way in which non-Daseinish entities are. That is, it is a

factual feature of a rock that it weighs one pound, but a factical feature of Green that she is obese.

21. These objections arose in oral presentations of earlier drafts of this paper.

22. David Finkelstein's observation!

23. Heidegger does not use (the German equivalent of) the word "challenge" (*herausfordern*) to articulate his concept of resoluteness, but I think it is a natural correlate of the anxiety he identifies as an incitement to resoluteness.

24. This is not to say that there can be practical demands without existential normativity. Far from it: if nothing were at stake in how I care for the grass in my yard, there would be no sense in which it would be appropriate or efficient or prudent for me to mow the lawn today, before the hurricane arrives.

25. Dreyfus 2000. Dreyfus's interpretation here diverges noticeably from his earlier approach in Dreyfus 1991.

26. The term "expert" derives from the analysis of "intuitive expertise" developed by Dreyfus and his brother Stuart. See Dreyfus and Dreyfus 1986. He borrows the term "virtuoso" from Pierre Bourdieu (1977).

27. Here is not the place to debate the proper interpretation of Aristotle's conception of *phronesis* (nor am I the one to do it).

28. Kukla and Lance 2013.

29. *Vorlaufen in den Tod* (Macquarrie and Robinson: "anticipation of death").

References

Blattner, W. 2000. Life is not literature. In *The Many Faces of Time*, ed. L. Embree and J. Brough, 187–201. Dordrecht, Holland: Kluwer.

Blattner, W. 2015. Essential guilt and transcendental conscience. In *Heidegger, Authenticity and the Self: Themes from Division Two of Being and Time*, ed. D. McManus, 116–134. London: Routledge.

Bourdieu, P. 1977. *Outline of a Theory of Practice*. Cambridge: Cambridge University Press.

Carman, T. 1994. On being social: A Reply to Olafson. *Inquiry* 37:203–223.

Carman, T. 2005. Authenticity. In *A Companion to Heidegger*, ed. H. L. Dreyfus and M. A. Wrathall, 285–296. Oxford: Blackwell.

Crowell, S. 2001. Subjectivity: Locating the first-person in *Being and Time*. *Inquiry* 44:433–454.

Dreyfus, H. L. 1991. *Being-in-the-World: A Commentary on Heidegger's Being and Time, Division I*. Cambridge, MA: MIT Press.

Dreyfus, H. L. 1996. Interpreting Heidegger on *Das Man*. *Inquiry* 38:423–430.

Dreyfus, H. L. 2000. Could anything be more intelligible than everyday intelligibility? Reinterpreting Division I of *Being and Time* in the light of Division II. In *Appropriating Heidegger*, ed. J. E. Faulconer and M. A. Wrathall, 155–174. Cambridge: Cambridge University Press.

Dreyfus, H. L., and S. E. Dreyfus. 1986. *Mind over Machine*. New York: Free Press.

Guignon, C. 1983. *Heidegger and the Problem of Knowledge*. Indianapolis: Hackett.

Haugeland, J. 1998. *Having Thought: Essays in the Metaphysics of Mind*. Cambridge, MA: Harvard University Press. (Abbreviated as *HT*.)

Haugeland, J. 2013. *Dasein Disclosed*. Cambridge, MA: Harvard University Press. (Abbreviated as *DD*.)

Heidegger, M. 1962. *Being and Time*. Trans. J. Macquarrie and E. Robinson. New York: Harper & Row.

Heidegger, M. 1979. *Sein und Zeit*, 15th ed. Tübingen: Max Niemeyer.

Kukla, R. 2002. The ontology and temporality of conscience. *Continental Philosophy Review* 35 (1): 1–34.

Kukla, R., and M. Lance. 2009. *'Yo!' and 'Lo!' The Pragmatic Topography of the Space of Reasons*. Cambridge, MA: Harvard University Press.

Kukla, R., and M. Lance. 2013. Leave the gun; take the cannoli! The pragmatic topography of second-personal addresses. *Ethics* 123 (3): 456–478.

Olafson, F. 1994. Heidegger à la Wittgenstein or "coping" with Professor Dreyfus. *Inquiry* 37:45–64.

Richardson, H. S. 1990. Specifying norms as a way to resolve concrete ethical problems. *Philosophy and Public Affairs* 19 (4): 279–310.

Richardson, J. 1986. *Existential Epistemology*. Oxford: Clarendon Press.

Withy, K. 2011. Moods change: Adding Aristotelian character to *Being and Time*. Paper presented at the Thirteenth Annual Meeting of the International Society for Phenomenological Studies, Kennebunkport, Maine.

2 Competence over Being as Existing: The Indispensability of Haugeland's Heidegger

Steven Crowell

> Heidegger calls his analysis of Dasein and disclosedness an *existential* analytic, not because the grounding of ontical truth is not transcendental, but because it can *be* transcendental only *as* existential.
> —John Haugeland, "Truth and Finitude" (*DD*, 220)

"No one is indispensable." This hard truth of the work world encapsulates the lesson of Division I of *Being and Time*. Engaged in its everyday activities, Dasein is not I-myself but the "one-self" (*Man-selbst*). Because I do what one does, one can replace me in anything I do. This is not because I am not unique, but because to *do* something is to function in a role, broadly understood, and thus to do something that someone else could do: raise my children, nail down that deal, make breakfast, write a lecture, dance with my wife. If it is said that not just anyone could do these things as well as I can, it is also true that there are those who could do them better; and if it is said that no one could do them precisely the *way* I do them, it is also true that I cannot do them precisely the way another does them. The fact is that they get done either way, and so I am not indispensable. Of course, my wife might feel differently about my replacement, but that is another matter altogether, one that has to do not with what I do but with *who I am*. That brings us to Division II, and to the indispensability of Haugeland's Heidegger; for what I will argue, in a roundabout way, is that John Haugeland has taught us what is indispensable about I-myself and in so doing has made himself indispensable to Heidegger studies and Heidegger indispensable to philosophy.

Of these three indispensabilities, the second—Haugeland's indispensability to Heidegger studies—is the least important (most dispensable?), yet it is important enough to warrant some mention in section 1. This will allow us to turn, in section 2, to Haugeland's demonstration of the indispensability of I-myself in ontology, a post-Cartesian brief for the first-person stance. I discuss the indispensability of this insight, and so of Haugeland's Heidegger for philosophy generally, in section 3, drawing on Haugeland's approach to natural science.

1   Haugeland's Heidegger and Heidegger Studies

By "Heidegger studies," I mean what Husserl called the "practical world," the vocational context, of scholars with a professional interest in Heidegger (Husserl 1970, 382; 1954, 462). A practical world combines two senses of the term "world" that Heidegger outlines toward the beginning of *Being and Time*: first, it means "any realm that encompasses a multiplicity of entities … as when one talks of the 'world' of a mathematician"; and second, it means "that '*wherein*' a factical Dasein as such can be said to 'live,'" where to "live" is to be engaged in the practices that are concerned with the "multiplicity of entities" in question (Heidegger 1962, 93; *GA*, 2:87).[1] In this sense, Heidegger studies is an *institution*, with insiders and outsiders, gatekeepers, protocols, and all the rest. Let us suppose that the point of such an institution is to amass expert knowledge of the works of Heidegger. Whether there is any further point to doing so is an issue we may leave aside; there is certainly no consensus about what such a further point might be. But already this is enough for us to identify the indispensability of Haugeland's Heidegger. For even if no one in Heidegger studies is indispensable, it is certainly the case that anything that *gets Heidegger right* is indispensable, and Haugeland has gotten Heidegger right on at least two crucial issues. That the issues are *crucial* is, well, crucial, for almost anyone who works in Heidegger studies can get something right about Heidegger, but it is surprising how infrequently these are the crucial things.

So what are these issues? The first concerns how to read *Being and Time*—in particular, how to understand the relation between Division I and Division II. At the start of his essay "Truth and Finitude: Heidegger's Transcendental Existentialism," Haugeland takes Dreyfus and Rubin to task for failing to note the relation between "the 'existentialist' portions of *Being and Time*—those having to do with authenticity, anxiety, falling, death, conscience, guilt, and resoluteness"—and the "overarching aim of *Being and Time*, which is to reawaken the question of being" (*DD*, 187). Later in the same essay, Haugeland makes a decisive claim: "Ultimately everything in *Being and Time* has to do with the question of being—and, with it, truth. The existential concepts are introduced for this reason and this reason only" (*DD*, 209). At first, this hardly seems like news. If one learns anything upon entering the world of Heidegger studies, it is that Heidegger had only "one thought," the thought of being, and that philosophy is at an impasse because it has forgotten the question of being.[2] Nevertheless, it is surprising how few commentators actually get any further than mere intonation of the term "being," at least when it comes to explaining just how the existential parts of the text contribute to answering the question of the meaning of being. An adequate explanation of that contribution requires that one see Heidegger as he was when he wrote *Being and Time*: a transcendental philosopher whose existentialism is part of a method for elucidating the conditions that make intentionality possible, that is, enable us to encounter entities as meaningful.[3] Haugeland gets this just right, whatever differences one might have with little details, such as the idea that General Motors is Dasein—an interpretation, however, that was recently upheld by the US Supreme Court.[4]

The importance of seeing Division I and Division II as integral parts of a transcendental project will emerge as we go, but we need to note the second crucial issue that makes Haugeland's Heidegger indispensable to Heidegger studies, namely, his understanding of what Heidegger does with the traditional distinction between *essentia* and *existentia*, essence and existence, *Was-sein* and *Dass-sein*. This distinction stood at the center of Sartre's reading of Heidegger, and though Heidegger repudiated that reading and is followed in this by most Heideggerians, it is surprising how many of those same Heideggerians tacitly appeal to something very much like the Sartrean view to distinguish Heidegger from Husserl and his phenomenological reduction. Haugeland is not among them.[5] The point here is a bit esoteric, but we may characterize it in broad strokes as follows.

In traditional metaphysics, any entity is some *sort* of thing, and the sort of thing it is is its "essence" (*essentia*). But an essence is the aspect of a thing that could be said also of other things of its kind; it is general, even if it is possessed in some metaphysical way by a particular thing. To designate the particularity of an entity, then, the tradition speaks of its "existence" (*existentia*).[6] The tradition devoted much attention to the problem of how to think the combination of these two ontological determinations of an entity, but, as Haugeland points out, Heidegger focuses on two: the ancient approach based on production (common to both Greek antiquity and the Christian Middle Ages), and the modern approach, represented by Kant's idea that "existence is not a real predicate" (*DD*, 171). In the ancient conception, existence is the material instantiation of an essence through a creative act; in Kant it is the positing (*Setzen*) of something within the horizon of space and time. What both approaches have in common is the view that *existentia* itself is *meaningless*. Stated otherwise, all the thinkability or intelligibility of a thing lies on the side of essence or *Was-sein*, what it is. *That* it is, in contrast, is something that can be noted or reported but not thought. With the exception of God—whose essence and existence are one and the same—the *existentia* of any entity (its producedness or positedness) is identical to that of any other. Sartre trades on this fact when he writes that with regard to human reality "existence precedes essence." I am not an instance of some pregiven kind; *what* I am (or rather *who* I am) is to be determined precisely through "existing." If we think of an entity's essence in normative terms, as designating what it is *supposed* to be, then human beings are not supposed to be anything at all; they are what they make of themselves, "without justification and without excuse" (Sartre 1956, 78).

When Heidegger writes, in *Being and Time*, that "the 'essence' of [Dasein] lies in its 'to-be'" and follows this with the claim that "its *Was-sein* (*essentia*) must, so far as we can speak of it at all, be conceived in terms of its being (*existentia*)," he is, as Haugeland will show, making an entirely different point—namely, that *existentia* is itself something thinkable or intelligible (Heidegger 1962, 67; *GA*, 2:56). What makes this possible (and here it is I, not Haugeland, who speaks) is the phenomenological reduction of being to meaning. In the world of Heidegger studies, however, this is rarely understood.[7] Instead, it is argued that Heidegger's ontological phenomenology is distinguished from Husserl's in its return to *existentia* conceived as imponderable facticity, contingency, thisness, particularity, *Jeweiligkeit*, and so on. The story goes something like this:

Husserl insists that phenomenology must employ an *epoché*, bracketing the being of entities to study them as pure phenomena. This is followed by an eidetic reduction of entities to their essences, which leaves their particularity out of account. Since Heidegger wants to approach being phenomenologically, he must reject both reductions. Though he never does so explicitly in *Being and Time*, a rejection is to be found (it is claimed) in a lecture course from the period, where he writes that in both "the transcendental and the eidetic" reductions, the "question of the being [of consciousness] in the sense of [its] existence … gets lost," and argues that "if there were an entity whose what is precisely to be and nothing but to be," then an "ideative regard" such as one finds in the reduction "would be the most fundamental misunderstanding" (Heidegger 1985, 110; *GA*, 20:151–152).

But this constitutes a rejection of "the" reduction only if the "what" in question is contrasted exclusively with "existence" in the Sartrean sense, and the problem with this is that neither Husserl nor Heidegger understands it in this way. Husserl does in fact bracket *existentia* in the sense of the particularity of a thing insofar as that particularity is the empirical instantiation of an essence, but this is not what Heidegger is reinstating when he insists that the "what" of Dasein must be thought in terms of its "to-be." As Heidegger repeatedly insists, this "to-be" (*Zu-sein*) is nothing ontical; it has nothing to do with the empirical occurrence of this or that case of Dasein. Rather, it is ontological; it has to do with what it is to *be* Dasein. As for ideation, he claims that his analysis of the to-be uncovers "not just any accidental structures, but essential ones" (1962, 38; *GA*, 1:23). Thus whatever else Heidegger may be doing phenomenologically, he avails himself of the same ideative regard that allowed Husserl to grasp essence as what-being. Further, bracketing *existentia* in the sense of particular instantiation (positing) is *all* that Husserl's reduction claims to do. Those who read Heidegger's passages as rejecting the reduction must therefore read them in just the Sartrean light that Heidegger disavows— namely, as restoring the priority of contingent existence over the instantiation of essence. Heidegger's own understanding of the reduction is, however, more nuanced.

Noting, in a lecture course from 1928, that "an ontology of the *modus existendi* is plainly apriori" (1984, 178; *GA*, 26:229)—just as an ontology of essence as *modus essendi* is—Heidegger goes on to claim that this "difficulty" has stymied traditional accounts of essence, including Husserl's. But he does not conclude from this that the reduction must be rejected; rather, it is the limits of Husserl's traditional understanding of the ontological problem that must be overcome. This overcoming is accomplished by recognizing the implications of the reduction itself: "By suspending what is actual (in the phenomenological reduction)," that is, by refraining from positing existence in the sense of *existentia*, "the what-character is set forth—but, in suspending the actual, the *actuality*, i.e., the *modus existendi*, and its intrinsic connection with the essential content in the narrower sense is *not* suspended. Essence here has a double meaning: it means the apriori of *essentia* and of *existentia*" (1984, 178; *GA*, 26:229). The reduction thus opens up the possibility of inquiring into the meaning of *both* ontological aspects of entities. *Existentia*, and not merely *essentia*, displays fundamentally irreducible modes.

Being and Time focuses on three such modes—*Existenz* (to-be), *Zuhandenheit* (availability), and *Vorhandenheit* (occurrentness)—and only *Vorhandenheit* is associated with the kind of meaninglessness that characterized the traditional concept of *existentia*. Thus Heidegger's ontological phenomenology has nothing to do with an "externalism" or "realism" that would make philosophy beholden to the particularities and contingencies of factical life, relativize it to the supposed facts of nature, history, culture, or whatever. But if it is impossible to understand the function of the existential topics in Division II as gestures toward this sort of ontic entanglement of philosophy, we must then give some other account of them, one that does justice to the transcendental, a priori, and methodological character of the inquiry.

John Haugeland has gone a long way toward meeting this challenge.[8] And to return to our theme, his indispensable contribution to Heidegger studies is to have explained *why* intelligibility is not exhausted by essence, why not merely *what* I am is thinkable but *that* I am is as well. The key insight of Haugeland's Heidegger is that existing is itself something at which I can succeed or fail. It is on this point that Dasein's *Existenz* contrasts most sharply with occurrentness, and so with the traditional notion of *existentia*. An occurrent entity either exists or it does not; it cannot *fail* at existing. For this reason, as we shall see in section 3, the occurrent represents a certain limit of intelligibility.

To say that Dasein can succeed or fail in its to-be is to say that it is assessable against a normative standard of that-being—the standard of authenticity (*Eigentlichkeit*)—just as the traditional concept of *essentia* points toward the normative standard for what a thing is, what a thing is *supposed* to be. Haugeland's great contribution to Heidegger studies is to have recognized the methodological implications of this fact. Briefly put, differences among *modii existendi* must correlate with different types of failure possible for the things that exist in these ways.[9] For instance, the *modus existendi* of a hammer is availability, and it fails to be how it is supposed to be when it is "unavailable" in some way: broken, missing, and so on. It remains a hammer (*modus essendi*) but does not live up to its *modus existendi*; it fails to function. Dasein, in contrast, whose *modus existendi* is *Existenz* or "care," is not available, does not have a function. Rather, Dasein fails to be how it is supposed to be when it acts *inauthentically*; it remains Dasein (*modus essendi*) but does not live up to its *modus existendi*.

These two possible ways of failing differ in important ways. In the first, the norm that determines what failure means is not intrinsic to the hammer but is defined by one of Dasein's projects; in the second, however, the norm that determines what failure means *is* intrinsic to Dasein as that being "for whom, in its being, that very being is essentially an *issue*" (Heidegger 1962, 117; *GA*, 2:113). As Haugeland shows, Dasein is *intrinsically* oriented toward measures of success or failure, intrinsically concerned with normative assessment of its own being in light of "what is best."[10] Dasein cannot "be" at all without being the way in which the normative, measure as such, is disclosed. But here Haugeland's indispensability for Heidegger studies passes over into the more important question of how Haugeland understands the indispensability of I-myself, an understanding that in turn makes Haugeland's Heidegger indispensable for philosophy generally.

2   The Indispensability of I-Myself: Competence over Being as Existing

I have said that one of Haugeland's indispensable contributions is his clear grasp of the methodological connection between Division I and Division II of *Being and Time*, the fact that "the grounding of ontical truth ... can *be* transcendental only *as* existential" (*DD*, 220). Using a vocabulary that strays a bit from Haugeland's own, I should like to phrase the "existential" part of this equation as the idea that the "I-myself" is a necessary condition for any understanding of being, thus including the kind of understanding of the being of entities on which truth claims rest. Phrased in this way, Haugeland's analyses of existential concepts such as anxiety, death, and guilt represent a recovery, *within* Heidegger's anti-Cartesianism, of the transcendental role of the first-person stance, the "I think," free of the Cartesian mentalism that identifies the first-person with *consciousness*.[11] Haugeland is able to find philosophical purchase here thanks to Heidegger's idea that the "being of the '*sum*'" (Heidegger 1962, 46; *GA*, 2:33)—the "I am"—is care (*Sorge*), that the *modus existendi* of the I-myself is to be an *issue* for myself, to be in question, *to give a damn*. "This," as Haugeland notes, "is the most basic definition of Dasein; all the others follow from it" (*DD*, 216).

Being an issue for myself is evident in everything I do and think, but it becomes explicit—hence also evident as a theme for those of us who engage in the project of elucidating our Dasein phenomenologically—in the experience of breakdown (existential "death"). If in its everydayness Dasein discovers itself in what it does—i.e., in its possibilities in the sense of abilities, skills, and practices—in breakdown it discovers itself as "primarily being-possible" (Heidegger 1962, 183; *GA*, 2:191). Such being-possible, I-myself, is not a competence over this or that but what Heidegger terms "competence over ... being as existing" (ibid.). As Haugeland shows, such competence is a form of "originary *self*-responsibility—one that cannot be public but can only be taken over by an individual" (*DD*, 209). Thus, for Haugeland, a being capable of saying "I" is not one who is able to perform some distinctive cognitive act of self-identification but one who is able to *commit* to its own existing, to be responsible for it. "Existential commitment ... is no sort of obligation"; it is "not a communal status at all but a resilient and resolute first-person stance" (*HT*, 341).

Haugeland offers penetrating analyses of what is involved in existential death, but he is more interested in what can *result* from it, namely, commitment as a form of owned (or authentic) existence. If we are to grasp the broader philosophical significance of Heidegger's ontology, however, we need to tease out a fundamental feature of existence as it is manifest in breakdown itself. Once we have gotten clear on this point, we will be able, in section 3, to appreciate Haugeland's insight that existential commitment is no voluntaristic leap of faith but the essence of mind.

The most economical approach is to begin where Haugeland himself begins: with the phenomenon of intentionality. For Haugeland, Heidegger's break with Husserl lies in rejecting his claim that consciousness is originary intentionality.[12] The semantic quality (or aboutness) of our experience cannot be understood as a function of synthetic relations among conscious acts, or *Erlebnisse*; the modalizations that belong to meaningful experience involve *normative* distinctions that reach into the temporal stream of consciousness from elsewhere. Laboring in the Sellarsian groves of Pittsburgh, Haugeland leveraged the idea that intentionality involves a normative moment—the idea that mental content is not information but meaning—into what might seem to be a very non-Sellarsian position: it is not our rationality that is responsible for the normative ground of meaning; it is our responsibility for normativity that is the meaning of our rationality. Because being responsible is not an act of consciousness but competence over being as existing, a phenomenologically adequate account of intentionality requires clarifying the link between responsibility (commitment) and the first-person stance, the I-myself.

To better understand the role of the I-myself in the constitution of intentional content, we may contrast Haugeland's account with one to which it is greatly indebted but with which it parts company on just this point. Hubert Dreyfus (1991) argues that the intentionality of experience—the intelligibility of things—cannot be grounded in consciousness; meaning just ain't in the head. Instead, following Heidegger, Dreyfus argues that the way things show up depends on Dasein's practical coping—the skills, practices, and roles in which everyday Dasein finds itself engaged. It is my ability to make a birdhouse, for example, that allows entities to show up for me as hammers, nails, and lumber; and it is my facility in the practices of the academic world that allows things to show up as papers to be written or as colleagues to be flattered. The intentionality of practical coping cannot be understood in terms of mental representation, however; its meaning does not derive from a complex of propositional attitudes. Heidegger groups such skills and abilities under the collective term "understanding," where understanding is not "one kind of cognizing among others" but a kind of *competence*, "being a match" for something, being up to it (Heidegger 1962, 182–183; *GA*, 2:190–191). As a category of Dasein's being, understanding is not something that Dasein exercises now and then; it is, rather, definitive of its to-be (*modus existendi*).

Thus, for instance, to understand oneself as a carpenter, philosopher, father, or friend is not to represent oneself under a certain description but to be *able to be* those things; that is, to act in greater or lesser conformity with the norms or rules that constitute them as the practices or roles they are. It is perhaps misleading to speak of rules here, since the norms in question need not be formulated or anywhere articulated.[13] We might better speak of the "measures" that are embedded in the practices themselves as the way "one" does such things. In Wittgensteinian fashion, Dreyfus holds that conformity to a way of doing things *just is* the ground of the normative appropriateness, suitability, or serviceability in light of which we can be intentionally directed toward things *as* this or that. The "ultimate 'ground' of intelligibility is simply shared practices" (Dreyfus 1991, 157).

Haugeland starts from this picture and does not so much reject it as refine it in an absolutely crucial way, without which it is not a *Heideggerian* picture at all. Heidegger himself alerts us to this. Commenting on the novelty of his idea that intentionality is based on "transcendence"—that is, on what *Being and Time* calls "ability-to-be" (*Seinkönnen*) or understanding— Heidegger notes that such transcendence cannot be conceived as a kind of "intuition" in Husserl's sense. It is not some sort of cognitive achievement. However, he continues, "Even less can it be packed into practical comportment, be it in an instrumental-utilitarian sense or in any other" (Heidegger 1984, 183; *GA*, 26:235). Haugeland helps us to see why this is so.[14] An understanding of being (*Seinsverständnis*) cannot be explained in terms of the sort of competence exhibited in skills and practices alone, because such behavior can be characterized *as* skilled practices or abilities-to-be, in the relevant sense, *only* if it is carried out by an agent capable of what Haugeland calls "first-person involvement." I will call such involvement "trying." For instance, for something to be "playing chess," it must be "trying to win" (*HT*, 340); it is not enough to move pieces around in a legal way. Further, "If you are to keep playing—you who are *involved* in this game as a player—then you must *insist* that both your own and your opponent's moves be legal. This insistence on legality is a kind of first-person involvement that is even more fundamental to game playing than is trying to win" (*HT*, 340).

This last point is key: trying to be a carpenter, philosopher, or father presupposes the possibility of acting in *light* of norms rather than acting merely in *conformity* with norms, even though it need not involve explicitly representing such norms to oneself. As I read it, Haugeland's concept of existential commitment captures just this point. But it entails a further point about the *modus existendi* of the entity so endowed: "trying" in the sense of first-person involvement—that is, practical coping, or trying to be *something*—is possible only for a being who is also able *not* to be *anything*, one who can cease to act without thereby ceasing to exist, one whose trying extends to competence over existing itself. If practical coping is a way of establishing what I am (carpenter, philosopher, physicist), then the distinctively first-personal involvement that makes commitment what it is concerns not "what" (*kein Was*) but what it is to be as such (*sondern Sein als Existieren*) (Heidegger 1962, 183; *GA*, 2:191). Here we find the post-Cartesian meaning of I-myself.

Division I of *Being and Time* begins with an analysis of how things show up in my everyday being-in-the-world—namely, as embedded in holistic equipmental contexts governed by in-order-to relations. *Zuhandenheit* (availability), then, is the *modus existendi* of entities whose possible ways to fail depend on what they are good *for*, in the sense of what they are *used* for. The norms that measure such success or failure are thus not intrinsic to what is available but derive from what Heidegger calls the "work" (1962, 99; *GA*, 2:94). In my current situation, for instance, things are showing up as available in relation to the work to be done, that is, suitable for writing a philosophy paper. The computer keyboard, the paper, pen, eyeglasses, books, desktop, and so on, are meaningful because they more or less successfully enable this work. In contrast, the plug on the far wall, the chairs in the corner, my wife's desk, and other such things in the environment (*Umwelt*) are not "available" in this sense. Just as it would be absurd to hold that the meaning of the computer keyboard, pen, paper, and other things with which I work is constituted by specific acts of consciousness aimed at them, so it would be absurd to say that *everything* that can be located in my vicinity right now is available—*if*, with Heidegger, we understand availability as a *modus existendi* whose intelligibility is governed by the work to be done. But what determines what the work to be done is?

For Heidegger, the work does not suffice to determine the normative in-order-to relations constitutive of availability because it is itself merely one more thing on hand, itself meaningfully there by the grace of some more encompassing in-order-to, some more encompassing work (1962, 99; *GA*, 2:94). Nothing about my wielding the pen, typing on the keyboard, looking up a citation in a book, printing out pages, and so on, makes it the case that I am writing a philosophy paper—exercising that particular skill, aiming toward that particular result or work—and not, say, making a work of art or just playing around. Rather, as Heidegger notes, meaning derives from the fact that the equipmental totality (including the work to be done) itself belongs to a *Bewandtnisganzheit*, or "relevance totality," whose character as a *determinate* totality derives from that "for-the-sake-of-which" (*Worumwillen*) everything is being done. And that for the sake of which something is being done—what is at stake in what I am doing—is not of the same *modus existendi* (availability) as what is being done; it "always pertains to the being of *Dasein*"; it is "for the sake of a possibility of Dasein's being" (1962, 116; *GA*, 2:112–130). Methodologically, this means that the normative order on which practical comportment (doing) depends involves a possibility of Dasein's being that is not, in turn, a kind of doing. This makes up what Heidegger calls Dasein's understanding of its *own* to-be, competence over being as existing. When this particular competence is isolated phenomenologically, as it is in breakdown, we learn why it is that "transcendence"—the ground of intentionality—cannot be "packed into practical comportment" (Heidegger 1984, 183; *GA*, 26:235).

For instance, what I am doing with my pen, keyboard, and books is writing a philosophy paper rather than making an artwork not because any of the skills I am exercising necessarily belong to one rather than the other, but because I am trying to be a philosopher (what it means to be a philosopher is at stake in what I am doing), and I am not trying to be an artist. Trying must not be understood as a kind of mental intention, which would reintroduce the cognitivism Heidegger wanted to escape.[15] No amount of intending to be a philosopher can make it the case that I am trying to be one if I am unable to exercise the relevant abilities and skills, though I can play at being one.[16] But what is essential to trying—to exercising competence over being a philosopher, lover, father, or any other such particular *Seinkönnen*—cannot be reduced to the exercise of any set of practical skills or abilities. Merely acting in *accord* with such norms cannot constitute me as trying to be a philosopher, any more than the mere intention could. I can act in accord with them quite accidentally, as when I am playing, or when the behavior they call for just happens to be identical to that called for by being an artist of a certain sort.

But if this is so, then the intelligibility that beings exhibit in the analyses presented in Division I of *Being and Time* rests on a condition that is not analyzed in Division I, a condition without which it is not possible to understand how behavior that happens to be in accord with norms can yield a world, a *determinate* totality of significance (*Bedeutungsganzes*) in which things do not merely appear but show themselves *as* the things they are. What, then, is competence over being as existing, if not a kind of mental intention or practical comportment?

Here the importance of Haugeland's methodological understanding of the existential elements of Division II comes into view. The problem that Division I leaves us with is not the problem of accounting for the norms that govern the practices in which the meaning (being) of entities can show itself. Division I suffices to show that such norms belong to those very practices, to the "public," to the historically variable forms of *das Man*. Rather, the problem lies in accounting for the normative *force* of such norms, how they can get a grip at all such that anything—including Dasein itself as philosopher, father, or physicist—can be measured by them and *itself* be said to live up to those norms or fail to do so. If Dasein is the being through whom behavior takes on the normative shape of a practice, this must be because beholdenness to norms is *intrinsic* to its being, because Dasein cannot be Dasein without being concerned with success or failure at being, without being an issue for itself. But nowhere in Division I has this been demonstrated, since Dasein shows up there only as the "one-self" (*das Man-selbst*), a self who is always engaged in some specific ability-to-be, some publicly recognizable role, always *doing* something. And though in doing something it is always an issue for itself, Dasein can easily "be" without being an issue for itself *in just that way*, since it can choose not to be a father, philosopher, or physicist. Such roles are ultimately as extrinsic to Dasein as the property of being suitable is to the pen I am writing with. And for all that Division I tells us, beholdenness to norms (commitment to holding oneself and others to the constitutive standards of the practices in which one is engaged; concern for successfully negotiating what is at stake in the practice) may *not* be intrinsic to Dasein considered as I-myself. What must be shown is that *I-myself* cannot be without being beholden to norms; that there are stakes merely in *being* at all; that, in addition to possessing competence over existing as a father or a philosopher (competence that I might equally not possess), I always exercise a certain competence over existing *as such*.

Haugeland understands how this possibility is attested in the phenomenon of existential breakdown, but to fully appreciate the deepest methodological point, we need to move a step beyond Haugeland's text. In existential death—with its attendant mood of anxiety, and as articulated in conscience—my practical comportment collapses.[17] Because I no longer gear into the world—that is, have become affectively "disengaged" from my involvements—the norms to which I have been beholden hitherto, which have thereby defined who I am, become inert, lose their normative force; in *Angst*, I am no longer moved by them. Nevertheless, I am "given to understand" something about myself in such a condition, thanks to the way my being is "articulated" in the call of conscience—namely, as "guilty" (*schuldig*) (1962, 326; *GA*, 2:373). Heidegger glosses this idea with the claim that, in conscience, I am called to "take over being-a-ground" (1962, 330; *GA*, 2:377), which, as I understand it, means that I am called to responsibility or answerability (*Verantwortlichkeit*). As Haugeland puts it, this sort of "*being* responsible ... must be something over and above that invariable responsibility that always characterizes Dasein—something that is a *possibility* for it but not necessary" (*DD*, 209). Taking over being-a-ground is *itself* a project, an ability-to-be, which can succeed or fail right along with the other projects in which I am engaged (father, teacher, physicist): the project of being I-myself. This is what I discover about myself in breakdown or existential death as the "possibility of no longer being able-to-be-there" (1962, 294; *GA*, 2:333). That such an "absolute impossibility of Dasein" is *possible*[18]—that it is a way for me to be—means that even when I am no longer *something*, I still *am*, and in such a way that a normative distinction between success and failure can still get a grip. How so?

Haugeland's analysis of existential death entails that even when I am not engaged in doing anything, I am still an issue for myself, still responsible for being. In existential death, I am for-the-sake-of my ownmost self. But what is at stake in such selfhood is no particular "what"; rather, it is my ability to take over my factical situation in light of the normative, that is, in light of the idea of measures of better and worse. This is what makes responsibility for the norms and stakes of my roles and practices—commitment as a "resilient and resolute first-person stance" (*HT*, 341)—itself possible. It is not that my ownmost self is normatively prescribed in terms of some particular role or roles; rather, it is *answerable* for what presents itself to me as normative in such particular roles. In being-a-ground I commit myself to the norms of some practice (say, physics or fatherhood), making them my reasons; and because I do so, what is at issue in that practice can show itself as assessable in normative terms. But commitment to a specific set of norms is a way of life, not a form of existential death, and while such resoluteness (authentic selfhood) may follow on existential breakdown, all such ontic commitments are grounded in the unavoidable (Heidegger says "*unüberhörbar*") call to "be responsible" as such—what might be called commitment to being-a-ground as such.

Haugeland's Heidegger expresses this with the claim that resoluteness involves holding oneself "free for the possibility of taking it back" (1962, 355; *GA*, 2:408), of abandoning altogether the practice in which I am engaged, that is, of recognizing that the world constituted in terms of the norms of that practice is *not sustainable* (Haugeland, *DD*, 216–217). But this is just to recognize that the norms of that practice lack normative authority; that is, their normative force depends on my commitment to them, on the project in which I am responsive to them as my reasons. Phenomenologically, this means that orientation toward the normative (measure) is intrinsic to existing as such, not merely intrinsic to being a father, physicist, and the like. And with that result, *Being and Time* completes its account of the transcendental significance of first-person responsibility by identifying the indispensability of the I-myself.[19]

One way of seeing the indispensability of the I-myself, of the "resolute first-person stance" that is the outcome of Haugeland's interpretation of Heidegger, is to see what happens when it is overlooked or dismissed. I will argue that in the debate between John Haugeland and Joseph Rouse, just this is at issue, and in so doing, I will try to make good on my claim that Haugeland's interpretation shows why Heidegger is indispensable to philosophy more generally. The debate focuses on a corner of philosophy that Haugeland cultivated quite productively, the philosophy of science. There Haugeland takes up the challenge of showing how our normative commitments not only let *equipment* and cultural things be what they are, but also let the beings of *nature* be what they are—beings that have no need of Dasein to be occurrent. The issues here return us to a certain paradox that emerges from Heidegger's seminal idea that *existentia*, and not merely *essentia*, has a plurality of meaningful modes. I mentioned in passing that only occurrentness retains the characteristic of meaninglessness traditionally ascribed to *existentia*. It is now necessary to reflect on the implications of that fact.

3   Essential Unintelligibility

We may introduce the problem with a line from Haugeland's essay "Letting Be." There he writes that "it is one thing to say that Dasein 'lets' its own equipment 'be,' ... but it's quite another to say any such thing about protons, planets, or prehistoric lizards" (*DD*, 173). Nor is this merely a matter of letting things *show up* in one way or another; rather, Haugeland's Heidegger holds that Dasein's "letting be" enables entities "to be—or, be objects—at all" (*DD*, 174). When properly unpacked, this is a very strong— or, to use Haugeland's technical term, "weird" (*DD*, 168)—claim, and he does not pretend to have spoken the last word on the subject.[20] Nor shall I. Nevertheless, its very weirdness offers an angle from which to test our reconstruction of Haugeland's methodological standpoint, since the claim is at issue in a set of objections directed at the supposed implications of this standpoint for understanding natural science. Having argued for the exegetical importance of Haugeland's Heidegger, I want now to examine those objections in order to defend the philosophical significance of the transcendental existentialism that I, following Haugeland, see as the enduring legacy of *Being and Time*.

In his penetrating and wide-ranging book *How Scientific Practices Matter*, Joseph Rouse gives extensive consideration to Haugeland's position on normativity and existential commitment, ultimately rejecting the idea that the I-myself—that is, existential commitment—can be the ground of normative force. Rouse bases his arguments on a residual voluntarism and a residual scientism that he purports to find in Haugeland's position. Regarding the first, Rouse argues that Haugeland's focus on commitment—his claim that binding oneself to constitutive standards is a condition for being able to hold entities to normative constraints—leaves us in a position with regard to normative force that is "like the authority of a monarch in a country where revolution is legitimate" (Rouse 2002, 247). The voluntaristic character of Haugeland's position makes it hard to see "how [he] could articulate a sense in which one ought to be (and remain) existentially committed" (256). For Rouse, there can be no answer to this question as posed, since it comes at the problem of normativity in the wrong way. "What needs to be understood," according to Rouse, "is not how we can commit ourselves to something, however insistently, but how something already at stake in our situation can have a hold over us" (247).

Of course, what Rouse looks for here is just what Haugeland thought he had explained with his account of existential commitment, namely, how in our scientific practice we can be normatively beholden to things—how, *despite* the fact that a thing "is not itself something normative" (Rouse 2002, 257), it can serve as a measure for that practice. And as I read him, Haugeland's transcendental existentialism shows that there is *nothing* "already at stake in our situation"—no context of ontic concern, no specific practice or ability-to-be—that can "have a hold over us" in itself, as Rouse puts it (247), nothing that has *intrinsic* normative force.

Heidegger marks this point by distinguishing between one's "circum-stances" (*Lage*), which are roughly what a third-person inventory of the things and practices in which one is engaged would contain, and the "situation" (*Situation*), which is that inventory from the first-person perspective of one who is resolutely involved therein.[21] The key point is that only this involvement constitutes the circumstances as a situation—i.e., as a deter-minate, normatively ordered space of meaning—precisely because reso-luteness (commitment) has taken over being-a-ground by *letting* things *be* normatively authoritative. Thus, on the one hand, Rouse is right that in the situation something "already at stake [has] a hold on us" (2002, 247), but there is no mystery about how this can be. Nor is there any volun-tarism here: *in* the situation I have precisely embraced claims that present themselves as normative; they are not mine to make up; I *acknowledge* them, feel their affective significance, insist on them. On the other hand, the situation is ontologically dependent on the I-myself disclosed in exis-tential death, a condition in which I can make no appeal to something that antecedently has a (normative) hold on me, since all such "holds" have lost their grip.

Nevertheless, even here I am not a monarch in a land where revolution is legal; I do not voluntaristically establish the normative *validity* of some-thing by means of my choice. Rather, I am in the position of having to recognize my responsibility (answerability) for the capacity of measures to *measure* anything—that is, for the normative force of norms. The validity or authority of these measures is not thereby fixed and established voluntaris-tically by my commitment to them; it is at stake in such commitment. The *meaning* of the norms whose normative force claims me remains in ques-tion. Thus if we recall my addendum to Haugeland's view—that is, insist on a distinction between the condition of breakdown and the situation of resolute commitment that may arise from it—we can see that Rouse's criticism elides the phenomenon of existential breakdown. His rejection of Haugeland's supposed voluntarism proves finally to be a denial of the transcendental significance of death.[22]

This point is connected to Rouse's second objection to Haugeland as well: the charge of residual scientism. According to Rouse, Haugeland fails to recognize the kind of normative force possessed by what is "already at stake in our situation" because his "appeal to objectivity to explicate what our commitments bind us to" involves a "problematic scientism" (2002, 257). The problem with Haugeland's conception of objectivity, Rouse argues, is that it takes nature to be what John McDowell calls the "realm of law," that is, something that "would be normatively constituted [but] not itself something normative" (257). For Rouse, in contrast, "what is at stake" in the practice of science—that for-the-sake-of-which one *is* a scientist—"is normative through and through" and so possesses the normative authority to hold us to our practices (257). According to Rouse, "What is authoritative over such practices is not the modally robust 'natures' of things, but the *normatively* irreducible stakes that emerge in intra-active practical configurations of the world"—which include social, political, affective, and other arenas in which practices of science "*matter* in specific, contestable ways" (261).

Thus both Rouse and Haugeland seek an alternative to the baldly naturalistic idea that "all meaning and understanding must be shown to be explicable in non-normative terms." But while Haugeland's strategy is to show how "the modally robust objectivity of nature could be constituted as significant within scientific practices"—a nature that remains in itself anormative—Rouse's approach is "more radical," since it seeks to head off scientistic naturalism from the outset by insisting on the "irreducibly normative significance of nature" (2002, 262). Here Rouse stands firm on a kind of normative holism for which Haugeland's "objectivity" seems to ignore scientific practices in favor of their propositional yield, granting it a kind of epistemic authority to transcend the limits of the practical situation.

Rouse's proposals deserve extensive examination on their own terms, but this is not the place for that. Instead I want to step back from the details to take a broader view of what motivates these objections, to see whether a different story can be told about the significance of the connection between Haugeland's supposed voluntarism and scientism. The crux of Rouse's criticism is that Haugeland's concept of objectivity, despite its advances over the "epistemic" view—which seeks either to eliminate subjectivity in favor of a view from nowhere or to constrain it with procedures that provide a meta-level surrogate for truth-as-correspondence (2002, esp. chaps. 1–2)—still fatally embraces the self-defeating idea that objectivity is a form of subject positioning that demands that the subject transcend its own epistemic or conceptual limits (336–337). For Rouse, such objectivity cannot be understood as the defining stakes of scientific practices. Objectivity is "not one generic virtue" of science but "a contested, historically specific field, in which scientific practitioners are accountable to what has shown itself to be at stake in their reflexive engagements with the world and their own practices, and thus to the way in which objects matter to the practice" (246).

Rouse thus opposes to Haugeland's existential approach a pragmatic approach to objectivity, one in which the various matters that postpositivist science studies have shown to be at stake in the discourses and practices in *and around* science provide the historically evolving context for establishing "phenomena" of relevance to science and for evaluating the success or failure of such establishments.[23] From a Heideggerian point of view, this proposal appears to embrace something like the nontranscendental reading of the relation between Division I and Division II of *Being and Time* that identifies existence with factical particularity, historical situatedness, embodiment, and all the ontic messiness of life. But while such a perspective does make it appear that Haugeland's concept of objectivity reflects a scientistic reduction of scientific practices, if we understand facticity in Haugeland's transcendental-existential way things appear otherwise: the ontological core (the "generic virtue") of the notion of objectivity is seen to arise from the existential constitution of Dasein itself. And here the distinctive feature of the *modus existendi* of the occurrent we noted earlier—namely, that it cannot succeed or fail at being—becomes relevant. The emergence of the occurrent is not a function of some modern, epistemically distorted conception of nature; it belongs to the being of Dasein.

Breakdown, existential death, evidences a peculiar *modus existendi* of entities. In *Angst* the world and things within the world "have the character of completely lacking significance" (Heidegger 1962, 231; *GA*, 2:247); I am confronted by the sheer fact "that they are beings—and not nothing" (Heidegger 1993, 103; *GA*, 9:114). Thus it is no accident that Heidegger describes natural science as having "strayed into the legitimate task of grasping the occurrent in its essential unintelligibility" (1962, 194; *GA*, 2:203). However, it would be a mistake simply to identify the beings that show themselves in this way with nature, which is not the totality of entities in the world but "itself an entity within the world" (1962, 254; *GA*, 2:280). When Heidegger says that "only on the basis of understandability ... is there a possible access to something which is in principle incomprehensible, that is, to nature" (1985, 258; *GA*, 20, 356), he is making the point that the *modus existendi* of what shows up in the world as (meaningful) nature is the occurrent.[24] Elsewhere he remarks that, unlike the being of the available, "intraworldliness does not belong to the essence [sc. *modus existendi*] of occurrent things as such. ... Rather, world-entry ... is solely the presupposition for occurrent things announcing themselves *in* their *not* requiring world-entry regarding their own being" (1984, 194–195; *GA*, 26:251, italics mine). Thus, as I will now try to show, the ontological picture is more complicated than Rouse makes it out to be: the mode of being of what is at stake in natural science *allows* for normative-epistemic determinateness but does not *demand* it, just as I-myself *exist* in my practical commitments but am not ontologically *defined* by them.

According to Rouse, the practices of natural science are concerned with entities in "nature," which show up as components of intra-active "phenomena." But according to Heidegger, what makes science distinctive is that the "essential unintelligibility" of entities is *simultaneously* at stake in the practice. Haugeland's appeal to objectivity as the norm of scientific life tries to do justice to this point. Objectivity is not, as Rouse suggests, the correlate of a subject position that insists on transcending its own conceptual or epistemic limits; rather, it is the marker, *within* our conceptual and epistemic space, for what belongs to that space in a distinctive way, namely, as possessing the "essentially unintelligible" *modus existendi* of the occurrent. Viewed from a historical and pragmatic approach to the institutions of science, nature will never appear other than a normatively ordered whole at stake in a variety of negotiated ways. But Haugeland's concept of objectivity attempts to acknowledge the ontological character that entities exhibit in the breakdown of such practices, their essential unintelligibility. Thus he agrees with Rouse that nature is "normatively constituted," but

Haugeland also recognizes that at least *part* of what is at stake in that constitution "is not itself something normative"—a position that Rouse rejects (2002, 262).

Put otherwise, Haugeland certainly agrees that scientific practices are caught up in a wider social and historical context in which nature matters to us in various ways, and that the evaluation of scientific success cannot *entirely* be abstracted from these other concerns. Yet he also argues that those practices possess a distinctive aspect, captured in his conception of objectivity, that acknowledges a kind of *limit* to the normative constitution of intelligibility, a limit reflected in the *contested* character of what is presented in such practices *as* objective. This becomes clear only if, in addition to approaching science as a set of historical practices, we also try to identify its meaning as an existential project, a specific for-the-sake-of.

When Heidegger devotes his 1929 *Antrittsrede* "What Is Metaphysics?" to a reflection on science (*Wissenschaft*), his concern is neither with the institutional structure of science nor with epistemological or methodological questions in a narrow sense. Rather, he wants to describe science in its character as a "freely chosen attitude of human existence," a "distinctive relation to the world in which we turn toward beings themselves"; that is, as a project, something for-the-sake-of-which Dasein is (1993, 94; *GA*, 9:104). Nevertheless, he does *open* with a consideration of science as a cultural institution, and he clearly acknowledges Rouse's view that the stakes of science in this sense are plural: political, psychological, social, and so on. More specifically, it is "the technical organization of universities and faculties" that alone "consolidates this burgeoning multiplicity of disciplines" and negotiates what is at issue in what is organized in this way (1993, 94; *GA*, 2:104). Behind all this scientific culture, however, Heidegger purports to uncover what makes science, as an attitude in which we turn toward beings themselves, "exceptional." In contrast to the political, social, technical, and other attitudes, science "gives the matter (*Sache*) itself explicitly and solely the first and last word"; it is "a peculiarly delineated submission to beings themselves … in order that they may reveal themselves." At the same time, Heidegger notes that this submission to beings is also an "irruption" of one being ("man") into the whole of beings such that beings "break open." The "irruption that breaks open … helps beings above all to themselves" (1993, 94–95; *GA*, 9:104–105). Here we cannot but recall the "weirdness" (Haugeland) of the claim that Dasein—precisely in giving beings the first and last word—enables them to be.

If we may call this scientific attitude of giving beings themselves the first and last word a concern for "objectivity," then Heidegger here finds the stakes of science qua science to be objectivity in Haugeland's sense. This notion does not emerge from a scientistic reduction of science as a practice, however; it does not entail the aspects of objectivity to which Rouse objects: the view from nowhere, self-defeating subject positioning, and so on. Indeed, Heidegger says virtually nothing about scientific practices. Rather, in the remainder of the essay, he attempts to identify the transcendental ground of one distinctive stake involved in this attitude or project, an existential sine qua non, even if in practice I don't "know" what "objectivity" amounts to, even if it is "a contested, historically specific field in which scientific practitioners are accountable to" the many ways "in which objects matter to the practice" (Rouse 2002, 246).

As we have seen, this distinctive stake shows up as a dimension of my being in existential death or breakdown. In "What Is Metaphysics?" breakdown is given pointed expression as the experience of the "receding of beings as a whole." In this receding there is *just enough* intelligibility for us to think what is happening: as "all things and we ourselves sink into indifference"—as the norms that constitute the meaning of things in our practical comportments lose their power to motivate us—this "'no hold on things' comes over us and remains," and with them "we slip away from ourselves." When in this way we are no longer claimed by our roles, no longer project ourselves into possibilities, "pure Da-sein is all that is still there"—namely, I-myself as "being possible" (1993, 101; *GA*, 9:111–112).

Heidegger goes on to describe how *das Nichts* is given in this experience, but that is not our concern here. Rather, we need to understand how this phenomenon lies at the basis of the distinctive attitude of science, which gives beings the first and last word. The crucial point is that breakdown discloses beings "in their full but heretofore concealed strangeness," namely, "that they are beings—and not nothing" (1993, 103; *GA*, 9:114). In other words, in this experience, the paradoxical *modus existendi* that both can and cannot be thought is disclosed, is let be. And because this total strangeness of beings can overwhelm Dasein, the "why" can "loom before us," and so also the demand for a project that "in a definite way ... inquire[s] into grounds, and ground[s]" beings (1993, 109; *GA*, 9:121)—namely, the demand for scientific objectivity in Haugeland's sense.

Stated otherwise, scientific explanation—the project of answering the "why" question by giving beings the first and last word—is the sort of thing it is because beings finally *do not matter*, that is, because there is something with the *modus existendi* of the occurrent. Even if nature (an entity in the world) is normatively constituted, occurrent entities are not normative through and through. But this means that the project of science, however ramified, will have something paradoxical about it, since the entities that are constituted in this project can be constituted only as *exceeding* that project in the direction of their essential unintelligibility. That such unintelligibility can be projected in the normative form of a demand that I give beings the first and last word (and thus can *fail* to do so, as when I let my prejudices, politics, and so on, determine what is or is not an acceptable scientific explanation—i.e., when I am not being "objective" in the ordinary sense) rests on the fact that I encounter unintelligibility within *my own* being as being-in-the-world. Thus, in being an issue for myself, the *modus existendi* of nature becomes an issue for me. Whether we call the anormativity of entities "nature" or not, this aspect of the occurrent shows up for me in existential breakdown and must be acknowledged in any ontological understanding of science, *if* it is true that science aims to give entities the first and last word.

In a certain sense, then, Rouse is right: beings *do* make a normative claim on us, bind us to them as demanding the first and last word. But this is only because, paradoxically, they can reveal themselves in what he calls—rejecting the very idea—an anormative aspect. Of course, it is always possible for us to disregard this demand. We do so, for instance, when we deny the strangeness or anormativity of beings and insist on the "irreducibly normative significance of nature" (Rouse 2002, 262) or mistake for "voluntarism" what Haugeland understood about the indispensability of the first-person stance and attempt to soften the responsibility of commitment by appeal to sociality, embodiment, and the always already practically situated character of everyday Dasein. But if one pursues this path, one will not find a less scientistic concept of objectivity; one will have changed the subject. If objectivity is characterized as the stakes of a project in which beings are given the first and last word, then certainly the *last* word, at least, of such beings lies precisely in this strangeness itself: science is the attempt to understand the occurrent in its essential unintelligibility. Understood phenomenologically, this thesis neither expresses skepticism nor arises from an incompletely carried-out rejection of the "epistemic" reduction of the stakes of scientific practices.

Of course, this leaves Haugeland with a problem, for it looks as though the project of science were fundamentally contradictory. On the one hand, scientific practices provide a normative context in which things can show themselves *as* protons, planets, and prehistoric lizards—that is, as meaningful, intelligible, thinkable. To be identifiable *as* anything at all is to belong within such a normative context. On the other hand, the existential roots of science lie in an experience that reveals entities in a *modus existendi* that is anormative, incapable of success or failure, and thus lacking a necessary condition for meaning. This is just what we should expect from an account that bills itself as a transcendental *existentialism*, since it is a hallmark of existentialism that the universe is fundamentally *absurd*. If Haugeland really did hold the sort of scientistic position attributed to him by Rouse, such absurdity should not be a problem: the real or objective would just be whatever science says it is. As I read him, however, Haugeland is far from scientistic. Such supposed scientism stems not from Haugeland's picture of scientific cognition but from his clear recognition of the implications of existential death.

But if that is so, how should we understand Haugeland's scientific realism, the claim that prehistoric lizards and protons have no need of Dasein in order to "be"? This is hardly a simple question, but our reflections suggest, in conclusion, one possible way out. If we could not avail ourselves of Haugeland's insight that *both* the what-being and the that-being of entities are thinkable, the problem that emerges for transcendental existentialism would be insoluble. We would have no way of explaining what science gets right when it gets it right about beings whose *modus existendi* is unintelligible. But if unintelligibility is itself merely one among a plurality of such *modii*, it provides something like a positive space for distinctions to be made. For instance, we can say that when Newton's laws are true, science has gotten the "what" of beings right (i.e., science has let them be *what* they are—can be—in themselves), but we can also recognize that such entities exist in such a way that it makes no sense to say that their "what" is "eternal" or that the different entities picked out by Einsteinian and Newtonian physics are or are not "simultaneously" occurrent. As Haugeland comments at the conclusion of "Letting Be," it is "a difficult and vexing problem to say just what the relationship is between the respective sets of entities [in Newtonian and Einsteinian physics], but simple identity seems ruled out" (*DD*, 177)—since identity depends on (re)identification, which depends on the different normative commitments of Einsteinian and Newtonian physics. But we can hold that these entities differ in *what* they are without asserting that "different" entities *exist*—since existence, for all such entities, is unintelligible and therefore the same.

If we leave such speculations aside, however, Haugeland's reading of Heidegger provides a framework for his *own* powerful alternative to prevailing dogmas in the philosophy of mind. His insight into the nature of existential commitment as the ground of intentionality sets us on a path that leads beyond the dead ends of internalism and externalism, toward what Heidegger might call "the Open": being in the *world*.

Notes

1. *GA* refers to Heidegger's collected works, *Gesamtausgabe* (Heidegger 1975).

2. See Heidegger 1968, 50; 1971, 20: "Every thinker thinks only one thought." Heidegger, who makes this remark in other texts as well, is speaking of Nietzsche here but obviously has himself in mind.

3. Heidegger asserts that, strictly speaking, "only Dasein can be meaningful [*sinvoll*] or meaningless [*sinnlos*]" (1962, 193; *GA*, 2:201). This does not mean, however, that it is wrong to speak of entities themselves as being present to us in terms of meaning, namely, *as* this or that. Indeed, Heidegger makes this point on the same page when he says that "Dasein only 'has' meaning, so far as the disclosedness of being-in-the-world is 'fulfillable' [*erfüllbar*] through the entities that are discoverable in that disclosedness." Disclosedness makes it possible for entities to show up *as* the things they are; to the extent that ("so far as") a mode of disclosure proves actually to be "fulfillable"—to the extent that the entities whose being (meaning) it prescribes can actually show themselves in it as they are (ontic truth)—we can say that these entities too are meaningful.

4. Actually, Haugeland did not discuss corporations in "Heidegger on Being a Person," the article in which he first put forth his controversial interpretation of Dasein. But he did mention "Cincinnati" as a case of Dasein (*DD*, 9), an interpretation that he later rejected after he had refined his conception of Dasein as a "living way of life." See Joseph Rouse's introduction to *Dasein Disclosed* for an illuminating discussion of this change.

5. I refer here to views—espoused in different ways by thinkers as diverse as Theodore Kisiel (1993) and Taylor Carman (2003)—that tend to identify Heideggerian "facticity" with the contingent particularity of social, historical, natural, and linguistic conditions.

6. I leave out of account Duns Scotus's notion of *haecceitas*, which for Heidegger was an important conceptual bridge to the position I am investigating here. See the discussion of "singularity" in Heidegger 1978, 251–254.

7. See the discussion of this regrettable state of affairs in Crowell 2001a and Sheehan 2001. More recently, Sheehan (2015) has expanded his argument into a full-scale interpretation of Heidegger's *Denkweg*.

8. An important precursor, whose work is still essential, is Carl-Friedrich Gethmann (1974).

9. This is the burden of what Haugeland (*HT*, 331–333) calls "the excluded zone," the details of which cannot concern us here.

10. This last is my way of putting the matter, not Haugeland's. I draw it from Heidegger's identification of Dasein's "transcendence" with Plato's *idea tou agathou* (see Heidegger 1998, 124; *GA*, 9:160). For further discussion, see Crowell 2007, 2013.

11. For the details of the argument, see Crowell 2001b.

12. Haugeland does not discuss Husserl specifically, but see his discussion, in "The Intentionality All-Stars" (*HT*), of Searle, out in right field, and the "right-wing phenomenology" of thinkers like Jerry Fodor.

13. Haugeland's "Truth and Rule-Following" (*HT*) provides one of the best accounts of the various sorts of measure or norm that come into play in an account of intentionality.

14. Though I cannot go fully into Haugeland's account of intentionality here, I have tried to develop it in Crowell 2013 as it relates to the Heideggerian category of "transcendence."

15. On the relation to what Searle calls "intention in action," see Crowell 2012. Of course, there are other uses of the term "intending" that do not entail cognitivism. But I would argue that any such use is best captured, ontologically, by Heidegger's analysis of what it is to try.

16. Of course, since which abilities and skills *are* relevant is relative to the understanding (in Heidegger's sense) of what it is to be a philosopher, the whole issue of whether I *can* exercise them, *am* in this case exercising them, is not a matter of empirical fact, not strictly determinable. On the one hand, there are typical ways in which one does things as a philosopher, and there must be such (public) ways of doing things, ways that are normative in the sense of what is normally done. But

because Dasein is not just one occurrent entity among others—because what it means for it to be is always at issue—the standards of what counts as being a philosopher are also always at stake or in question. It is not possible for Dasein to make them up whole cloth; but neither can they be seen as fixed and immutable. I return to this point in section 3.

17. It is central to my argument that Heidegger's analyses of death, anxiety, and conscience delimit a *unitary* phenomenon—the care structure in breakdown—and are not free variables that can function independently. If the care structure is made up of the equiprimordial aspects of affectedness (*Befindlichkeit*), understanding (*Verstehen*), and discourse (*Rede*) (Heidegger 1962, 224; *GA*, 2:239) and are found together in *every* mode of being-in-the-world, then breakdown is the particular mode in which affectedness takes the form of *Angst*, understanding takes the form of death, and discourse takes the form of conscience. Heidegger marks the difference between my sense of myself in breakdown and my authentic sense of myself by characterizing the latter not as anxious but as *ready*-for-anxiety, not as hearing the call of conscience but as "*wanting*-to-have-a conscience" (1962, 342; *GA*, 2:392), not as dead but as "*anticipatory* being-*towards*-death" (1962, 356; *GA*, 2:409). Note also his claim that "authentic 'thinking about death' is wanting-to-have-a-conscience" (1962, 357; *GA*, 2:409). For further argument, see Crowell 2013.

18. As Haugeland puts it, "Death is a way to be … ; in other words, death is a way of life" (*DD*, 210). For a fine account of the difference between existential death and our ordinary conception of biological or biographical cessation (which Heidegger calls "demise"), see Blattner 1994.

19. See Heidegger 1984, 188–189 (*GA*, 26:242–244), where this point is discussed in terms of what he calls "metaphysical egoicity [*Egoität*]." Here Heidegger insists on its methodological centrality, explaining why the "extreme model" of existential breakdown was analyzed in *Being and Time* to demonstrate "existing" as "being towards oneself."

20. Haugeland's closing paragraphs (*DD*, 176–178) express a good deal of tentativeness.

21. Heidegger writes that "the situation has its foundation in resoluteness. … It is not an occurrent framework in which Dasein appears [*vorkommt*], or into which it might even bring itself. Far removed from any occurrent mixture of circumstances and accidents which we encounter, the situation *is* only *through* resoluteness and in it" (1962, 340; *GA*, 2:397).

22. One might gloss this claim in the following way: Rouse's view, like Hegel's, elides the significance of breakdown in favor of a kind of pragmatic holism. Haugeland's conception of commitment represents a certain "tear" in any holistic fabric of intelligibility, which will look like voluntarism and decisionism from the point of

view of a certain conception of the space of reasons. Robert Pippin (2007) makes a similar point in rejecting the idea that breakdown can be "total."

23. Rouse's concept of "phenomenon" is central to his analysis, but we cannot pursue it here. "Phenomena ... are not regularities, but differentially repeatable patterns involving norms of correct or incorrect repetition" (2002, 22). Regarding the issue at hand, "It is easy to forget that objects only exist within [intra-active] phenomena" (2002, 274).

24. For an excellent recent account of the ambiguities in Heidegger's usage here, see Golob 2014.

References

Blattner, W. 1994. The concept of death in *Being and Time*. *Man and World* 27:49–70.

Carman, T. 2003. *Heidegger's Analytic: Interpretation, Discourse, and Authenticity in Being and Time*. Cambridge: Cambridge University Press.

Crowell, S. 2001a. *Husserl, Heidegger, and the Space of Meaning: Paths toward Transcendental Phenomenology*. Evanston, IL: Northwestern University Press.

Crowell, S. 2001b. Subjectivity: Locating the first-person in *Being and Time*. *Inquiry* 44:433–454.

Crowell, S. 2007. Conscience and reason: Heidegger and the grounds of intentionality. In *Transcendental Heidegger*, ed. S. Crowell and J. Malpas, 43–62. Stanford: Stanford University Press.

Crowell, S. 2012. Reason and will: Husserl and Heidegger on the intentionality of action. *Heidegger-Jahrbuch* 6:249–268.

Crowell, S. 2013. *Normativity and Phenomenology in Husserl and Heidegger*. Cambridge: Cambridge University Press.

Dreyfus, H. L. 1991. *Being-in-the-World: A Commentary on Heidegger's Being and Time, Division I*. Cambridge, MA: MIT Press.

Gethmann, C. F. 1974. *Verstehen und Auslegung: Das Methodenproblem in der Philosophie Martin Heideggers*. Bonn: Bouvier.

Golob, S. 2014. *Heidegger on Concepts, Freedom, and Normativity*. Cambridge: Cambridge University Press.

Haugeland, J. 1998. *Having Thought: Essays in the Metaphysics of Mind*. Cambridge, MA: Harvard University Press. (Abbreviated as *HT*.)

Haugeland, J. 2013. *Dasein Disclosed*. Cambridge, MA: Harvard University Press. (Abbreviated as *DD*.)

Heidegger, M. 1962. *Being and Time*. Trans. J. Macquarrie and E. Robinson. New York: Harper & Row. Published in German as *Sein und Zeit* (*Gesamtausgabe* 2), ed. F.-W. von Herrmann (Frankfurt: Vittorio Klostermann, 1976).

Heidegger, M. 1968. *What Is Called Thinking?* Trans. J. Glenn Gray. New York: Harper & Row. Published in German as *Was Heißt Denken?* (Tübingen: Max Niemeyer, 1971).

Heidegger, M. 1975. *Gesamtausgabe*. Frankfurt: Vittorio Klostermann.

Heidegger, M. 1978. *Frühe Schriften (Gesamtausgabe 1)*. Ed. F.-W. von Herrmann. Frankfurt: Vittorio Klostermann.

Heidegger, M. 1984. *The Metaphysical Foundations of Logic*. Trans. M. Heim. Bloomington: Indiana University Press. Published in German as *Die Metaphysische Anfangsgründe der Logik im Ausgang von Leibniz* (*Gesamtausgabe* 26), ed. K. Held (Frankfurt: Vittorio Klostermann, 1978).

Heidegger, M. 1985. *History of the Concept of Time: Prolegomena*. Trans. T. Kisiel. Bloomington: Indiana University Press. Published in German as *Prolegomena zur Geschichte des Zeitbegriffs* (*Gesamtausgabe* 20), ed. P. Jaeger (Frankfurt: Vittorio Klostermann, 1979).

Heidegger, M. 1993. What is metaphysics? In *Basic Writings*, ed. D. F. Krell, 93–110. San Francisco: Harper Collins. Published in German in *Wegmarken* (*Gesamtausgabe* 9), ed. F.-W. von Herrmann (Frankfurt: Vittorio Klostermann, 1976).

Heidegger, M. 1998. On the essence of ground. In *Pathmarks*, ed. W. McNeill, 97–135. Cambridge: Cambridge University Press. Published in German in *Wegmarken* (*Gesamtausgabe* 9), ed. F.-W. von Herrmann (Frankfurt: Vittorio Klostermann, 1976).

Husserl, E. 1970. *The Crisis of European Sciences and Transcendental Phenomenology*. Trans. D. Carr. Evanston: Northwestern University Press. Published in German as *Die Krisis der europäischen Wissenschaften und die transzendentale Phänomenologie* (*Husserliana* VI), ed. W. Biemel (Den Haag: Martinus Nijhoff, 1954).

Kisiel, T. 1993. *The Genesis of Heidegger's Being and Time*. Los Angeles: University of California Press.

Pippin, R. 2007. Necessary conditions for the possibility of what isn't: Heidegger on failed meaning. In *Transcendental Heidegger*, ed. S. Crowell and J. Malpas, 199–214. Stanford: Stanford University Press.

Rouse, J. 2002. *How Scientific Practices Matter: Reclaiming Philosophical Naturalism.* Chicago: University of Chicago Press.

Sartre, J.-P. 1956. *Being and Nothingness: A Phenomenological Essay on Ontology.* Trans. H. E. Barnes. New York: Washington Square Press.

Sheehan, T. 2001. A paradigm shift in Heidegger research. *Continental Philosophy Review* 34:183–202.

Sheehan, T. 2015. *Making Sense of Heidegger: A Paradigm Shift.* New York: Rowman & Littlefield.

3   Ostension and Assertion

Rebecca Kukla

To the extent that man is drawing that way, he points toward what withdraws. As he is pointing that way, man is the pointer. Man here is not first of all man, and then occasionally someone who points. ... His essential nature lies in being such a pointer.

—Martin Heidegger, *What Is Called Thinking?*[1]

In making an assertion a speaker lets what is being talked about show itself from itself, by pointing it out—putting it on exhibit, so to speak.

—John Haugeland, *Dasein Disclosed* (*DD*, 67)

This essay is motivated by a comment that John Haugeland made in a graduate seminar well over twenty years ago, in the early 1990s. Philosophy, he told us, was just a particularly sophisticated and elaborate form of ostension; we use philosophical discourse to direct one another's attention to how things are. Traditional arguments are philosophers' favorite ostensive tools, though they are neither unique nor universally appropriate tools. Contained in this comment was the idea that ostension is generally not a mute pointing but an elaborate, structured activity. There are many means of ostending, and they reveal different things in different ways. In *Dasein Disclosed*, Haugeland puts the point in Heideggerian terms: formal indication, he claims, is the method of philosophy. As Heidegger uses the term, a formal indication is a pointing or showing (indication) that does not (merely) have a determinate descriptive content (hence "formal" as opposed to contentful). Philosophy, then, according to Haugeland, is a kind of ostending or indicating that does not function (merely) by describing. As I took the point, philosophy is in the first instance a skilled collaborative activity wherein we draw one another into contact with an articulated world.

My goals in this essay are as follows. I will develop an account of ostension that builds on Haugeland's and Heidegger's comments but goes into substantially more detail about what ostension is and how it functions.[2] I will argue that while Haugeland was onto a deep insight about ostension and discourse, his account needs correcting in a couple of ways. First, the role he assigns to ostension is actually too small. Heidegger thought that *all* discourse, and not only philosophical discourse, had an essential ostensive component, and I agree. In particular, even descriptive assertions function ostensively in an important sense, although their descriptive content taken independently does not. Second, Haugeland misses the structure and significance of the essentially social character of ostension. My account of the role of ostension in assertion is Heideggerian, at least in flavor. Once I have offered it, I will end with some comments on Heidegger's account of assertoric truth in *Being and Time*. When we understand the character of ostension and its place in discourse, I claim, some of Heidegger's comments on truth that can seem either self-contradictory or vapid become much clearer and in many ways compelling.

1   The Nature and Function of Ostension

Pointing is our stock, minimal example of an ostensive act. But in fact we can ostend in a wide variety of different ways. In describing philosophy as ostension or formal indication, Haugeland is pointing out that speech acts can themselves be ostensive acts, and this includes not just unstructured exclamations like "Lo!" or "Hey look!" but grammatically sophisticated speech acts that make highly theoretical claims.

In *Philosophical Investigations* and elsewhere, Wittgenstein makes clear that the epistemology of ostension is complex. When I (for instance) point at something, I might, as far as the direction of the point goes, be calling your attention to any of an infinite number of objects and properties. But my interest here is not with such Wittgensteinian (or Kripkensteinian) puzzles. I take it as given that we somehow manage to ostend successfully most of the time. And notice that the success conditions for ostension are in part epistemological, in an inherently social sense: I ostend successfully only if the person for whom I am ostending does manage to figure out what I am indicating or ostending. This will be crucial later. My goal is not to explain how this epistemological success comes about but rather to explore what ostension *is* and what exactly its success conditions involve.

Successful ostension directs another's attention so as to allow her to encounter and grasp an object as it is, on its own terms. I propose that ostension, as I am using the word, has three necessary and interlocking dimensions. One, ostension does not merely pass along theoretical knowledge or entitlements to assertions; rather, it brings someone to a substantial practical encounter with a thing. It seeks to allow someone to grasp how entities are, to "get it," as Haugeland sometimes puts it. In Heideggerian terms, when an ostension is successful, the one for whom something has been ostended now comports herself toward the ostended thing in a way that counts as understanding it, coping with it, or grasping it—even (but not only) if this just takes the form of making sense of what she sees. Two, ostension is a collaborative and second personal activity—I ostend something *for you* so as to draw you into an act of mutual seeing *with me*. Notice that we need this piece to understand why the success conditions for ostension include epistemological success of a certain sort. My ostension is successful only if the person for whom I am ostending latches onto the correct object. Three, successful ostension involves practical skills on the part of both the ostender and the addressee. As Wittgenstein argues, even the most basic act of pointing counts as ostending only insofar as it draws on shared standards of salience and a concrete performative context, as well as both parties' complementary skills at drawing attention and having it drawn. In ostending something for you, I comport myself toward the entities ostended in a specific way, as do you in picking up on my ostension.

Calling attention to an object is a form of embodied engagement with it—one that requires a complicated understanding of the object itself, the addressee, the practical norms for drawing attention, and the rich background context of shared salience that makes drawing attention possible. Many philosophers have discussed how our capacities to see are not mere receptive capacities. They involve skills that range from the mundane (how to hold my head, where to stand) to the rarefied (how to use scientific instruments, interpret information, make fine discriminations that can only be made after extensive training, etc.).[3] Likewise, ostension is a practical act, which draws on its own set of skills; these are essentially related to but distinct from our skills at seeing.

Ostension puts us into real, practical contact with the thing ostended—a kind of contact that involves attention to the thing. Cashing this out precisely is difficult, but at a minimum it involves some sort of first-person encounter with the thing that includes skills for coping with it, even if the kind of coping called for is quite minimal and mediated. Notice, as will be important, that we can make true, justified claims about something without having this kind of first-person attentive encounter. I might memorize a sentence and have good reasons to believe it is a true sentence—because I read it in a textbook published recently by a respectable press, say—without encountering the part or features of the world that the sentence is about.

Ostension brings about this kind of attentive encounter through a social process. Unlike some other skills at seeing, ostensive skills are inherently social; they are the skills that enable us to draw others into acts of mutual seeing. They require that we be able to capture and focus the attention of others in the right way, enabling us to share undistorted contact with specific features of entities. Furthermore, ostension will succeed only insofar as the one addressed by the ostension has her own different but corresponding skills at seeing objects, interpreting ostensive gestures, and catching on to shared saliences. While I can perhaps, in some derivative sense, ostend something to myself, clearly this act is parasitic on the more typical case in which I ostend second-personally. We ostend using language, gesture, touch, signs and symbols, and anything else that can be communicative.

I have just tried to give an account of *what sort of act ostension is*. A related but separate question is what kinds of things ostension can pick out, and how. What we manage to ostend and whether our ostension is successful depend on a vast array of social conventions and our skills at marshaling them, as well as the character of the world itself. Three points about *what can be ostended* are important here.

First, there is no such thing as mute or simple ostension. Ostension is a structured act in at least three senses: The act of ostension draws on structured skills, as does the response to the ostension. And less obviously, but importantly, all ostensions as such uncover something as having a determinate character of some sort. We cannot pick out a mute "this" but rather must draw on our joint skills at noting salience and communicating to bring someone to grasp something in a specific way—even if it is as minimal as "that material thing in that location over there." Were we to try to pick out something as—what?—a mere presence or something like that, our structured skills at seeing, attending, and drawing attention would have nothing correspondingly structured to grab onto.

Second, there is no reason to restrict ostension to physical entities. Successful ostension results in grasping or seeing what is ostended in a way that is undistorted and enables practical coping with it, and this will be embodied. But there is no need for us to assume any simple sensory contact as the medium of this grasping. Indeed, Heidegger is explicit that "seeing" and "sight" are his generic terms for concretely and explicitly grasping how entities are, whether or not they are directly available to sensory perception (2010, 187). Ostension draws someone's attention to something as having some determinate character that it actually has (while leaving other features of it indeterminate), in a way that involves a practical grasp of the thing. There is no reason why direct perception needs to be the medium of this attention, or that the practical grasp need involve direct physical manipulation. Wittgenstein's main example of ostension is of how to go on following a rule. This is an excellent example for my purposes. The ostender tries to get the addressee to attend to a norm of behavior in a way that involves practical grasp of how to continue it. Regardless of any puzzles concerning how we manage to do this, we are routinely successful at this kind of ostension. Hence we can indicate things that are nothing like midsize physical objects. We can ostend states of affairs, moral complications, responsibilities, abstract trends, and things with any number of other metaphysical characters.[4]

Third and finally, once we acknowledge that we can ostend things other than physical objects, it becomes clear that the physical proximity of the object of ostension is not required. When something is ostended successfully for me, I encounter it in a way that lets it show itself as it is, but since this showing need not be a literal physical seeing, and since my comportment toward it can be indirect, there is no reason to think that something needs to be physically proximate to show up for us in the relevant sense. This is so even when what is ostended is in fact a midsize physical object. Physical proximity is not a requirement for attentive engagement. Think, for instance, of the way that a close friend can be acutely, even viscerally, present to you over Internet chat. You have much more than a theoretical knowledge that your friend is the one out there somewhere typing back to you; often, in the flow of conversation, you have a direct, attentive grasp of your friend's presence, albeit not her proximate physical presence. In arguing that discourse is ostensive, Heidegger claims that in discourse with someone we are "already *with him*, in advance, alongside the entity that the discourse is *about*" (2010, 207; italics mine). Ostension, as I am using the term—and I will return shortly to the claim that discourse is fundamentally ostensive for Heidegger—brings us alongside that which is ostended. But "alongside" does not mark a physical distance and orientation here. A dance performance may enable me to grasp the pain of losing a child; a theoretical discussion may bring me "alongside" the plight of the Palestinians in Israel; a painting may give me practical access to a way of life.[5] Discourse is especially (though not uniquely) well suited to bringing about this kind of attentive encounter with something that isn't in one's perceptual path.

If ostension does not require physical contact or even a physical entity as its object, what makes it distinctive? How do I ostend something for someone else, as opposed to just teaching her something about it, passing on a claim to her, or giving her new knowledge? I will return to this question later, and I also do not think that any sharp line exists where ostension stops and the mere passing on of knowledge begins. Yet, clearly, we can inherit beliefs and entitlement to claims without directly grasping or being alongside their object in the way ostension enables—without having the thing "in view." For instance, I have excellent evidence that my friend is to be trusted when it comes to claims about money management. He has convinced me that that short-selling stock is a wise financial strategy at the moment.[6] I am willing to repeat this claim to others and can draw a large number of inferences from it. But I have little or no concrete sense of what it means. He convinced me of it, but not by ostending—not by directing my attention toward something that I can now grasp and comport myself toward skillfully, except in a minimal, degenerate sense. I trust him because of other pieces of knowledge I have about him, but his words put me in no meaningful practical relationship with short-sold stocks.

An ostension succeeds only when it brings someone to see something as it really is, and this involves success in at least two dimensions: we must succeed at directing attention, and we must succeed at displaying things correctly. An ostensive act can fail for a number of reasons: it can misdirect or fail to capture someone's attention; it can call on someone to attend in a way that he does not have the requisite skills to accomplish (e.g., because it relies on his making discriminations he is not in a position to make); it can attempt to direct attention to something that is not actually there; it can direct attention to different features of a situation from the ones the ostender meant to point out; or it can distort the addressee's vision in directing it. Distorted ostension can be quite mundane: I shout, "Lo, a rabbit," when actually in the presence of a cat. Or it can be highly rarefied; in conversing with my friend about his dying relationship, I may gradually bring him to see it as a case of tragic lost opportunity, whereas it is in fact an affair that burned itself out because it was only based on both partners' fantasies of escape from other problems in their lives.

To summarize, *successful ostension is a communicative, coordinative, structured social activity that functions to bring about shared, accurate attention to the determinate character of some object(s) or feature(s) of the world.*

2 Ostension and Assertion

Haugeland does not give us much by way of an explicit account of ostension or formal indication. But he does make clear that he associates the kind of ostension distinctive of philosophy with a special kind of grasping of things as they are in themselves that thrusts us past the everyday, fallen understandings that mark daily life.

Ordinary language and common sense are relentlessly *homogenizing*. They absorb into themselves and redigest all contrary conceptual innovations, spitting back homonymous domesticated simulacra. ... Therefore, all viable philosophical concepts must remain *directly* grounded in the phenomena themselves—as opposed to those comfortable public accommodations. (*DD*, 74–75)

Whether Haugeland is right or not about the role of philosophy is a question I put aside here. What I want to push on is his quick opposition of grounding in the phenomena themselves and "comfortable public accommodations." Even very routine assertoric discourse can function ostensively and be "directly grounded in the phenomena themselves," without being some kind of special ontological discourse that thrusts us out of our average, everyday understandings.

For his part, Heidegger clearly took assertion in general to have an essentially ostensive, indicative function. An assertion, for Heidegger, is a communicative directing of shared attention through a pointing out, and its distinctive method of pointing out is predication: "Every predication is what it is, only as a pointing-out" (2010, 199). Assertions, in essence, call others to shared attention and bring them into common contact with their objects *by way of a predication* that assigns a definite character to the object. We have seen that ostension does not constitutively require physical presence or perceptual contact with its object. Thus the fact that we can make assertions about things that are not there does not preclude assertion having an ostensive component. Heidegger writes:

> The basic structure of assertion is the exhibition of that about which it asserts. ... I am making an assertion not about ideas but about what itself is meant. All further structural moments of assertion are determined by way of this basic function, its character of display. ... The primary character of assertion as display must be maintained. (1988, 207)

Furthermore, the way that assertion displays is in the first instance *communicative*; in the most fundamental case, assertions display determinate features of the world *to others*:

> "Assertion" means "*communication*," speaking forth. ... Letting someone see with us shares with the other that entity which has been pointed out in its definite character. That which is "shared" is our *being towards* what has been pointed out—a being in which we see it in common. ... We may define "*assertion*" as "*a pointing-out which gives something a definite character and which communicates*." (Heidegger 2010, 197–199)

Indeed, if assertion were not essentially communicative, it would become oddly pointless, in a Heideggerian context. Heidegger repeatedly argues that although assertion exhibits or displays beings as they are, it can do so only if they are already available for display. Hence assertion does not make something available but points to what is already available, and this pointing has its paradigmatic purpose in a communicative context. Assertion is primarily an interactive activity that requires us to comport toward entities and display them *for* and *with* others.

> As display, ... [assertion] always relates to some being that has already been unveiled. What thus becomes accessible in determinative display can be *communicated* in assertion as uttered. ... Communication does not mean the handing over of words, let alone ideas, from one subject to another, as if it were an interchange between the

psychical events of different subjects. *To say that one Dasein communicates by its utterances with another means that by articulating something in display it shares with the second Dasein the same understanding comportment towards the being about which the assertion is being made. …* Communications are not a store of heaped up propositions but should be seen as possibilities by which *one Dasein enters with the other into the same fundamental comportment toward the entity asserted about, which is unveiled in the same way.* (1988, 210)

Thus assertions do not merely exhibit something *so that* others may see it. Rather, they have a more specific performative structure. They serve to let us "enter with another into the same fundamental comportment toward the entity," and to "let someone see with us" or "see in common." In assertion, one person draws another (or others) into shared comportment toward entities, through an act that shapes and binds the attention of the other.

3   Telling, Communication, and Reference

Heidegger's account of assertion as in the first instance a special kind of ostension has its home in his larger story of understanding, interpretation, and discourse, and it is worth briefly recapping this story to motivate and clarify his account. For Heidegger, we grasp entities in the world first in our practical dealings with them; when we grasp them, they present themselves to us as being certain kinds of entities to be coped with in specific ways. We do not hear pure noises or see patches of colors that need interpreting separately before we can cope with them but rather hear and see objects that already present themselves as calling for responses and interactions of specific sorts. In Heidegger's terminology, this first encounter is already an *understanding* of entities, even though this understanding may not be accompanied by an explicit linguistic description. This understanding always embeds an "as-structure," an *interpretation*; we always take things *as* being a certain way or demanding a certain kind of response, and this as-structure can either be rendered explicit or remain "circumspective": "In dealing with what is environmentally ready to hand by interpreting it circumspectively, we 'see' it *as* a table, a door, a carriage, or a bridge; but what we have thus interpreted need not also be taken apart by making an interpretation which definitively characterizes it" (Heidegger 2010, 109).

Equally fundamental to, and co-constitutive with, circumspective understanding and interpretation, for Heidegger, is *Rede*, typically translated as
"discourse," but translated by Haugeland as "telling."[7] *Rede* articulates the
intelligible as-structure of that which we understand, in a public, communicative way. Understanding may remain practical and circumspective, but
because it involves comportments toward objects *as* entities of specific
sorts, this understanding always contains within it the possibility of being
made explicit and public in telling. Haugeland prefers the translation "telling" to "discourse" because of the term's flexibility: to tell, he points out,
means both to speak and to discriminate (as when we tell the time, e.g.).
Telling, for Haugeland, is a matter of comporting toward objects in a way
that discriminates how they are, and as such it is an essential activity of
Dasein. Telling is broader than speaking in language. It is the "articulation
of intelligibility," broadly construed—that with which we communicate an
interpretive understanding.

If we stop here, it is difficult to see how to draw boundaries around *Rede*.
It can look as though telling is just the ever-present public face of our practices of comporting toward objects. William Blattner bites this bullet:

> As Reiss sits in her mountain retreat writing a novel, she makes known or manifest
> the workshop of the author. She writes with the computer, thus, as it were, stating
> publicly that computers are to be written with. ... Making publicly known (i.e., com-
> municating) does not require another person to receive the communication; com-
> munication is not, after all, transportation. Rather, communication requires a public
> domain in which one can, in principle, make known things that are interrelatedly
> differentiated as one takes them in one's comportment to be. (1999, 73)

I take Blattner's point that communicative acts need not have an audience
in each case, but I worry about this sweeping reading of telling. It is not
clear what makes something a specifically *communicative* act as opposed to
a merely public act as long as we just read telling as discriminating in action.
We can also raise a more narrowly interpretive worry about Blattner's reading: while Heidegger does not think that *Rede* needs to be specifically linguistic, he clearly uses the notion to make the transition from his discussion
of circumspective understanding and interpretation to his discussion of
language and assertion, and we need to understand how and why.

But Haugeland's term of choice, "telling," contains a crucial clue. Telling
indicates not just discriminating and talking but *talking-to*. The communicative dimension of *Rede* does not just lie in its publicity. Rather, when we

tell, we engage in an activity that, at least in its central cases, has a specific second-personal dimension. Blattner's Heidegger does not impute any pragmatic complexity to communication beyond public availability. But I have emphasized that communication requires the engagement of special sets of skills, many of them intersubjective and inherently second personal. Typing on one's computer does not constitute a second-person telling. Language, gestures, and other forms of communication do. Telling includes much more than assertion; it can include all forms of speech, along with gesture, facial expression, and anything else that calls on others to see and interpret the world the way that we do.

In drawing together discriminating and articulating in the notion of telling, Haugeland gives us a tool for understanding the communicative dimension of *Rede*. However, he does not attend to the second-personal structure of this communication. As far as his account goes, telling can just involve putting the determinate character of something on display, in an impersonal sense. As we saw at the start of this chapter, he wants ostension to be fundamental to discourse. But we can only give that intuition precise cash value once we add in the second-personal pragmatics of this kind of communicative display. Heidegger writes:

> Discoursing or talking is the way in which we articulate "significantly" the intelligibility of Being-in-the-world. Being-with belongs to Being-in-the-world, which in every case maintains itself in some definite way of concernful Being-with-one-another. Such Being-with-one-another is discursive as assenting or refusing, as demanding or warning, as pronouncing, consulting, or interceding, as "making assertions," and as talking in the way of "giving a talk." (2010, 161)

Heidegger clearly wants it to be essential to *Rede* that it is a communicative act, in the pragmatically full-bodied sense that it has the structure of an address to another person or other people (even if in particular instances it is to no one in particular). His examples of speech acts—demanding, consulting, warning, interceding—are all acts that are *essentially* addresses to specific others, and not speech acts that incidentally reveal something to others merely in virtue of being public events. Merely using equipment on a mountain or even in a public square does not meet this criterion.

The other clear distinguishing character of discourse, as opposed to other forms of engagement with objects, is that discourse is *about* something. It has an intentional object. When we type on a computer, what we type is (we hope) about something, but presumably the act of typing is not about the computer. Heidegger writes: "Talking is talk about something. ... Even a command is given about something. ... What the discourse is about is a structural item that it necessarily possesses; for discourse helps to constitute the disclosedness of Being-in-the-world" (2010, 161–162).

Now, typically, philosophers reason thus:

1. Discourse is about things; it has intentional objects.
2. This is because it has semantics.
3. If we want to understand how discourse hooks onto the world, we'd better study what semantics are and how discourse manages to have any.
4. {Hijinks ensue.}[8]

Heidegger enables us to construct a picture of discourse, or telling, that is at root ostensive—one that begins with basic pointing and gestures and builds up to assertions and other kinds of structured speech acts. I think one of the most powerful offerings of such a picture is that it changes this path in a way that removes the need to address resilient puzzles about semantics. In this picture, we reason thus:

1. Discourse is about things; it has intentional objects.
2. This is because it is ostensive, and ostensive practices direct us toward objects; that's just what they do, by definition and in a concrete sense.
3. Discourse has semantics because it is ostensive; "semantics" is just a name for what you get when you have the kind of complex ostensive practices that let you indicate things that are not there.
4. {No hijinks ensue.}

I want to spend some time spelling out this second path and trying to make it seem plausible. In discourse, we draw one another into shared comportment toward objects. This involves practical engagement with those objects, even if from a mediated distance. Discourse is not a tool for trading representations around; it is a set of second-personal practices that orient us toward a shared world. Telling is something we *do*; it is a special kind of comporting activity. "Communication is never anything like a conveying of experiences ... from the interior of one subject into the interior of another. ... In discourse, Being-with becomes 'explicitly' shared" (Heidegger 2010, 205). All understanding embeds the possibility of discursive articulation, but not all understanding is telling. Hammering a nail is not about nails. Making a declarative claim about oranges is about oranges, but also,

catching my friend's eye at a party and raising my eyebrow just so may well
be about his ex-girlfriend. It may serve to draw him into shared attention
toward her, uncovering her for both of us *as* … (having just walked in,
being distraught, looking hot tonight, whatever). Note that the eyebrow
raise need not have a determinate propositional content to have a specific
ostensive function. My raised eyebrow succeeds in telling my friend something about his ex-girlfriend, not because it has a representational structure
that corresponds to his girlfriend, but because we have in place a wide array
of world-involving practical norms that enable this act to function ostensively in this way. Raising my eyebrow here serves as a coordinating activity
that succeeds in putting us both into attentive engagement with the same
features of the world.

A large part of what makes this story compelling to me is its potential to
Heidegger insists that we must understand assertion in this context, as a
special and rarefied form of *Rede* or telling. It is in the first instance an activity; the noun form of "assertion" is derivative. In assertion, we use the compositional structure of language to accomplish a highly specific form of
shared ostension: we draw attention to the determinate predicative character of things. Assertions are "exhibitive discourse," as Heidegger puts it
(1988, 180). They still constitute comportment toward entities, but of a sort
that has a kind of minimal form: assertion as such involves little complex
material engagement with objects; rather, it lets them be seen as they are,
in their determinate character.

A large part of what makes this story compelling to me is its potential to
demystify "aboutness," or reference. Philosophers have struggled to articulate the special relationship between representations and referents that
constitutes intentional aboutness—a relationship that always seems in
danger of being unnaturalizable. According to the Heideggerian story I am
suggesting here, practices, rather than representations, are what most para-
digmatically have aboutness, although representations (or signs, as Heidegger likes to put it) can perfectly well have aboutness as equipment
within such practices. And what gives them that aboutness is their
function: insofar as the function of my practice is to call attention to something, to make it show up in a way that allows it to be grasped, then that
thing is what it is about. This can be accomplished in an indefinite variety
of ways, and so there is no special relation that is "the" reference relation.
But again, while each of us can perhaps, in a slightly odd and derivative
sense, point things out to ourselves, the primary home of aboutness is in

communicative, social ostension. Trying to induce shared attention is just a kind (or actually a wide, loose network of kinds) of coordinative activity that we engage in together. Hence reference is at root a social accomplishment.

This gives us a way of understanding how assertion and other discourse direct us to specific parts or features of the world without a prior or independent account of semantic content as the mechanism for doing so. As I alluded to earlier, philosophers often ask what the relationship is between assertions (or sentences, beliefs, etc.) and the bits of the world they are about, which begins a notoriously philosophically frustrating quest for a self-standing account of semantic content that can be used to glue together bits of language or mental states and the world. Following Heidegger, I am suggesting that we start by understanding "aboutness" as a feature of ostension, where ostension is a coordinative social activity used to calibrate and share attention and engagement with the same entities. The most straightforward and basic forms of ostension are concrete uses of signs and pointing and the like. We can then build up to assertion and other discourse with traditional "semantic content." Instead of focusing on *assertions* as *entities* whose content requires explanation, we focus on *asserting* as an *indicative practice* that happens to have an elaborate and powerful recursive structure. This structure allows flexible and detailed forms of indicative coordination. The content of the assertion is an abstraction from this practice; it is what is indicated. But the indication practice explains the content, rather than our needing an independent account of content that makes it capable of explaining the hookup to the world.

For the most part, this is, I think, an account that Haugeland would embrace. But again, the social dimension of the story is critical to its making any sense, and Haugeland himself did not give attention to the social pragmatics of asserting. It is because ostension is a coordinative activity aimed at establishing a *shared* object of attention that it makes sense to think of it as having aboutness. Without this sharing, there is no particular reason to count an engagement with the world as an indication relation. When I hammer a nail, I cope with the nail, and my actions are directed toward the nail, but we do not think of the hammering as having a content that makes it about the nail. On the other hand, if I were using a hammer to get you to attend, with me, to the nail—banging on the nail to draw you to it—then the hammering would in fact be about the nail. Haugeland, like me, takes asserting—and more generally truth telling, of which more later— as a practice that *displays* through articulation how things are. But unless one has on board the idea that displaying is characteristically a social transaction, in which I show something *to* someone (or even to everyone or anyone), Haugeland does not seem to have the resources to distinguish between an activity that is merely directed toward and engaged with an object and one that is (also) about an object. Yet surely we want to be able to say something about the sense in which language, art, and other representational practices have content or aboutness whereas other engaged practices like normal hammering do not. I am working to give an enhanced version of Haugeland's story that lets us explain this distinction purely through attending to social pragmatics, without any extra metaphysics of meaning.

I recognize that we typically think of utterances and thoughts as having semantic aboutness even when they are not addressed to anyone. I would like to remain agnostic, here, as to whether we are wise to do so, but in any case I want to insist that such content is at most of a derivative sort that would need to be cashed out counterfactually somehow. For Wittgensteinian reasons, I think that we can retain the identity of inherently social practices when we engage in them all on our own in an oddly degenerate, problematic sense at best.[9] But since, as I have described the situation, semantic contents are an abstraction that do not themselves play any explanatory role, I needn't have much invested in what we decide to say about this matter.[10]

Heidegger insists repeatedly that assertion is a derivative form of uncovering, because we can only use assertion to ostend that which is *already uncovered* and thereby available for this type of ostending: assertion displays an as-structure that is already present in interpretation. It expresses this as-structure, "and this is only possible in that it lies before us as something expressible" (2010, 190). It is only because we are already in the midst of circumspective understanding and interpretive comportment, because entities are already uncovered for us within a shared structure of salience, that we are able to point one another's attention to these entities and their character. "Assertion cannot disown its ontological origin in an interpretation which understands" (Heidegger 2010, 201). Now, I think that Heidegger overstates the case here; I see no reason why assertions, as well as any other ostensive tools, cannot sometimes serve to reorient the addressee to the world in a way that makes new things available for attention rather than just drawing attention to what was already available. But the deeper point is that assertion emerges as a possible ostensive tool only against the background of an elaborate set of shared interpretive comportments toward entities.

If one thinks that the key to understanding semantic content is figuring out what features of bits of language make them hook onto corresponding bits of the world, then Heidegger can sound as if he is just giving a vapid restatement of the problem when he insists that assertions "bring us along-side" entities. Furthermore, since what assertion uncovers is, according to Heidegger, already uncovered, the claim that assertion directs us to objects can sound pragmatically as well as semantically vapid. If entities are already uncovered for me, then what exactly happens when an assertion points or directs me to that entity? What is the practical cash value of this new act of directing?

But this act of directing seems odd and redundant only when construed individualistically. There is nothing mysterious about the idea that I can use words, gestures, and other equipment as tools in aid of drawing others into shared attention to something that may already have been available and uncovered, in the sense that our shared public interpretation gives it salience and significance. And we all understand that we should not expect points, eyebrow raises, expletives, and so forth to have an internal compositional structure that explains or determines what they point at. We understand that these actions are *about* the world because of how they are used within our developed practices of ostension and practical coping. As Taylor Carman aptly puts the point, "Semantically referring linguistic terms, then, are refined modifications of preassertoric interpretive gestures whose communicative function is to highlight salient expressible aspects of the world as it is already intelligible in practice" (2007a, 220).

If we try to understand how assertions have content first, via some sort of account of content and reference, and then add on that they can be used to communicate and ostend, we misconstrue their basic character, I think. We also render their relationship to the world puzzling, because we then need to find internal linguistic features of assertions that suffice to somehow attach them to the world in a structured way. Instead, follow-ing Heidegger, I suggest that we should see them as being of a piece with, and drawing on, the resources already provided by the broader context in which we comport ourselves toward objects and draw on practical norms for how to uncover things for one another. Assertions manage to ostend in a particularly sophisticated way that draws on the compositional structure of language (of which more later), but we need not find their content "inside" them somehow any more than I need find my friend's ex-girlfriend within my eyebrow.

4 Assertion as Mere Passing-On and "Idle Talk"

As I mentioned earlier, if we are going to read assertion as fundamentally an ostensive act, we need to make room for the fact that not all communication that imparts knowledge or entitlements to claims is meaningfully ostensive. Clearly my speech can have a topic, an intentional object, and it can be taken up and used appropriately by others, even if it does not serve to bring them to grasp the object itself attentively, as my friend's financial advice about shorting stocks exemplified. In fact, Heidegger both insists on the ostensive essence of assertion and at the same time uses assertion to explain how this derivative kind of *Rede*, which has an object and communicates but is not really ostensive, is possible. One of the special things about assertions is that they are particularly suitable for passing on claims, whether or not doing so brings those who inherit the claim into a proper encounter with their object.

> As something communicated, that which has been put forward in the assertion is something that Others can "share" with the person making the assertion, even though the entity which he has pointed out and to which he has given a definite character is not close enough for them to grasp and see it. That which is put forward in the assertion is *something which can be passed along in "further retelling."* There is a widening of the range of that mutual sharing which sees. But at the same time, what has been pointed out may become veiled again in this further retelling, although even the kind of knowing which arises in such hearsay ... always has the entity *in view.* (Heidegger 2010, 198)

In other words, unlike commands, raised eyebrows, and various other forms of telling, assertions—because of their minimal call for practical engagement with their objects and their compositional structure—allow for a special kind of breadth of use. They can be merely passed on and used by others as entitlements to make further claims. This kind of passing-on is double sided, as Heidegger describes it. On the one hand, this kind of mere passing-on still has a minimal ostensive core; it is still a "mutual sharing which sees." On the other hand, the retelling can "veil" the object, even while it is still in some sense "in view." Similarly:

> What is expressed [in assertion] becomes, as it were, something ready-to-hand within-the-world *which can be taken up and spoken again*. Because the uncoveredness has been preserved, that which is expressed (which is thus ready-to-hand) has in itself a relation to entities about which it is an assertion. Any uncoveredness is an uncoveredness of something. Even when Dasein speaks over again what someone else has said, it comes into a Being-toward the very entities which have been discussed. But it has been exempted from having to uncover them again, primordially, and it holds that it has thus been exempted. ... Dasein need not bring itself face to face with entities in an "original" experience; but it nevertheless remains in a Being-toward these entities. In large measure uncoveredness gets appropriated not by one's own uncovering, but rather by hearsay of something that has been said. (Heidegger 2010, 224)

Heidegger here emphasizes the double-sided character of assertion that is merely passed along. It at once allows us to communicate without face-to-face encounters with entities, via hearsay, but at the same time it involves an attenuated form of comportment toward entities. This is ostension stretched to its limit, and in it the uncoveredness of entities is mostly "appropriated" but not entirely lost. According to Heidegger, this kind of mere passing on of a claim involves mere talking without real grasping, even if we don't notice that as we go.

This is at least plausibly what Heidegger means by *Gerede*, or "idle talk." Haugeland writes, "A pretty good colloquial characterization of idle talk would be as follows: 'blathering on without knowing what you are talking about'" (*DD*, 163). This is the common but attenuated kind of talk that keeps us all on the "same page," as it were, but does not really involve substantial understanding or contact. We can have genuine ostensive discourse that uncovers even when we talk about something that is not physically proximate. But proper *Rede* directs others to objects themselves and uncovers them, whereas *Gerede* does so only in a minimal, degenerate form and mostly just adds to our repertoire of repeatable claims. Perhaps the most obvious example of idle talk is the political sound bite: the kind of platitude that gets passed on nearly verbatim by large numbers of people who do not reflect at all on their lack of substantial grasp of its objects. "A public health insurance option would lead to socialized medicine," for instance. But as we saw earlier, it can also take the form of perfectly legitimate acceptance and repetition of a claim we have good reasons to accept but no ability to grasp, such as that short-selling stocks is a wise financial strategy right now, or that in gauge theory, the Lagrangian is invariant under a continuous group of local transformations. Idle talk is far from merely pernicious. *Gerede* provides an inferentially fertile framework that fills in and gives shape to our shared interpretive space within which genuine uncovering and ostending are possible. It provides a public fabric that helps to hold us in a shared world. But at the same time, it exists only as a degenerate and derivative form of ostensive discourse, which lets us fall away from a direct shared grasping of objects.

Heidegger claims, "That which is put forward in the assertion is *something which can be passed along in 'further retelling'*" (2010, 197). Haugeland reads this as a degenerate kind of asserting, albeit one whose possibility is built into the structure of assertion. I have further glossed this as a draining out of the ostensive character of assertion, a way of communicating and passing along claims that no longer involves shared attention to things themselves or the direct grasping of the object of speech. In contrast, Brandom, reading Heidegger on assertion, (unsurprisingly) assigns this "passing on" a nearly opposite significance: for him, it is the central function of assertion to enable reasserting, or passing on through retelling. This fits with Brandom's deeply nonostensive, noninteractive reading of assertion, according to which "the output of perception is assertion" (2002, 78). Brandom focuses on the point that assertions can be passed along and retold, even in the absence of direct exhibitive contact with the object. But while Heidegger does indeed emphasize this point, I think its place in his argument is quite different from where Brandom locates it. Heidegger's comment about passing along and further retelling comes in the midst of a paragraph in which he says, "Assertion … is letting someone see with us what we have pointed out" (2010, 198). Heidegger's central point is not that assertions license reassertions that are cut free from direct contact with the object. Rather, he is making a point that should be antithetical to Brandom: first, that even in these cases of merely passing along claims, we are broadening the range of shared seeing and keeping the object in view; and second, that this kind of hearsay is a derivative, even if pervasive, use of assertion. Heidegger never loses the idea that an assertion is most fundamentally an *act of exhibition* that serves to draw others into the mutual shared seeing of a common world.[11]

5 Ostension and Assertoric Truth

In "Truth and Rule-Following," Haugeland gives an account of truth as *letting be* (*HT*, 325). The core idea is that truth is objects displaying how they are, which they can only do insofar as we engage with them in norm-governed ways. By committing to taking objects as obeying various consti-tutive norms, and taking violations of those norms as problems we are responsible for addressing, we let objects show themselves. Haugeland is most interested in how this plays out at the level of whole bodies of norms, which disclose entire "worlds," in his parlance. Sets of normative practices can die out under the pressure of anomalies, and with that their ontologies can be undermined—consider phrenology, or ancient animism, for instance. Likewise, new sets of practices can let new ontic domains be, and thereby display new truths. The proliferation of gender identities over the last two decades might offer a nice example.

I have argued that Haugeland's notion of display, of letting objects show themselves, needs to be filled in by a social pragmatics of ostension in the context of second-person interactions, if it is to have bite. Without this, it is hard to give a naturalistically acceptable story about the difference between engaging with something in a way has meaning and counts as let-ting it display itself, on the one hand, and regular old causal interactions of the sort that robotic arms might have with objects they manipulate or worms might have with fruit they eat. Haugeland's picture of letting be often comes off as quite individualist, involving some kind of internal moment of commitment, and as a result it can come off as unacceptably magical. For instance, he writes:

The governing or normative "authority" of an existential commitment comes from nowhere other than itself, and it is brought to bear in no way other than its own exercise—that is, by self-discipline and resolute persistence. A committed individual holds him or herself to the commitment by living in a resilient, determined way. Thus, its authority is *sui generis* in a stronger sense than just "of its own genus": it is of its own genesis, self-generated. (*HT*, 341–342)

My version of letting be as ostension need not involve any such existential-ist melodrama but can otherwise retain the core of Haugeland's picture.

Particular assertions let-be in a more modest and less creative sense than the adoption of networks of constitutive standards that disclose new worlds. Indeed, as I discussed earlier, Heidegger insists that assertions disclose truths that are *already available*. So the kind of display involved in assertion is not a constitutive display; it's not a case of innovation. This makes it even clearer that we need some distinction between displaying and just interacting or engaging—a distinction that I have tried to draw by way of the pragmatics of ostension. Without this distinction, if assertion merely displays a truth that was already available, it is hard to see it as doing anything substantive at all.

Heidegger was acutely aware of this problem. He gives a rather notorious account of assertoric truth in §44 of *Being and Time*.[12] Commentators have been divided as to whether he is rejecting a correspondence theory of truth in that section or explaining the ontological preconditions that make a correspondence theory possible.[13] It is also easy to read this section as more or less empty or antitheoretical—as Heidegger just stamping his foot and saying that true assertions are ones that get the world right. This seems unhelpful. I want to suggest, in this final section, a way of thinking about assertoric truth that is at least consistent with §44. I will argue that the apparent vapidity of Heidegger's discussion dissipates if we keep our focus on the ostensive function of discourse.

Comportments toward objects can be successful or unsuccessful, and success comes in degrees. Comportment can be less than successful in many ways; it might, for instance, be simply clumsy. What interests me here is comportments that are compromised by a *misunderstanding* of their objects. If I misunderstand an entity, then I will not be able to cope with it appropriately, and my comportment toward it will be distorted. If I don't understand how to drive a car with a manual transmission, then my attempt to cope with it will, literally, backfire. All understanding, as we have seen, embeds an interpretive as-structure. However, we saw that assertions are a special kind of comportment. They are about their objects, and they are communicative. A successful assertion reveals objects as they are by way of a description. As we saw earlier, this requires success along two dimensions: they must succeed in drawing attention, and that attention must be undistorted. Haugeland writes: "There is a distinctive kind of success that descriptions *as such* must aim at—one that depends on the described entity itself. ... A description as such undertakes to be correct or *true"* (*DD*, 199). But this is only half the story. A descriptive assertion has to be not only *true* but also *socially effective* to be a successful act. This depends not just on the described entity but on the social context and practice in which the description is situated. But of course, an assertion can be *true* even if it is not socially successful as an ostension. We can use the language of truth to artificially isolate the descriptive accuracy of an assertion from its overall social success. Being able to do so is often strategically useful. A true assertion does not distort or occlude its object in virtue of its predicative structure, regardless of whether it actually manages to function socially as an assertion ought to.

According to Heidegger's version of the traditional picture of truth, an assertion agrees with its object, and is true, in virtue of some sort of likeness between a linguistic or mental entity that forms the content of the assertion—the proposition, mental representation, and so on—and the object itself. This raises two problems: on the one hand, we have no substantive theory of what kind of thing this linguistic or mental entity *is*; and on the other hand, it is not clear what kind of similarity or agreement relationship the mental or linguistic entity and the world are supposed to have that counts as the relevant correspondence relation. If they are not identical to each other, then how are we to isolate a relevant similarity relationship; and even if we pick one, haven't we then lost the idea that the assertion is true in virtue of getting the world exactly right? Heidegger writes: "With regard to what do the *intellectus* and *res* agree? ... If it is impossible for the *intellectus* and *res* to be equal because they are not of the same species, are they then perhaps similar? But knowledge is still supposed to 'give' the thing *just* as it is. This 'agreement' has the relational character of the 'just as'" (2010, 258–289). This worry nearly exactly mirrors a worry of Frege's:

A correspondence ... can only be perfect if the corresponding things coincide and so just are not different things at all. ... If the first did correspond perfectly with the second, they would coincide. But this is not at all what people intend when they define truth as the correspondence of an idea with something real. For in this case it is essential precisely that the reality shall be distinct from the idea. But then there can be no complete correspondence, no complete truth. So nothing at all would be true; for what is only half true is untrue. (Frege 1977, 3)

Heidegger responds to the problems with the traditional picture by offering an extended analysis of an example in which someone says "the picture is askew" with his back to the wall and then turns around to confirm the *truth* of this assertion. "In carrying out such a demonstration," Heidegger says, "the knowing remains related solely to the entity itself. In this entity the confirmation, as it were, gets enacted. ... It shows that it, in its selfsameness, is just as *it* gets pointed out in the assertion as being" (2010, 261). Heidegger's point here is that confirming truth is *not* a matter of comparing the assertion with the world and checking whether they have some kind of proper correspondence relation, and in fact our attention is not directed to any kind of propositional content of the assertion at all; instead we confirm truth by directing our attention to the picture itself, thereby confirming that the picture is just as the assertion points it out as being.

Now, this example can be mysterious and confusing in a number of ways. It is difficult to see how we can tell, by looking "solely" at the picture, whether it is just as it is pointed out in the assertion as being; this would seem to require that we compare the picture to the assertion. Clearly Heidegger wants to avoid assigning a role in the judgment of truth to some special kind of propositional entity that we compare with the object. But what, then, is the alternative type of judgment of agreement we are making? It can sound here as though all he is doing is refusing to give a theory of what sort of correspondence between assertion and object we are looking for by putting his foot down and insisting that somehow the object itself does all the theoretical work we need in grounding truth. It is not surprising that commentators such as Mark Wrathall have concluded that Heidegger has not really moved away from a correspondence theory here. In *Basic Problems of Phenomenology*, Heidegger reminds us that if we want to figure out "where" assertoric truth is, "it is necessary to go back to the determination of assertion that was given, that it is *communicative-determinative exhibition*. ... The *hearer* is directed from the very beginning in his understanding of [the assertion] toward the entity talked about" (1988, 215; italics mine). But this leaves us no reading of what Heidegger means by saying that the demonstration directs us *to the object itself*. What kind of directing is this, in practical terms?

The original assertion directs us to the picture. Its fundamental existence is as an *act* of ostension. Hence in confirming the truth of the assertion, we are, as it were, checking on the success of a comportment: having already directed our attention to the picture, albeit not through direct sensory contact, we now attend to it anew, to confirm that our initial comportment toward it was not distorted or misdirected. Once we focus on the *act* of assertion as an act of ostension, we notice that confirming the success of this act does not require that there be some entity that "matches" or is "similar to" the object, any more than when we confirm the success of our hammering job by checking the smoothness and grip of the nail, we need to compare that to some likeness of the result in the original hammer. In confirming truth, we move from one kind of attention to the object itself to a new kind, although in the first case the attention did not happen to involve direct sensory contact.

It is interesting that after all his emphasis on how assertion is a social, communicative act, Heidegger exemplifies the notion of assertoric truth with an example that is not social but involves only one person checking truth on her own. Here the assertion is not addressed to anyone other than the asserter herself. Although I admit this is a strong reading, Heidegger's choice here makes sense if the point is to focus our attention on the accuracy and the undistorted character of what is revealed in an ostension by assertion, as opposed to the communicative and coordinative success of the assertion, thereby conceptually separating the two components. In moving from his extended discussion of *Rede* and assertion to a focus on truth, I submit that he is pointedly bracketing the social function of assertion so as to focus on a different success condition of ostension, namely, undistorted and accurate uncovering.

This is not to introduce a substantive theory of truth as some special sort of correspondence relation or anything similar, but rather to use truth as a tool for isolating and drawing attention to this dimension of the phenomenon of ostending with language. The appearance that Heidegger is reiterating the correspondence theory but just refusing to theorize the correspondence relation comes from the hard-lost tendency to understand the assertion as in the first instance an entity rather than an act of pointing, in which case it seems that it can "point out" the object as being a certain way only in virtue of its inherent properties, which lands us back at correspondence square one. When Heidegger says that confirming truth is a matter of checking whether things are as the assertion pointed them out as being, this act of pointing can sound just as mysterious as the original correspondence relation we felt we needed to explain. It is tempting to read the phrase "pointing something out" as just a proxy for "having a particular content," rather than as actually ostensive. But we need no more metaphysical baggage here in saying that the assertion is true than we need in saying that an act of hammering was successful. While we have no unified theory of how acts of ostension manage to ostend what they do, we are not tempted to think that an ostensive act—a point, an eye roll, an exclamation "Lo, a rabbit!"—needs to have some structure in common with what it ostends.

Now, it may seem as though I am cheating somewhat. In the case of assertion, surely the compositional structure of the linguistic act is central to determining what it ostends, what entities and features of the world it is pointing us toward. This is of course right. Nothing in Heidegger's account undercuts the point of doing compositional semantics to understand how particular assertions draw on elaborate conventions to point out particular ways that things are. It is a central and interesting fact that we manage to engage in rich and elaborately structured acts of ostension through our mastery of a compositional language. But that we have this toolbox available to us does not mean that there is some entity, the content of our assertion, that is the truth bearer. Often our assertions in some sense structurally mirror the states of affairs to which they point us, and this shared structure is obviously helpful in making it possible for us to understand one another. At other times, a mere sign, gesture, or code word with no internal structure can serve the same function. Whether an act succeeds in being a true assertion depends on how it manages to direct attention to objects, and this in turn depends on our participation in a rich and tight web of social norms.

I have tried to show that ostension is a complex and multifaceted set of social practices that are essential to our ability to talk about the world. Furthermore, I have argued that the pragmatics of truth-talk depend on the pragmatics of ostension. In doing so, I have offered an account of assertion and truth that is Haugelandian and Heideggerian in spirit, but at the same time reveals that both their pictures have a large and important lacuna: neither attends to the constitutively communicative, second-personal character of either ostension or assertion, or to the social pragmatics of either practice. As a result, neither can offer a naturalistic, metaphysically spare account of how assertions are *about* their objects, or of how truth-talk is pragmatically substantive. Attention to the nature of ostension and its place in assertion can dissolve or clarify multiple puzzles surrounding the relationship between language and the world. It also lets us understand how the content of language is derived from its communicative function, rather than the other way around.

Acknowledgments

I presented previous, quite different versions of this essay at the "Mind, Meaning, and Understanding: The Philosophy of John Haugeland" symposium, University of Chicago, May 2010, and at the annual meeting of the International Association for Phenomenological Studies, Asilomar Conference Center, Pacific Grove, CA, July 2010. I am grateful to both audiences for extremely helpful feedback. In revising the paper, I have benefited greatly from conversations with Bill Blattner, Andy Blitzer, Bryce Huebner, Eric Winsberg, Kate Withy, and especially Oren Magid.

Notes

1. The Heidegger epigraph is taken from the Harper & Row edition of *What Is Called Thinking?* translated by J. Glenn Gray. I have modified Gray's translation slightly for emphasis. See Heidegger 1968, 9.

2. Or equivalently to "ostension" for my purposes, "indication," which is Heidegger's preferred term.

3. For fuller discussions of this point, see, e.g., Kukla 2006; Noë 2006; and Rouse 2002, esp. chap. 6.

4. What is less clear is whether there is anything that is real but not ostendable. Private sensations are Wittgenstein's famous example of things that could not be ostended, and they seem an excellent candidate. Since ostension involves shared attention, it does seem impossible by definition to ostend something private. For Wittgenstein, this is reason to doubt their reality; I remain agnostic about them here. Types of private sensations, such as types of pain, seem to be decently ostendable; it is only particular private tokens that seem by definition immune from ostension.

5. See Heidegger 2008.

6. At the moment at which this sentence was first drafted, that is. I have no idea whether the advice still holds.

7. See, e.g., "Reading Brandom Reading Heidegger" and "Truth and Finitude," both in *DD*.

8. That is to say, we have a century of complicated and technical theorizing behind us trying to account for the nature of meaning and how it can manage to hook discourse onto objects, including computational accounts, teleological accounts, causal accounts, and many others.

9. "Why can't my right hand give my left hand money?—My right hand can put it into my left hand. My right hand can write a deed of gift and my left hand a receipt.—But the further practical consequences would not be those of a gift. When the left hand has taken the money from the right, etc., we shall ask: 'Well, and what of it?' And the same could be asked if a person had given himself a private definition of a word; I mean, if he has said the word to himself and at the same time has directed his attention to a sensation." Wittgenstein 2010, §268.

10. But see Kukla and Lance 2015 for a detailed discussion of the relationship between speaking and thinking and whether and how thinking can be said to have semantic content; and see Kukla and Winsberg 2015 on whether we should be deflationists about semantics altogether.

11. Since Brandom takes what Haugeland and I read as idle talk as the essence of assertion, he needs a different account of idle talk. Haugeland points out that Brandom interprets idle talk in epistemological terms: for Brandom, "idle talk" is talking without having good evidence for what you are saying. This is quite different from being merely passed on, as merely passed-on, retold assertions are often perfectly justified. I can be justified in claiming something of which I have no direct grasp, either because I had direct grasp of it at another time, or because I have perfectly good reasons for trusting the source of the claim, although I don't really have a practical grasp of its content myself; we saw examples of this earlier. In Haugeland's reading, the distinction between proper assertion and idle talk is not an epistemological one, at least not in the narrow sense of a distinction at the level of justificatory status. The kinds of memes that travel through a community and form part of its discursive background may or may not be justified. They simply serve a different sort of purpose from traditional knowledge building. Likewise, in the picture I have been developing, assertion is not defined in epistemological terms, either. Assertion is a social tool for engaging shared attention via descriptive predication, where the epistemological question of how we would establish that it is doing so properly remains a separate question.

12. As opposed to full-blown *aletheia*, or the disclosing of ontological domains.

13. See, e.g., Carman 2007b; Wrathall 1999.

References

Blattner, W. 1999. *Heidegger's Temporal Idealism*. Cambridge: Cambridge University Press.

Brandom, R. 2002. *Tales of the Mighty Dead*. Cambridge, MA: Harvard University Press.

Carman, T. 2007a. *Heidegger's Analytic: Interpretation, Discourse, and Authenticity in Being and Time*. Cambridge: Cambridge University Press.

Carman, T. 2007b. Heidegger on correspondence and correctness. *Graduate Faculty Philosophy Journal* 28 (2): 103–116.

Frege, G. 1977. *Logical Investigations*. Trans. P. T. Geach and R. H. Stoothof. Oxford: Blackwell U.K.

Haugeland, J. 1998. *Having Thought: Essays in the Metaphysics of Mind*. Cambridge, MA: Harvard University Press. (Abbreviated as *HT*.)

Haugeland, J. 2013. *Dasein Disclosed*. Cambridge, MA: Harvard University Press. (Abbreviated as *DD*.)

Heidegger, M. 1968. *What Is Called Thinking?* Trans. J. Glenn Gray. New York: Harper & Row.

Heidegger, M. 1988. *Basic Problems of Phenomenology*. Trans. A. Hofstadter. Bloomington: Indiana University Press.

Heidegger, M. 2008. The origin of the work of art. In *Basic Writings*, ed. D. F. Krell. New York: Harper Perennial.

Heidegger, M. 2010. *Being and Time*. Trans. J. Stambaugh. Albany: SUNY Press.

Kukla, R. 2006. Objectivity and perspective in empirical knowledge. *Episteme: A Journal of Social Epistemology* 3 (1): 80–95.

Kukla, R., and M. Lance. 2016. Speaking and thinking. In *Sellars and His Legacy*, ed. J. O'Shea, 80–99. Oxford: Oxford University Press.

Kukla, R., and E. Winsberg. 2015. Pragmatism, deflationism, and truth. In *Meaning without Representation: Essays on Truth, Expression, Normativity, and Naturalism*, ed. S. Gross, N. Tebben, and M. Williams, 27–46. Oxford: Oxford University Press.

Noë, A. 2006. *Action in Perception*. Cambridge, MA: MIT Press.

Rouse, J. 2002. *How Scientific Practices Matter: Reclaiming Philosophical Naturalism*. Chicago: University of Chicago Press.

Wrathall, M. 1999. Heidegger and truth as correspondence. *International Journal of Philosophical Studies* 7 (1): 69–88.

4 Love and Death

Joseph Rouse

Love is the mark of the human.
—John Haugeland (*HT*, 2)

That is why Heidegger speaks of death—or rather, of resolute being-towards-death. Taking responsibility resolutely means living in a way that explicitly has everything at stake.
—John Haugeland, "Truth and Finitude" (*DD*, 216)

John Haugeland's philosophical work has justly been celebrated for many accomplishments. These include his astute analysis and assessment of "good old-fashioned artificial intelligence" (GOFAI) as doomed because computers just don't give a damn, his groundbreaking work on weak supervenience, representational genera, the ontological intertwining of patterns and pattern recognition, constitutive rules, and much more. This paper will not further celebrate these or any other of Haugeland's recognized accomplishments. I instead take up two of his most controversial claims. Both claims are central to his philosophical work and express the moral, existential, and ontological seriousness of his primary concerns. Neither has yet entered the common wisdom of the philosophical profession. Both claims are phenomenological, although not in any of the methodologically specific senses advanced by Husserl or his successors. They are phenomenological in seeking to identify and clarify the phenomena to which some influential philosophical discussions are (or should be) accountable.

Haugeland's first claim considers what philosophers are and should be talking about when we speak of "intentionality." Haugeland starts with exemplary cases of intentional comportments: playing chess, doing empirical science, walking a batter intentionally in baseball, or speaking a natural language. He then argues that not only do many prominent and influential theories of intentionality fail to account for these paradigmatic cases, but their apparent plausibility arises from describing some other phenomenon that mimics but falls short of intentional directedness. Philosophical reflection on intentionality has thereby missed the distinctiveness of human intentional comportments, even in these familiar examples. In constructive response to these failures, he offers the striking but puzzling slogan that "love is the mark of the human" to replace various proposed "marks of the mental" (*HT*, 2; *DD*, 274).

The second claim initially seems to involve textual interpretation, rather than phenomenology. Heidegger devoted the first chapter of Division II of *Being and Time* to a phenomenon he characterized as existential death, or "being-towards-death." In any interpretation, existential death is distinct from both the perishing of a human organism and the loss of recognition and the abdication of commitments that constitute a person's social demise. Almost all commentators nevertheless conclude that perishing, demise, and existential death, while distinct and separable, are closely related. Each phenomenon concerns different aspects of, or orientations toward, the end of an individual human life. Haugeland will have none of this. He instead locates existential death in the neighborhood of Kuhn's (1970) account of the persistent vulnerability of normal science to crisis and revolution, Popper's (1959) insistence on the empirical falsifiability of any legitimate scientific theory, and the "death" of a natural language when no one speaks or learns it anymore. In Haugeland's reading, existential death is the utter collapse of the intelligibility made possible by what Heidegger called "understandings of being." Such a collapse of intelligibility in our own engagement with the world inevitably seems impossible, and when confronted with the apparently impossible, we initially seek to explain it away. Rightly so. Haugeland argued that such efforts to sustain or restore intelligibility in the face of apparent impossibility also call for openness to the possibility that those efforts will fail. Heideggerian being-toward-death is our comportment toward the possibility of the impossibility of *any* intelligible intentional directedness, and the responsibility to own up to that possibility rather than ignore, deflect, or dismiss it.

These two sites of controversy, intentionality in the philosophy of mind and death in existential phenomenology, may seem far apart. Haugeland nevertheless understood being-toward-death and love as the distinctively human character of genuine intentionality as the same issue viewed from two directions. Perhaps only that well-known former NYU philosophy student Woody Allen has previously suggested a philosophical convergence of love and death, and then only facetiously. In this paper, I show what Haugeland was claiming in each case, and why these two discussions are reciprocal. I also indicate why Haugeland was onto something importantly right about both issues, without trying to argue for either wide-ranging claim in detail. In the paper's final section, I briefly suggest some directions in which Haugeland's treatment of intentionality in terms of love and death should lead us. This final consideration is especially important in the context of this volume. I do not just want to look back and celebrate Haugeland's philosophical accomplishments, as if these were settled and completed. My final section emphasizes how his work should be taken up and carried on. In pressing into the possibilities Haugeland has opened, I note points where our views differ, but less as objections than as strategies for building on what he achieved.

1   Intentionality and Love

Haugeland argued that intentionality has often been misidentified with one of two misleading simulacra: "ersatz intentionality" and what I shall call "lapsed intentionality." In some respects, each phenomenon mimics intentional directedness. Each nevertheless lacks a crucial dimension of even ordinary instances of intentionality. Since Haugeland thinks many prominent philosophical accounts of intentionality actually succeed only in explicating these simulacra, his descriptions have both critical and constructive import: they call attention to important aspects of intentionality that these philosophical theories obscure.

To grasp this line of argument, however, we need to consider briefly Haugeland's broader challenges to other philosophical theories of intentionality. His critical arguments, in *Having Thought* and elsewhere, typically address whole families of philosophical theories rather than specific instances, and we need to understand the relevant forms of philosophical kinship. His later work implicitly maps the field via two distinctions. The first distinction is between conceptions of intentionality as an operative process in cognition, and conceptions of it as a normative status. An operative-process approach seeks to identify processes or states that bring about or constitute intentional directedness toward objects. Salient examples include Fodor (1979, 1987) on representations that function in cognition, Husserl (1980) on correlated noetic acts and noematic senses, Searle (1983) on intentionality as a complex biological property, Millikan's (1984) teleosemantics, Dreyfus (1979, 1991) on practical coping, and Dretske (1981) on primary information-bearing features of cognitive states.

A normative-status approach, by contrast, identifies intentional comportments with performances and capacities that can be held accountable to relevant standards. They are intentional in virtue of whether and how they would stand up to such accounting, for example, if they are interpretable as mostly rational in context. On normative-status accounts of intentionality, not all intentional states or performances need actually involve the constitutive forms of accountability, such as reflective assessment or interpretation by others. Grandmasters playing blitz chess, for example, are intentionally directed toward a strategic configuration of chess pieces when making their rapid moves, even though they need not have any chess concepts "in mind." Their moves only need to be appropriately accountable to the regulative, constitutive, and strategic norms of chess play. Normative-status approaches include Brandom (1994) on the game of giving and asking for reasons, Davidson (1984) on radical interpretability, McDowell (1994) on conceptual understanding, Heidegger (1962) on care and "authenticity" (*Eigentlichkeit*), and Haugeland (1998) himself on existential commitment.

A second dividing line draws on a distinction from Husserl (1970) to illustrate two different strategies for how to explicate intentionality. Husserl distinguished "empty" intendings such as imagination or memory, which present objects in their absence, from "fulfilling" intendings such as perception or mathematical intuition, which make the objects themselves directly manifest. This distinction suggests two alternative strategic directions to take in accounting for intentional directedness. A common philosophical approach starts by explicating the contentfulness of empty intentional comportments and only then asks what it is for such content to be fulfilled. This approach is typically motivated by the need to understand intentional directedness toward objects that do not exist or are misconstrued, which seems to preclude starting with actual relations to existing objects. Alternatively, one can start with a system's actual relations to entities and ask what it would be for those relations to be meaningful or aspectual, and thus intentional. The most common motivation for this second strategy has been "baldly" naturalistic, in McDowell's (1994) phrase. The analysis starts with entities, states, or performances that are causally or functionally interactive with their surroundings, and asks what would make those interactions intentionally directed and aspectual, such that the system's actual dealings with the world could be in error. It must, after all, be possible for an intentional system to "mean" or "try to do" something other than whatever it actually interacts with or accomplishes. Dretske's (1981) appeals to information-bearing states or Millikan's (1984) teleosemantic functional norms are familiar examples of strategically beginning with "fulfilling" interactions. Not all versions of this strategy are naturalistic, however. Heidegger (1962) also begins with intentional fulfillment (an understanding of being exhibited in an ability-to-be) without construing fulfillment in causal or other baldly naturalistic ways. Dreyfus (1991) on practical/perceptual coping and McDowell's "direct realist" account of perception as rational second nature also start with a fulfilling intentional comportment, without espousing a "bald naturalism."

We can combine these two distinctions in a 2×2 array (table 4.1) that sorts approaches to understanding intentionality by their location on that grid.

Haugeland primarily argues against the B1 and A2 strategies (having relegated the A1 strategies to the "outfield" of intentionality in "The Intentionality All-Stars," *HT*, 127–170). Broadly speaking, the problem with the A2 conceptions is a failure of meaning. Although they characterize comportments that actually and effectively respond to their surroundings, Haugeland concludes that "there is nothing that the response can 'mean' other than what *actually* elicits it in [its] normal [functioning] in normal conditions" (*HT*, 310). The B1 approaches fail in the opposite direction, by not accounting for truth and error. Their systematically interconnected comportments seem to constitute a meaningful conception but cannot actually be conceptions *of* anything, because they are only accountable to their own further comportments. In the B1 accounts, actions, thoughts,

Table 4.1

A two-dimensional mapping of approaches to understanding intentionality.

Accounts of Intentionality	1: Primacy of empty intending/linguistic meaning	2: Primacy of fulfillment: causality, perception, being-in-the-world, etc.
A: Operative-process accounts	Husserl: essential structures of consciousness Carnap: logical structure of language Jackson: a priori partitions Searle: intentionality as biological Minsky et al.: GOFAI	Dretske: information-bearing states Millikan: teleo-semantics Fodor: cognitive representations Dreyfus: practical/perceptual coping Dennett: what "satisfies" the intentional stance
B: Normative-status accounts	Quine: radical translation Davidson: radical interpretation Rorty: conversation of mankind Brandom: game of giving/asking for reasons Dennett: the intentional stance?	Heidegger: Dasein's disclosedness McDowell: perception and action as rational second nature Haugeland: existential commitment Dennett: the intentional stance?

and utterances could at best exhibit "mere coherence" among their own performances, without being accountable to objects themselves.

Haugeland associates the groups I have identified under B1 and A2 with distinct simulacra of intentionality. The best A2 approaches aim to explicate intentionality but only succeed in accounting for what Haugeland calls "ersatz intentionality." Nonhuman animals, at least those with relatively flexible, nonsphexish behavioral repertoires, are the paradigm case, but Haugeland also thought that artificial intelligence (AI) robots exhibit a similar engagement with their surroundings. It is telling for his conception of "ersatz intentionality" that computer programs cannot achieve even ersatz intentionality without a body enabling them to move and act.[1] Ersatz intentionality is a form of causal or functional involvement in the world that superficially resembles genuine cases of intentional directedness. Both animals and sophisticated robots display behaviors that misleadingly resemble intentional comportments, sometimes needing careful analysis to tell the most interesting cases apart.

The best B1 approaches (Haugeland took Brandom 1994 and Davidson 1984 as exemplary) also characterize a real phenomenon, although Haugeland did not explicitly name it as such. "Lapsed" intentionality, as I call it, falls short of the real thing in a different way. Unlike ersatz intentionality, it presupposes an achievement of intentional directedness. Lapsed intentional directedness nevertheless systematically avoids, conceals, or undermines a crucial dimension of its own presupposed achievement, namely, its accountability to anything beyond its own interconnected performances. The B1 theories erase in principle the accountability to the world that lapsed-intentional performances seek in practice to avoid (with only partial success). These theories thus idealize a problematic tendency within intentional engagement. Heidegger (1962) presented one version of this tendency, "falling" into idle talk, curiosity, and ambiguity, as exhibiting an unavoidable tension within dasein's disclosedness.[2] Harry Frankfurt (2005) characterized a similar phenomenon, "bullshitting," as a lamentable and preventable lapse. Ian Hacking (1992) instead celebrated related tendencies within experimental practice as the "self-vindication" of the laboratory sciences.

Both descriptive terms, "ersatz" and "lapsed," are fighting words that would be rejected by the theorists they target. If various human activities are paradigm cases of intentionality, then it is no surprise that what nonhuman animals and sophisticated robots do resembles those capacities. We, after all, are animals. AI robots, in turn, reflect sustained, sophisticated attempts to match some of our intentional capacities in vitro. The challenge for Haugeland is to show in what sense animal and robotic behaviors nevertheless merely simulate intentional comportments. He likewise must show what is obscured by inadvertently identifying intentionality with something like idle talk or "self-vindicating" adjustments of experimental systems to their theoretical models.

Consider animals. Many animals exhibit a sustained and often highly flexible purposiveness in responding to their surroundings. They constitute themselves as entities by coordinated processes that differentiate their bodies in a nonarbitrary way from other things. Moreover, they are not merely differentiated from other entities by boundaries between them. As organisms, they also engage their surrounding environment, which is not just whatever is in their physical vicinity. An organism's environment is composed of the interconnected aspects of its surroundings that matter to its ongoing way of life. Organisms respond to things in their environments by eating them, avoiding them, mating with them, concealing other things beneath them, and so forth. Nor are such engagements haphazard;

organisms' behavior mostly responds to the features of their surroundings that they *need* to take into account as the kind of organism they are. Animals also typically engage their environments in a systematically self-directed way, by movements that track and respond to what matters to them. Some animals, in turn, devise flexible strategies united by their common end and adjust their behavior in response to failures to attain it. Robots achieve similar selectivity via functional design and a focused interface between their capacities for detection and response. Goal-directed biological or technological teleology and adaptation clearly resemble human intentional directedness and so not surprisingly have figured prominently in many A2 theories of intentionality.

In what sense could we dismiss these sophisticated, flexible, purposive, self-corrective responsive repertoires as merely "ersatz" simulacra of intentionality? Several interconnected features ground Haugeland's insistence on sharply differentiating human intentionality from nonhuman animal or robotic behavior. First, nonhuman organisms respond to their surroundings "narcissistically," in Kathleen Akins's (1996) telling term. Evolved perceptual/practical systems (our own included) do not discern or register objective features of things but only respond to how those things matter to their own functioning. Thus organisms' thermoreceptor systems do not register or respond to the ambient temperature, but only to a more gerrymandered set of features defined by their possible effects on, and significance for, the organism's way of life. Such narcissism may also characterize successful AI systems: the computer scientist Eric Aaron (Aaron and Mendoza 2011) has achieved some robust successes in dynamic navigation programming by also directing robotic agents toward such narcissistically specified ("agent-centric") features.

Second, nonhuman animal behavior remains closely tied to its actual circumstances, with no space for symbolic displacement. A vervet monkey's distinctive warning cries in response to different predators, for example, direct attention toward an actual location and prompt a characteristic response (Cheney and Seyfarth 1990). Absent such circumstantial ties, the warning cries would be pointless. Such comportments are thus complex, high-level patterns of differential responsiveness to actual configurations of the world. One can then assess individual organisms' abnormal responses with respect to what is normal for organisms of that kind, but that approach leaves no standard against which to assess their overall pattern of normal responsiveness.

Third, organisms' purposive directedness outward is circumscribed by its own way of life as a relatively fixed "end." Here we must be careful, because many organisms do have significant abilities to alter their behavior in response to things. Cats, for example, must *learn* to discriminate the indigestible shrews from the comestible mice.[3] This pattern of changing responsiveness to their surroundings still only exhibits what Haugeland calls "first-order self-criticism" (*DD*, 265): cats and other animals can revise their own behavior in the face of recalcitrant experiences but can only do so with respect to their normal functioning as the kind of organism they are. They cannot engage in the "second-order self-criticism" that could revise those functional norms in response to how things show themselves. Their first-order skills for learning are thus assimilable within their normal functioning. Normal cats do learn not to eat shrews after a small number of trials, but there is no further standard with respect to which one might classify and evaluate this capacity for learning. Perhaps there is some sense in which it would be "better" for cats if shrew avoidance were genetically assimilated rather than having to be learned anew by each cat. Such a possibility nevertheless has no normative force for cats themselves, and hence there is no sense in which the current capacity for relatively rapid empirical uptake could be understood as a failing. It could only become a "failure" in a limited sense, if cats as we know them could not reproduce themselves successfully in a different environment in competition with other organisms that did not need to learn to tell that difference. Apart from successful self-maintenance and reproduction in changing circumstances, there is no standard with respect to which the normal functioning morphology, physiology, or behavior of cats could be assessed or could fail.

Why isn't first-order self-criticism good enough to constitute intentional directedness and accountability? In several crucial respects, it is better than good enough. Nonhuman organisms do not *lack* intentionality, just as fish notoriously do not lack bicycles.[4] Nor does moral regard for animals require their philosophical assimilation to our own self-understanding as intentional entities. For all the flexibly goal-directed appropriateness of a great deal of animal behavior, however, I think we should endorse Haugeland's claim that no known nonhuman organisms comport themselves intentionally toward other entities or themselves. The goal directedness of an organism's physiology and behavior is not articulated into intermediary or subordinate goals.[5] We are inclined to think that other organisms take some things in their environment as food, for example, because they mostly eat what is edible for them, but there are many edible things they do not eat, even when hungry. Sometimes individual organisms make mistakes relative to their normal pattern. Some edible things may not fall in their normal pattern of recognition and response, perhaps (from an evolutionary perspective) because the requisite discriminative capacity would be too energetically or cognitively costly. In that case, failing to eat such things when hungry would not be a mistake but instead a successful strategy. These animals thus respond to what they do eat not "as food" but "as energetically and cognitively accessible food." That category may also have exceptions, however, which must in turn be added to a more complex description of what the animal is directed toward. There is no principled stopping point to the process of qualifications to the supposed as-structure of the organism's behavior, short of its entire normal behavioral pattern in response to its normal environmental range. Similarly, we cannot rightly say that a lion was trying to catch a gazelle but failed. That "failure" would instead be part of a larger pattern of behavior that balances energy inputs and outputs, vulnerability to predation or injury, feeding and reproductive behavior, and so forth. There is no principled way from within an animal's behavioral repertoire to distinguish what it is *trying* to do from what it typically *does* do when functioning normally. Evolutionary analysis of an organism's behavior can of course *explain* its reproductive success in its environment, by discriminating components of its functioning that differentially contribute to evolutionary success. These classifications invariably treat some aspects of the organism's normal behavior as "noise" or ceteris paribus violations, however, which are only exceptions with respect to the *analyst's* categories and norms. For the organisms, the exceptions belong just as much to what they normally do.

The difficulty of projecting our explanatory distinctions into the animal's own behavioral repertoire is actually even greater, because even if its actual behavior perfectly mapped onto the analyst's classifications in the organism's actual historical range of environments, it would almost surely diverge under counterfactual conditions not part of their evolutionary history. Such counterfactual accountability is nevertheless crucial in understanding intentionality. That was why Hubert Dreyfus's (1979) objection that Roger Schank's (Schank and Abelson 1977) restaurant script programs could not handle questions that were "off-script" was so telling against that iteration of artificial intelligence research. Haugeland (*HT*, chap. 10) similarly asked how dogs would respond to "impossible" permutations in the facial features of its human family. These examples illustrate Haugeland's (*DD*, 267) point that intentional directedness must be modal to express a "nonaccidental" directedness and accountability toward entities. Circumstantial coincidence, even evolutionary coincidence, is not enough. Of course, intentional systems also sometimes diverge from the responses toward entities that are appropriate for them, often rather more so than well-adapted organisms do. What nevertheless lets a genuinely intentional system be directed toward an aspect of an object, rather than toward a "narcissistic" projection of the system's environmental dependence, is its *own* capacity for what Haugeland (*DD*, 269–271) called second- and third-order criticism. First-order learning is governed by the organism's own goal directedness toward what matters to its own self-maintenance and reproduction. To be directed toward *objects* (rather to its own gerrymandered, "narcissistic" environment), an organism must be capable of second-order criticism of those norms and goals, and third-order criticism of its entire way of life. Only then could it behave in ways that are right or wrong about the object, rather than about its own successful self-maintenance and reproduction in its environment. Even flexible, learned animal behavior thus shows itself to be a form of "derivative" intentionality, whose accountability *to objects* depends on our capacities and norms for theoretical explanation.

These considerations explain why Dennett's intentional stance must be placed ambiguously on the chart in table 4.1. If Dennett's account identifies intentionality with the gerrymandered properties that allow a system to be sensibly interpreted from the intentional stance, then it describes a form of "ersatz intentionality" and belongs in B1. If instead he identifies intentionality with the pattern of rationality-in-context *ascribed* to those systems by an interpreter, then we need to know more. Depending on how those explanatory ascriptions are themselves normatively accountable, the theory may belong in A2 or B2, but it then takes our biological understanding rather than nonhuman animal behavior as exemplary of "original" intentionality.

I will treat lapsed intentionality more briefly. This phenomenon is the articulation and development of intentional comportments so as to render them accountable only to other intentional comportments, and never to their objects. Such systematically interconnected performances are likely only possible for systems whose overall behavior is genuinely intentional. Yet the very efforts to improve and refine intentional directedness in systematically consistent ways can cut them off from their objects. Haugeland emphasizes how *talk* about things can become hermetically closed off from accountability to anything other than more talk. Sophisticated scientific theories, for example, risk rendering further theoretical articulation accountable only to other theoretical models. Haugeland (*HT*), McDowell (1994), and I (Rouse 2002) have each argued that Davidson's or Brandom's theories of intentionality that highlight linguistically articulated beliefs and desires mistakenly take such hermetic self-enclosure to be constitutive of the intentional domain.

It matters to Haugeland's arguments that intentionality on such B1 accounts is all-encompassingly holistic. To be an intentional comportment is to belong to this overall structure, whether understood as the conversation of mankind (Rorty), the game of giving and asking for reasons (Brandom), or a token-reflexive practice of truth-theoretical interpretation (Davidson). Beliefs are then only accountable to other beliefs, and their purported objects can only have a causal impact without justificatory significance. The normative significance of objects is confined to how they should be *taken* to affect our beliefs, and what one *takes* as reliable differential responsiveness to them. One of Haugeland's (*HT*) most important constructive contributions, as counterpoint to the B1 strategies, has been to insist that skillful perception and action must be normatively integrated all the way down within a theory of intentionality. Intentional states are not just essentially "interrelated" with their objects but "intimately" embedded in the world, through mundane and constitutive skills and meaningful equipment. Perception and action cannot rightly be taken as external causal impingement on or by a "belief" system, a token identity between mind and body, or a merely instrumentally successful interpretive scheme.

Despite the prominence of linguistically oriented theories in B1, however, the phenomenon of a hermetically self-enclosed "lapsed intentionality" is not confined to idle talk. Experimental practices in the sciences are also intentional systems, and by refining and stabilizing their implementation, and making them the proximal target of scientific theory, a discipline can be severed from accountability to anything beyond its own concepts and procedures. Ian Hacking saw this phenomenon as a source of the stability of much of scientific knowledge:

> [An instrumentarium] evolves hand in hand with theories that interpret the data that they produce. As a matter of brute contingent fact, instrumentaria and systematic theories mature, and data uninterpretable by theories are not generated. There is no drive for revision of the theory because it has acquired a stable data domain. What we later see as limitations of a theory are not data for the theory. (1992, 55)

The problem then is that, in supposedly securing the correctness of theories within their coadapted domains, Hacking renders the theories empty. His own account undercuts his presumption that such self-vindicating constructions are nevertheless theories *of* some empirical domain. That exemplar of conceptual stability, geometrical optics, tellingly illustrates his claim:

> Geometrical optics takes no cognizance of the fact that all shadows have blurred edges. The fine structure of shadows requires an instrumentarium quite different from that of lenses and mirrors, together with a new systematic theory and topical hypotheses. Geometrical optics is true only to the phenomena of rectilinear propagation of light. Better: it is true of certain models of rectilinear propagation. (1992, 55)

That final qualification is telling. Hacking helps himself to a presumption that his own account undermines, namely, the understanding of geometrical models of rectilinearity as a *theory* about *optics*. It is one thing to say that a theory only accurately describes some phenomena in its domain and thus has limited range or accuracy. It is another thing altogether to confine its domain to those phenomena for which the theory seems to work. The fine structure of shadows *is* directly relevant to geometrical optics and thereby displays the theory's empirical limitations. Only through its openness to empirical challenge does the theory even purport to be about the propagation of light. Such constructions are only theories *of* anything through accountability to a broader domain that leaves them open to empirical challenge. Restricting a theory to the range of its empirical success often seems a tempting way to secure the theory's accuracy, or to restrict one's epistemic aspirations to those claims one can securely justify. Such seeming security is nevertheless purchased by emptying the theory of empirical content.

We can now better understand, at least indirectly, why Haugeland characterizes intentionality as a mode of love. Ersatz intentionality is directed beyond itself in a fundamentally narcissistic way. It encounters and responds not to other entities but only to its own needy dependence beyond itself. It is then a derivative mode of directedness: only the "narcissist's" analyst (the biologist or computer scientist, in these cases) can recognize the independence and autonomy of its purported object and thereby constitute any objective purport. Lapsed intentionality instead invokes an all-too-creative, self-enclosed, imaginative imposition. Any reader of Jane Austen's *Emma* can recognize such directedness toward self-fulfilling projections as a pathological simulacrum of love. According to Haugeland, intentionality involves letting oneself be open to entities in their own range of possibilities, rather than merely those we project on them or need from them. Love, like intentionality, is integrally connected to the freedom to let the beloved change oneself and one's sense of possibilities. It also involves a mode of "existential" involvement that outruns merely functional neediness or social obligations and responsibilities. As we shall see, this existential dimension of intentionality, in which one's whole way of life is at stake in its accountability to the world, provides the philosophical link that Haugeland discerned between love and death.

2   Existential Death

Haugeland's account of existential death is closely bound up with his interpretation of what Heidegger referred to as *Dasein*. The standard view of this concept is cogently expressed by Taylor Carman: "The analytic of Dasein is an account of the existential structure of concrete human particulars, that is, individual persons" (2003, 42). Haugeland instead identifies dasein as a way of living that embodies an understanding of being. That way of living can only be lived by individual persons and thus only occurs so long as persons live it. Moreover, their individuation within that overall way of life is constitutive of dasein's way of being, which is "in each case mine." Nevertheless, dasein is the way of living that its "cases" take up together, and not the individuals themselves (whom Heidegger refers to as "others").

On the standard reading of dasein, existential death concerns how one comports oneself toward one's own possible nonbeing as an individual person: what else could "death" refer to? For Haugeland, by contrast, death concerns the possible collapse of an understanding of being, in which it discloses entities in ways that show its own and their impossibility, that is, unintelligibility. Haugeland's revisionist interpretation does agree with other accounts of existential death at many points. Existential death is not an actual event but a comportment toward the ever-impending possibility of dasein's own impossibility. Moreover, dasein's *ontological* character centers on this ownmost possibility. Its own being is at issue for it because in understanding itself as an entity that might not be, it thereby does not have to be what it "is." Dasein's ordinary response is to flee from its own responsibility for that and how it is, and for the disclosedness of entities made possible by its being-in-the-world. Each of these claims is differently inflected, of course, depending on which construal one gives to dasein and its possible "nonbeing."

I will not try to unravel the textual basis for these competing interpretations, at least not directly. I begin my explication and defense of Haugeland's reading by recognizing that standard readings have seemed not only correct, but obviously correct, to most readers of Division II, chapter 1, and related texts. Haugeland's reading depends primarily on its place within his careful, systematic attentiveness to key features of the overall argument in *Being and Time*: the referential-individuative apparatus by which dasein is "the entity that we ourselves are," the methodological role of "formal indication," how the discussion of death develops the immediately preceding discussion of truth, and how it thereby sets the problem of understanding dasein "as a whole." Perhaps above all, Haugeland's account turns on how existential death matters not just to the existential analytic of dasein but as preparatory to the unwritten Division III of *Being and Time*. Haugeland's reading nevertheless confronts the interpretive problem of why Heidegger would characterize a possible collapse of intelligibility in terms of death. In addition to Haugeland's own justifications of his account, I offer three philosophical responses to this interpretive challenge.

My first response is that Haugeland's account can accommodate the orthodox readings while developing a deeper understanding of their ontological significance. Traditional conceptions of existential death as an existential orientation to the prospect of human mortality turn out to be an important special case of Haugeland's more general conception of existential death as the possible impossibility of an understanding of being. Everyday dasein has its own predominant understanding of its own being, as publicness. This mode of understanding is an *existentiale*, an essential structural dimension of dasein's way of being as constituted by existence and mineness. Any case of dasein must press into possibilities that are only possible for it through a certain level of conformity to social norms. In those everyday comportments, dasein also confronts a constant inclination to go further and let the public understanding of its social roles govern its self-understanding as a "for-the-sake-of-which." Understanding the anonymous norms of "what one does" as that for the sake of which one comports oneself is nevertheless incompatible with facing up to the ever-present possibility of one's own nonbeing. The possibility of the impossibility of continuing to press into possibilities disconnects one's own case from dasein's everyday anonymous modes of comportment, whose understanding of being as publicness is unaffected by the perishing or demise of any particular case of dasein. For this public understanding, dying is an event that happens to others, and to oneself at some future time that one needn't take into account now, except as anyone does, by writing a will, buying insurance, crossing the street carefully, exercising regularly, and so forth. Each of us cannot avoid dealing with this ever-present possibility, if only by thus fleeing from it. Haugeland could therefore recognize that the possibility of the impossibility of each of us pressing into possibilities, in each of our own cases, is the most proximate manifestation of a possibly impossible understanding of being. Traditional interpreters are right to recognize this specific case as especially important in Heidegger's analysis, since it introduces the shift in self-understanding from publicness to resoluteness that individuates dasein. The importance of this specific case also helps explain why Heidegger would pose the more general issue by thinking through and beyond the everyday understanding of human mortality. The more general issue, of which grasping the existential import of individual mortality is only a special case, would nevertheless still be what matters for posing the general question of the meaning of being.

A second consideration is relevant to Haugeland's own reasons for reading Heidegger as he does, and is also more germane to the connection I am drawing between his accounts of love and death. For Haugeland, Heidegger's account of owned being-toward-death as a resolutely finite commitment to an understanding of being is a telling description of the third level of self-criticism that I discussed in the first section. Resolute being-toward-death goes beyond the empirically responsive revision and repair of their own comportments that many nonhuman animals can do, and also beyond a second-order critical reassessment of the very norms with respect to which such revision and repair are undertaken. Haugeland argues that adequately understanding even the most ordinary human intentional directedness depends on attributing to us the (existentiell) possibility of a third level of self-criticism, as authentic, "loving" intentionality. "Authentic" intentionality is a stereoscopic involvement in the world (*DD*, 270-3). It must sustain a dogged, resilient effort to overcome any obstacles to the intelligibility of its constitutive engagement with the world, alongside a resolute determination not to cover over its failures and even to give up that understanding of being altogether in the face of unsurpassable failure. Nor can this reading be seen as an extraneous imposition on Heidegger's text: anyone familiar with Kierkegaard's influence on Division II will recognize his deep kinship with Haugeland's reading of Heidegger. Haugeland's account highlights the centrality of Kierkegaard's (2006) concern with the possibility of having to give up love as an existentially constitutive commitment in faithful responsiveness to its impossibility. In short, for Haugeland, the principle of charity requires reading Heidegger on existential death and owned resoluteness as coincident with an adequate understanding of intentionality as a form of existentially committed love.

My third and most important reason for accepting Haugeland's reading of Heidegger on death is that it offers an important and revealing new way to understand the project of *Being and Time*. One of the more perplexing interpretive problems confronting readers is why Heidegger begins the transition to the unwritten Division III through an extensive contrast between his account of dasein's originary temporality and Hegel's conception of the relation between time and Spirit in *Phenomenology of Spirit*. Haugeland's account of existential death and its place in his overall reading of *Being and Time* removes the perplexity and brings out new dimensions of Heidegger's project by allowing us to see the entire book as a critical philosophical engagement with Hegel. Haugeland himself did not recognize this consequence of his interpretation. He occasionally acknowledged Hegel's conception of Spirit as a precursor to Heidegger's account of dasein, but only in an attenuated sense. Spirit stands alongside Kant's Transcendental Unity of Apperception, Husserl's transcendental ego, Christian souls, Diltheyan historical communities, and other philosophical accounts of the human as alternatives to Heidegger's conception of human beings as dasein. Haugeland's account nevertheless brings out illuminating parallels between Heidegger and Hegel, which then confer greater significance to their telling divergences, including the contrast in their conceptions of time that heralds Division III.

Hegel and Heidegger each understand human ways of life as historically situated openings onto the world as a whole that allow entities to show themselves intelligibly. For Hegel, human life in its various historical forms constitutes Spirit's self-recognition in otherness; for Heidegger, dasein is its disclosedness of the being of entities. Hegel and Heidegger also reject the traditional individuation of human bodies, minds, or lives in favor of the complex intertwining of an individuated collectivity: Spirit as "the I that is We and the We that is I," and dasein as "*the* entity that *we* ourselves in each case are." Haugeland's reading of Heidegger brings out an even deeper parallel in the two books' subject matter and its manifestation. Both books are centrally concerned with the truthful disclosure of the intelligibility of what-is, as in and for itself for Hegel, or as the sense of being in general for Heidegger.[6] Both Hegel and Heidegger explicitly reject epistemological conceptions of that intelligibility as derivative modes of comportment whose intelligibility depends on a more-encompassing conception (Spirit's self-comprehension in otherness for Hegel; dasein's being-in-the-world for Heidegger) that they cannot accommodate in their own terms.[7] Moreover, this issue comes to the fore in each book through encounters with the impossibility of an entire mode of disclosure, which is how Haugeland interprets Heidegger on existential death. *The Phenomenology of Spirit* reconstructs world history as a logical succession of "forms of consciousness," each taking itself as a truthful uncovering of its characteristic object. In each successive formation, the "experience" of living in terms of its constitutive orientation discovers its intended object(s) as recalcitrant to its understanding of, and comportment toward, that object: the object *in itself* turns out to be radically opposed to its being *for* consciousness.

Recognizing this thematic parallel makes Hegel a telling precedent for Haugeland's Heidegger to talk about the collapse of an entire understanding of the intelligibility of entities in terms of death. This motif is omnipresent in *Phenomenology of Spirit*. Consider the following version in the preface:

The life of Spirit is not the life that shrinks from death and keeps itself untouched by devastation, but rather the life that endures it and maintains itself in it. It wins its truth only when, in utter dismemberment, it finds itself. (Hegel 1977, 19)

The book's two most pivotal junctures involve literal confrontations with death. Spirit recognizes itself in a radically new way, first in "Mastery and Servitude" through the "absolute fear" of death in which consciousness "trembles in every fibre of its being" and thereby discovers "the absolute melting-away of everything stable [as] the simple, essential nature of self-consciousness" (Hegel 1977, 117). This absolute fear is then collectively recapitulated in "Absolute Freedom and Terror," whereby the previous determinations of alienated social life "vanished in the loss suffered by the self in absolute freedom," whose "sole work and deed is death ... the coldest and meanest of deaths with no more significance than cutting off a head of cabbage" (362, 360). The significance Hegel accords to *fear* of death is implicitly challenged by Heidegger in his juxtaposition of fear and angst as disclosive moods—angst is the philosophically significant attunement to death.

Neither Hegel's nor Heidegger's invocation of death as synecdoche for a collapse of intelligibility is merely figurative. Both emphasize that inquiry and understanding involve vulnerability and risk, such that one's (way of) life is at issue and at stake in comporting oneself understandingly toward the world.[8] Hegel takes up this point most explicitly in the introduction's scathing criticism of the quest for epistemic security as a "fear of error [that] reveals itself as fear of the truth" (1977, 47), but the point is also central to *Being and Time*: in dasein's understanding of being, its own being is always at issue. *Phenomenology of Spirit* culminates in the striking image of the Golgotha of Spirit, the comprehension of Spirit's unfolding in time and history as the path to its crucifixion. Hegel argues that this self-understanding (as "being-towards-death" in a rather different sense!) is the achievement of Absolute Knowing, the comprehension and recapitulation of all prior formations of consciousness as its own partial appearances. From this standpoint, the collapse of each prior formation manifests its finitude, a self-undermining dependence on what eludes its own understanding. The completion of modern forms of experience, by contrast, permits recollection and recapitulation of Spirit's path to that self-recognition as a self-completing, infinite whole.

We can now recognize that each book culminates in reflecting on the temporality and historicity of being, and at that point, Heidegger for the first time explicitly engages and criticizes Hegel, specifically the concluding passages of the *Phenomenology* on Spirit as "emptying out into Time" (Heidegger 1962, sec. 82; Hegel 1977, 492). For Heidegger, this critical response to Hegel's views on time sets the stage for posing the question of the meaning of being in general, which was the task assigned to the unwritten Division III. These parallels between the two projects, in light of Haugeland's interpretation of existential death, not only show why the critical juxtaposition to Hegel sets the stage for properly posing the central question for *Being and Time*. Recognizing the centrality of Heidegger's confrontation with Hegel's conception of Spirit's "infinitude" also puts in a new light the undisputed importance for Heidegger of Kant's conception of the finitude of human understanding, now directly contrasted to Hegel on Spirit's self-comprehension as a self-completing, infinite whole.[9]

Read and extended in this way, Haugeland's account also provides new insight into the troubling question of the political significance of *Being and Time*. For Hegel, prior formations of consciousness are *merely* finite modes of understanding. Each formation overcomes its recurrent external dependence by giving way to a new formation, until, in Hegel's concluding adaptation of Schiller, "from the chalice of this realm of spirits foams forth for [Spirit] its own infinitude" (1977, 493). Haugeland's Heidegger then responds to Hegel with a call for an owned, resolute openness to the ineliminable *finitude* of any understanding of being. As a challenge to Hegel, that commitment would specifically require us to face up to the possibility of the impossibility of a modern, rationalized "ethical mode of life" (*Sittlichkeit*). The formidable abstractions of Hegel's *Phenomenology* were conceived as a necessarily philosophical response to the French Revolution.[10] For Hegel, the Revolution and its failure made possible a new and adequately nonfinite comprehension of the modern world. This new understanding is the culmination of Spirit's "externalizing self-sacrifice that displays the process of its becoming Spirit in the form of *free contingent happening"* (1977, 492). *Being and Time* then looks to be a comparably formidable and necessarily ontological response to subsequent events. The senseless slaughter of the First World War and what seemed to Heidegger to be the humiliation and political collapse of modern Germany put Hegel's understanding of modernity in serious question. We know all too well what path Heidegger followed from there. The disastrous and repugnant outcome of Heidegger's existentiell interpretation of his historical situation should nevertheless not deter us from taking seriously his onto-logical-existential reading of what was at stake in his confrontation with Hegel. In this reading, *Being and Time* neither suggests nor rationalizes Heidegger's political commitments or his later deceptions and apologetics. The book does, however, provide a clearer indication of what Heidegger later thought was at stake in his political forays. Moreover, we can utterly repudiate his political commitments and still think that he was right about the finitude of any understanding of being, including the understanding that Hegel saw embedded in modern ways of life. In any case, Haugeland's interpretation of existential death takes us much further toward adequately understanding the philosophical significance of Heidegger's challenge to Hegel on Spirit's infinitude.

3   Philosophical Life after Existential Death

In this final section, I will indicate too briefly some issues to take up in the wake of Haugeland's phenomenological reconstitutions of intentionality as love and of existential death as the possibility of the impossibility of intelligibility. I limit myself to four themes, all of which I address more extensively elsewhere (Rouse 2002, 2015).

I begin with something problematic about Haugeland's designation of the environmental responsiveness of nonhuman organisms as ersatz intentionality. This term serves a polemical point for his criticism of how seriously other philosophical theories have missed their common target. This formulation nevertheless mistakenly suggests that intentionality as a thoroughly human phenomenon then serves as a norm with respect to which the ways of life of other animals are somehow deficient. Haugeland's account should instead encourage us to repudiate such conceptions of our relation to other animals. Nonhuman animals are crucially not like us in this important respect, but more importantly, they do not thereby "fail" to be like us. Recognizing a fundamental difference between intentionality and nonhuman animals' often-flexible responsiveness to their environments need not be attached to anthropocentric presumptions of human superiority. We should acknowledge and respect their own characteristic and astonishing ways of engaging and responding to the environments constituted by their ways of life, without interpreting such engagements as "failing" to achieve our self-critical intentional directedness.

My second theme is reciprocal. While nonhuman animal lives are not "genuinely" intentional, we are nevertheless animals and must understand ourselves as such. Here Haugeland himself has in his own terms been uncharacteristically irresolute. He noted in passing that "by ['norms of objective correctness can be understood in a spirit of] naturalism, appropriately construed,' I mean the thesis that people are, though distinctive, still naturally evolved creatures (somehow implemented in whatever physics tells us about)" (*HT*, 317, 358n15). That commitment is appropriate, but its viability cannot just be taken on faith. If Haugeland's account of intentional normativity turned out to be irreducibly antinaturalist in this broad sense, then a resolute response to that discovery would give it up as impossible in a way that would render it unintelligible to us. That responsibility has special force for Haugeland, because his account makes a naturalistic conception of intentionality not only harder to achieve but harder to conceive as even possible. Redeeming Haugeland's merely promissory footnote requires a dogged, resilient effort to revise and repair our understanding of science, nature, intentionality, and philosophical naturalism. In any case, that is what I have been trying to do in *How Scientific Practices Matter* (2002) and *Articulating the World* (2015). Moreover, I think a more adequate naturalism adds a powerful argument for Haugeland's differentiation of intentionality from other animal behavior. The best work on the evolution of language and conceptual understanding, as a form of niche construction and coevolution, reinforces this discontinuity even while vindicating an evolutionary recognition of our thoroughgoing animality (Rouse 2015, chaps. 3–5). The evolution of the capacity for an articulated "as-structure" required a partial break from other animals' highly attuned and flexible responsiveness to their environments.

A third theme emerges in asking what intentionality directs us toward and makes us accountable to. Haugeland emphasizes that intentionality requires *accountability* to something beyond our own patterns of comportment, and that is why intentionality cannot be assimilated to its ersatz or lapsed simulacra. He still makes common cause with the philosophical tradition, however, in regarding intentionality as a directedness toward objects, and "objectivity" as the governing normative issue. This focus on objectivity is Haugeland's Kantian rather than Heideggerian moment. He construes objective directedness as a kind of existential receptivity, a "letting-be" of an object or an entity standing over against us. His view also has a voluntarist moment, of committing oneself not to impose one's will on the object, and thus letting it stand forth and be what it is. This formulation, although attractive in many ways, does not do justice to our involvement in the world. Far from standing over and against us, the world has us in its normative grip. Taking "objectivity" to express our accountability to something beyond our own comportments and commitments is a kind of existential Myth of the Given, something nonnormative at which existential normativity supposedly comes to a full stop.

The sciences, Haugeland's own primary example of "authentic intentionality," bring out the importance of this issue especially clearly. Sciences do not simply seek truths about natural objects, or even laws about objective domains. Most truths about nature are not scientific truths, and even natural laws only constitute domains that are governed by normative issues. Marc Lange (2000, 2007), who more than anyone else has recently advanced our understanding of scientific laws, still indexes their normativity to disciplinary "interests." That term obscures that what a scientific discipline is "interested in" is itself normatively accountable. Intentional comportments get their normative significance and content as part of a larger pattern of practice, and such practices thereby constitute something at stake in their ongoing performance. What is at stake in a practice is nevertheless usually at issue within the practice itself. What the practice is about and how it matters are contested and open to further transformation in the practice's ongoing differential reproduction. There is much more to say here, but suffice it to say that talk of what is "at issue" and "at stake" in our practices and performances is best understand as what Brandom (1994) would call an extension of logical vocabulary. Just as logic makes explicit what we do in inferring, so these anaphoric concepts allow us to talk about the accountability of any intentional entanglement in the world as normative "all the way down." Our accountability to our discursively articulated environment does not terminate in already-determinate objects but is accountable to normatively structured issues and stakes that can only be discerned from within ongoing patterns of conceptually articulated involvement in the world.

This consideration points directly toward my final theme, already raised in my juxtaposition of Hegel and Heidegger as Haugeland interprets him. Heidegger was horribly wrong about what was at stake in the cataclysm of the First World War and the political and economic turmoil in its wake. We should nevertheless take seriously Heidegger's recognition of the fragility and possible unintelligibility of the practices and self-understanding of our modern way of life. I have already mentioned one partial response to that possibility: the commitment to working out a viable philosophical naturalism that neither oversimplifies the task nor settles for optimistic hand waving. A blithe assumption of the inevitable triumph of a reasoned, secular, naturalistically intelligible way of life is irresponsible. Other loci for the possible impossibility of our modern ways of life should also be taken into account. For example, the sciences give all-too-clear indications that our way of living may irrevocably undermine the climatic, energetic, and other conditions of its own biological possibility. The postcolonial and post–Cold War conditions of a global political economy may also seem to render a just society inconceivable. Moreover, justice and biological sustainability may be utterly unrealizable together. Can we respond to these and other challenges to the intelligibility of the possibilities afforded by our ways of life, neither fearfully nor complacently but resolutely? That is, can we doggedly revise and repair our understanding of what is at issue and at stake in who we are and how we live, while also remaining open to the possible need to give it up as impossible? A call for a resilient, resolute responsiveness to these and other possible impossibilities in our self-understanding as worldly human beings is the final challenge I take from John Haugeland's placement of intentionality at the philosophical juncture of love and existential death.

Notes

1. Haugeland does not discuss whether a character in a video game or an "eater" in the game of Life (*HT*, chap. 11) has an ersatz-intentional directedness toward other objects in the on-screen "world."

2. I follow Haugeland's recommended practice of treating "dasein" as a naturalized English word and therefore capitalize it only when quoting others or citing Heidegger's German term.

3. Thanks to Mark Okrent for this useful example, although we take it in different directions.

4. Indeed, in the same spirit, normal cats do not lack a fully genetically assimilated basis for shrew recognition.

5. As I argue later, *we* (as biologists equipped with conceptual capacities) can distinguish functional subsystems within an organism's physiology and behavior, but the organism's behavior does not make such distinctions, even implicitly.

6. Haugeland did not live to write the part of *Dasein Disclosed* that aimed to interpret what Heidegger would have said in Division III, but his other writings make clear the direction of his account. Heidegger presented the lecture course *Basic Problems of Phenomenology* as "a new elaboration of Division III of part I of *Being and Time*" (1982, 1), focused on four traditional theses about being that constitute the "basic problems of phenomenology." Haugeland clearly intended to argue that Division III would aim to account for the unity of these various aspects of being in terms of the fourth and final thesis concerning the "truth-character" of being.

7. Heidegger (1962, sec. 13) explicitly claims that knowing is a derivative mode of comportment dependent on dasein's being-in-the-world. *Phenomenology of Spirit* makes a similar reordering central to its whole line of argument. The book begins by considering forms of consciousness in which consciousness takes itself to be sensation, perception, or understanding of its object. Each turns out to be deficient, and from the standpoint reached at the end of the book, one can recognize that deficiency as a dependence of their intelligibility on their belonging to a whole socially articulated way of life ("Spirit") that it cannot recognize in its own terms as consciousness of an object. "Knowing" as traditionally conceived in theories of knowledge is thus also quite literally a "derivative mode of comportment" in Hegel's account.

8. Thanks to Rebecca Kukla for calling to my attention the importance of this theme.

9. Heidegger (1962, 480) calls attention to the significance of this difference between Kant's and Hegel's views of time at the end of sec. 81, setting the stage for the extended comparison to Hegel in sec. 82.

10. For an extended discussion of this aspect of the *Phenomenology*, see Comay 2011.

References

Aaron, E., and J. P. Mendoza. 2011. Dynamic obstacle representations for robot and virtual agent navigation. In *Advances in Artificial Intelligence*, ed. C. Butts and P. Lingras, 1–12. Berlin: Springer.

Akins, K. 1996. Of sensory systems and the aboutness of mental states. *Journal of Philosophy* 93:337–372.

Brandom, R. 1994. *Making It Explicit: Reasoning, Representing, and Discursive Commitment*. Cambridge, MA: Harvard University Press.

Carman, T. 2003. *Heidegger's Analytic Interpretation, Discourse, and Authenticity in Being and Time*. Cambridge: Cambridge University Press.

Cheney, D., and R. Seyfarth. 1990. *How Monkeys See the World*. Chicago: University of Chicago Press.

Comay, R. 2011. *Mourning Sickness*. Stanford: Stanford University Press.

Davidson, D. 1984. *Inquiries into Truth and Interpretation*. Oxford: Oxford University Press.

Dretske, F. 1981. *Knowledge and the Flow of Information*. Cambridge, MA: MIT Press.

Dreyfus, H. 1979. *What Computers Can't Do*, 2nd ed. New York: Harper & Row.

Dreyfus, H. 1991. *Being-in-the-World*. Cambridge, MA: MIT Press.

Dreyfus, H., and S. Dreyfus. 1986. *Mind over Machine*. New York: Free Press.

Fodor, J. 1979. *The Language of Thought*. Cambridge, MA: Harvard University Press.

Fodor, J. 1987. *Psychosemantics*. Cambridge, MA: MIT Press.

Frankfurt, H. 2005. *On Bullshit*. Princeton, NJ: Princeton University Press.

Hacking, I. 1992. The self-vindication of the laboratory sciences. In *Science as Practice and Culture*, ed. A. Pickering, 29–64. Chicago: University of Chicago Press.

Haugeland, J. 1998. *Having Thought*. Cambridge, MA: Harvard University Press. (Abbreviated as *HT*.)

Haugeland, J. 2013. *Dasein Disclosed*. Cambridge, MA: Harvard University Press. (Abbreviated as *DD*.)

Hegel, G. W. F. 1997. *Phenomenology of Spirit*. Trans. A. V. Miller. Oxford: Oxford University Press.

Heidegger, M. 1962. *Being and Time*. Trans. J. Macquarrie and E. Robinson. New York: Harper & Row.

Heidegger, M. 1982. *Basic Problems of Phenomenology*. Trans. A. Hofstadter. Bloomington: Indiana University Press.

Husserl, E. 1970. *Logical Investigations*. Trans. J. Findlay. London: Routledge & Kegan Paul.

Husserl, E. 1980. *Ideas I*. Trans. F. Kersten. The Hague: Martinus Nijhoff.

Kierkegaard, S. 2006. *Fear and Trembling*. Trans. S. Walsh. Cambridge: Cambridge University Press.

Kuhn, T. 1970. *The Structure of Scientific Revolutions*, 2nd ed. Chicago: University of Chicago Press.

Lange, M. 2000. *Natural Laws in Scientific Practice*. Oxford: Oxford University Press.

Lange, M. 2007. Laws and theories. In *A Companion to the Philosophy of Biology*, ed. S. Sarkar and A. Plutynska, 489–505. Oxford: Blackwell.

McDowell, J. 1994. *Mind and World*. Cambridge, MA: Harvard University Press.

Millikan, R. 1984. *Language, Thought, and Other Biological Categories*. Cambridge, MA: MIT Press.

Popper, K. 1959. *Logic of Scientific Discovery*. New York: Harper & Row.

Rouse, J. 2002. *How Scientific Practices Matter*. Chicago: University of Chicago Press.

Rouse, J. 2015. *Articulating the World*. Chicago: University of Chicago Press.

Schank, R., and R. Abelson. 1977. *Scripts, Plans, Goals, and Understanding*. Hillsdale, NJ: Erlbaum.

Searle, J. 1983. *Intentionality*. Cambridge: Cambridge University Press.

II Embodiment

5   Language Embodied and Embedded: Walking the Talk

Mark Lance

Interrelationist accounts retain a principled distinction between the mental and the corporeal—a distinction that is reflected in contrasts like semantics versus syntax, the space of reasons versus the space of causes, or the intentional versus the physical vocabulary.

—John Haugeland, "Mind Embodied and Embedded" (*HT*, 208)

What could it mean to say that Dasein's world "lets entities be" ...?

—John Haugeland, "Letting Be" (*DD*, 94)

How should a Heideggerian—especially one who accepts the general contours of Haugeland's powerful development of Heideggerian ideas—understand the phenomenon of language in its epistemological and ontological significance? In particular, if we take language to involve the production of speech acts via the utterance of (repeatable) sentences that have a *content* that can be shared with or embedded within other sentences used in other contexts, how are we to understand the Being of such speech acts and sentences, and the relation between them? Put another way, how can a Haugeland-style Heideggerian make sense of a category of Being that incorporates a force–content distinction?

It is not uncommon to characterize the epistemologically central Heideggerian notion of engagement, or embodied coping, by *contrast with* reasoning, representation, calculation, linguistic articulation, and related notions. For some—notably Dreyfus—little if anything is said about what concepts, reasoning, sentence meaning, or propositional knowledge are, other than the elements we should not take to be fundamental to understanding. Dreyfus's efforts to distinguish his own approach to understanding from that of traditional artificial intelligence often leave one with the impression that he is abandoning the linguistic ground to the representationalist, accepting that there is little more for the Heideggerian to say about language than that it arises in breakdown by way of distancing oneself from a world.

Brandom is, in one sense, nearly the opposite of Dreyfus. Whereas Dreyfus neglects language, it is Brandom's primary focus. And in *Making It Explicit: Reasoning, Representing, and Discursive Commitment* (Brandom 1994), he begins his elaborate discussion of language in a place familiar to Heideggerians: suggesting that any understanding of language must be in terms of the practical role that speech acts play within a context of engaged skillful understanding. Thus whereas Dreyfus contrasts linguistic practice with engaged coping, Brandom purports to understand it as a special case. Sentences in Brandom's view, that is, are ready-to-hand.

But by the end of *Making It Explicit*, Brandom has severely transformed this idea. An account emerges that ultimately abandons its Heideggerian roots in all but the most rarefied formal sense, leaving linguistic practice a thin and abstract shadow of engaged coping—linguistic practice is merely scorekeeping, and one is engaged with no more than scorecards—and connecting linguistic and nonlinguistic modes of engaging the world only causally. The practice of scorekeeping, out of which Brandom constructs contents, that is, has no essential connection to embodied and embedded coping in a material world. It is itself a practice, to be sure, but it is merely the practice of moving abstract tokens about in a rule-governed manner. Any embedding in the environment is explicitly deemed contingent. Thus, for Brandom, language gets interesting when it ceases to be understood in terms of the richly embodied and embedded coping of a typical Heideggerian world, and we are left again with a Dreyfusard dualism of practical and linguistic understanding.

John Haugeland rejected this separation of the contentful and the richly practical. In his masterful "Truth and Rule-Following," he argued that one cannot retrieve a genuine notion of semantic content from the interrelation of a normatively structured practice of moving linguistic tokens about—syntax—and various causal interactions with an environment. If we are to make sense of empirical content—of the very idea that claims have contents that are true or false insofar as they get the world right—we must understand the normative role the world plays in constituting linguistic correctness. It seems to me, however, that Haugeland does not address the central problems inherent in such a project. In particular, his talk of allowing the world to constrain our normative practice remains frustratingly metaphorical.[1] Further, a strange individualism lurks in Haugeland's discussion of these issues. Certainly one must attend to the right sense of worldly embeddedness if one is to understand genuinely empirical content. But one would think that the central ontological context of language would be Being-with. Language is, after all, most fundamentally the vehicle of communication, and yet this dimension is simply not a significant moving part in "Truth and Rule-Following."[2]

Rebecca Kukla and I have tried to offer another approach. In Kukla and Lance 2009, we offered a more systematic account of the normative functional space of speech acts, and in Kukla and Lance 2014, we offered an account of normative accountability to the world that emphasizes the essentially intersubjective ground of such accountability. But *semantic content* was not the topic of either of these works. And even if it is a mistake to begin one's theorizing by helping oneself to a dichotomy between syntax and semantics, or the space of reasons and the space of causes—as Haugeland, I believe rightly, claimed—these are distinctions that surely must emerge. Even the most semantically nihilistic philosopher must admit that there is something pragmatically significant to the question of whether a given pair of sentences "share content." Meaning talk is legitimate talk, with a legitimate place in the language game, whether or not it is the theoretical key to understanding language as a whole. It is just as obvious that one can express a content in multiple syntactic structures, or with different pragmatic performances, as it is that justifying and causally explaining an event are different performances.

While I certainly offer here nothing that can be called a "theory of content" (and, indeed, have grown increasingly skeptical of the very project over the years),[3] my goal is to clarify the place of "semantic content" (or, better, something very much like the speech act of judging that two performances have the same semantic content) within the broadly Heideggerian conception of understanding developed by Haugeland. I begin with some phenomenological observations about the role of simple signaling performances among creatures engaged in simple social practices. I look at the ways that engagements with the world are functionally augmented by the addition of various layers of "protolinguistic complexity"—more and more structurally complex, normatively constrained signaling behavior—until we see the emergence of something recognizably like semantic content, specifically the force-content distinction.[4]

Proceeding in this manner will have two points. First, looking at the way that protolanguage works will point us toward a more phenomenologically plausible idea of the function of language (and move us away from what I take to be the seriously misguided idea that language arises only in breakdown of engaged comportment). Second, beginning with toy examples and developing them carefully will allow us to frame the question of what it would be for (embodied and embedded) social norms to institute norms related to the content of, rather than the particular contextual performance of, a given speech act. The answer to this question will add a surprising social-structural element to the ontological constitution of objective purport.

1   Skills and the Circumspective Understanding of Changing Circumstance

As Heidegger emphatically reminds us, the fact that we do not explicitly represent our circumstances does not mean that skills are "blind." Involved skillful engagement has its own mode of uptake, which Heidegger calls "circumspection." (It is important that the skillful soccer player has her eyes open while playing, after all.) My goal is not to say anything specific about circumspection, about what changes in us—or maybe in us-in-the-environment, or us-cum-environment—when something changes in our world.[5] But however one understands our responsiveness to environmental events in the midst of circumspective practice, it is clear that we do skillfully modify our engagement to deal with environmental input.

Suppose that I'm a defensive midfielder, and the midfielder on the other team passes the ball wide on the left flank. Situation S_1 is the totality of all that is relevant to my skillful engagement with the game before the pass and situation S_2 after the pass. From the outside, we can characterize me as seeing "that the pass went left" and adapting appropriately, but of course we should not read this as implying that I form a representation of the discrete event or think the thought "the pass went left" and *infer* what to do from this. That would precisely not be a circumspective mode of engaging with the environment.

With this temporal adaptability in mind, we might use the following diagram: with S_1 being our original situation, S_{11} being the result of the pass going left, S_{12} being the result of the pass going right, S_{111} being the result of my following the ball, S_{112} being the result of my staying with my man, and so on.

Let us now distinguish two sorts of changes that we skillfully cope with in the course of a practical engagement with a world: agential and environmental. For now, we will treat the agential changes as merely things I might do, and the environmental as anything that happens that isn't under my control. So if our tribe of primates is on a hunt, I might throw a rock into the bushes, on the one hand, and a rabbit might dart out of the bush, on the other. Regardless of whether one thinks of rabbits as performing actions, I will call the former agential and the latter environmental, limiting "agential" to changes brought about by actions of the critters I am interested in—the potential language users. So consider the second diagram. S_1 is a moment in a boar hunt. E_1 is the event of a young boar rushing out of the bushes; E_2 stands for a mature boar rushing out toward us. A_{111} and A_{121} stand for my throwing my spear at the boar, A_{112} stands for my chasing after the boar, and A_{122} stands for running away toward shelter. (Mature boars are more dangerous than young ones, let us stipulate.) Thus this diagram represents a branching system of possible ways a skillful event might unfold from a given moment. Presumably, as a skillful hunter, I am capable of "tracking" the changes that result from any of these events and adjusting my engagement with the situation in light of them, again, with

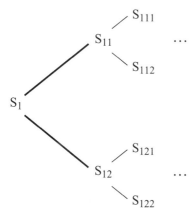

Figure 5.1

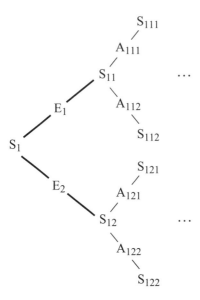

Figure 5.2

no suggestion that any of this happens consciously, or even as a matter of subconscious processing of contentful information.

So in the initial situation S_1, either a young boar rushes out (E_1) or a mature boar does (E_2). Either results in a new situation. In situation S_{11}, with a young boar now present, the hunter either throws her spear (A_{111}) or charges toward the boar (A_{112}), and so on.

2   Signaling, or Really-Really-Proto-Speech Acts

A distinctive feature of human practice is that it is social. We hunt as packs, play games as teams, eat meals and raise children as extended families. And we do this socially not only in the sense that there are lots of us doing things in proximity to one another but also in the sense that our skills are coordinated, beholden to each other, and often functional only in the context of others. Further, we are *normatively* embedded in our communities in the sense that our actions not only tend to serve various individual and collective ends but also are subject to appraisal and correction by others. We hold each other to the social norms and conventions that allow us to deal with our worlds in this distinctively human manner. And this holding also requires signaling.

Much of our ability to function as groups—not to mention much of the utility that accrues to social practice over individual behavior, and the very possibility of mutual holding to norms—depends on our ability to skillfully respond to signals from others, at least in the sense that communicative acts by others can function to change our situation in ways that make a difference to our ongoing comportment. So what, in a suitably minimal sense, is such a normatively and behaviorally significant signal?

Suppose that we are on a boar hunt. There are four of us, and we are looking in different directions as we near the large oak tree. Boars being dangerous, we have a coordinated hunting pattern that involves three of us surrounding the boar and holding it at bay with our spears while the fourth throws a net over it. If we were all to spot a boar—circumspectively give skillful uptake to its presence—we would carry out this coordinated attack. But suppose we are looking in different directions, and only one of us sees the boar behind the tree. Since the others are looking away, they can give no circumspective uptake, and so the coordinated hunting pattern cannot continue.

How useful it would be if our party came to have a social practice of shouting some distinctive sound at such moments. Just so it is easy to remember, imagine it sounds like this: "Boar by oak!" But this is, for current purposes, just a distinctive noise, no more. (We presume no compositional grammar, no inferential role, no truth conditions, no meaning, and no determinate pragmatic force.) We can now imagine a straightforward revision of our usual hunting practice: we give uptake to such a signal by moving directly into our surround-and-trap practice. Signaling, that is, functions as a sort of surrogate for whatever sort of perceptual uptake goes on in circumspection, leading to the same functional output in our broader practice of hunting. And the benefit of signaling is obvious: we can react with more collective competence if anything anyone's eyes are pointed at provides environmental input to each of us as if our own were.

There are two points to notice about these simple signaling moves. First, they share functional structure with a wide range of mature pragmatic performances. Consider the declarative, or perhaps better observative, "There is a boar over by the oak!" or the imperative "Move into position to attack a boar by the oak!" or the normative "We ought to move into position to attack a boar by the oak."[6] All of these, in the context of competent group hunting, lead to the same immediate behavioral consequence, and at this point the more complex normative structures that distinguish the functions of these speech acts are not assumed to be in place at all. In particular, as I discuss in Lance 2015, there is, at the level of simple signaling, no clean distinction between normative and descriptive claims. This signal involves an implicit prediction about what others will do, and an implicit endorsement of relevant behavior, together with an inclination to hold them to that behavior. If I say "boar by oak" and you do not perform the expected action in response, I will be both surprised and disapproving.

Second, this signaling act will have differential behavioral implications across the various members of the community to whom it is directed. I might move to the right of the tree, Ned to the left, while Nob deploys his net. The signaling act is inspired by a sighting of the boar and leads to all these differential responses. This is its functional role in the collective hunt. It is a signal the uptake of which is differential across those participating in the practice in a way that corresponds to their usual skillful response to a given perceptual uptake of a situation. Note that we are not understanding the act as the moving of some discrete bit of information from a speaker to various hearers and then understanding them to deploy it in various diverse ways given background knowledge. The act is, rather, defined by the function that moves from production to variable output. Just as the skillfully engaged agent responds to the holistic context that includes specific changes in environment, so the skillful social agent can respond to various signals that take place in such environments.

3   Very-Very-Proto-Speech Acts Are Not Speech Acts

Such simple signaling obviously falls short of anything that is genuinely linguistic. Many (related) features of language are missing from such a system—compositional grammar, systematic situational variation, stable conventions of use, inference, and so on—but the most fundamental (for our purposes) is that there is nothing like a context-independent content, no force-content distinction. No matter how complexly differentiated the performances become (perhaps imitating a grammar that allows for the production of a huge variety of signaling performances), no matter how many "signaling-signaling" moves come to be licensed, the significance of any signaling performance is thoroughly tied to a particular context of skillful engagement: this hunt, with these hunters, at this point in the day, for this community, in this condition, and so on. And there does not seem to be any obvious way to remove that context dependency by adding more of the same sort of practical signaling. A linguistic act, by contrast, has a meaning. If I assert that P, this is something that someone else can assert, or disagree with, when not hunting, on a different day, in a different social context.

Imagine that the same utterance (side question: what does that mean? For now, we can suppose that it is a matter of objective sonic similarity) is used in a range of situations. Maybe a group of gatherers/cooks who will be preparing dinner when we return from the hunt are watching from the safety of a nearby hill. They are gathering herbs, vegetables, and firewood. Perhaps their collective activity differs depending on the likely result of the hunt—some animals will be grilled for a long time, others boiled, some cooked with these herbs, some with those—thus explaining why they keep a watch on our activity. In such a context, one gatherer might call out "boar by oak," and this may set the group off on the specific boar-meal gathering pattern. *The point is that adding this usage by others does nothing to bring us closer to a public meaning for a declarative utterance.* In fact, if anything, it takes us further from a common meaning because this pattern of use is so very different. The import of the utterance for the gatherer/cook group is utterly different from that for the hunting group. It is true that the utterance is elicited by the same event in the world, but this is not of use to the pragmatist—we are trying to explain aboutness and content in terms of proprieties of use, not assuming them to characterize use—and if we are looking at the more specific mode of skillful engagement that precedes the utterance, then we see no obvious similarity between the two contexts. Thus either we take this to be a functionally different signal with superficial sonic similarity to the previous—in which case it adds nothing—or we stipulate that it is the same signal, and now have two radically different functions for the same signal.

The problem can be put abstractly as a tension between two core ideas of the sort of contemporary Heideggerian pragmatism one finds running along the Berkeley-Chicago-Pittsburgh axis. The first is that skillful understanding is tied to local contexts of involvement, and the second is that language is to be understood as a particular sort of tool, defined by the way it is taken up in skillful use. Taken together, these two ideas seem to imply that there is no such thing as the noncontextual meaning of a claim, and if this assertion is right, the whole idea of objectivity and truth seems threatened. A proposition is not true-in-this-context, true-for-me, true-for-us, and so on. It is either true or false simpliciter, and the same goes for meaning. A content is something that anyone can endorse or challenge as true or false, whatever their contextual goals. If I can think that P, then you can think "that's wrong; ~P." So the meaning (of a proposition) can't be something that is defined by its function in a particular context, individual skill, mode of comportment, or anything else. But, as we emphasize in Kukla and Lance 2014, this means that the norms of correctness governing signaling acts are simply the norms of practical utility in that context, not norms of truth. There is no sense in which the utterance has a constitutive goal of getting the world right.[7] I return to this issue in section 6.

4   Simulation

Our skills are not perfect. This is a simple point that is often lost in discussions of the phenomenology of engaged coping, no doubt because it is so obvious as to be thought not worth discussing. (What is usually at issue and deemed worthy of explanation in the current philosophical context is that unthinking skillful behavior is so successful.) Obvious or not, however, the point has important implications for the development of social practice. Suppose our hunters come upon a young boar hiding in a bush. In an epistemically ideal situation, they would have a clear view of the surrounding area and understand how to proceed. In a less-ideal situation, the heavy vegetation would make it unclear whether a mother boar is around. This makes a difference, of course, since mother boar are formidable adversaries, whereas young ones are easy dinner. (Again, to say that this distinction matters to our hunters' practice is not to assume that they are engaging in propositional representation. So long as there are behavioral implications to the difference between the two circumspectively ideal situations, then we will expect to see an intermediate form of cautious engagement in nonideal situations.)

A signaling-simulator is a critter capable not only of responding to signaling performances by others in the same way that they would in the presence of receptive uptake of changes in their environment, but also of simulating such responses without acting. That is, signaling-simulators are capable of experiencing and acting on the basis of internal processes that run for a time in a way similar to processes that form a part of their actual engaged behavior.[8]

Imagine a signaling-simulator on a hunt of the sort just described; he spots a young boar but is in the mode of circumspective uncertainty between the-boar-is-alone mode and the-boar's-mother-is-nearby mode. Our signaling-simulator can at this point fruitfully engage in something like protoreasoning—again without assuming that there is anything like an explicit grasp of propositional contents, or of propositional contents at all, for that matter. He might—out loud or sotto voce—token the "there is a mother boar" signal and run through a simulation of the next few minutes of the hunt, simulating how he would behave had that signal been received. Perhaps this takes the form of a sensory experience, a visual, emotional, and auditory presentation that goes the way that a normal involved engagement in a hunt would go; or perhaps it involves something else. The engineering of the inner life of signaling-simulators is not my concern here.[9] What is true of such creatures, by definition, is that they are capable of running through something like a skillful engagement in a hunt, one occasioned by the signaling move "there is a mother boar," where this likeness is sufficient for them to experience the likely emotional or evaluative upshot of that process.

But this upshot is not certain until it happens. Sometimes our hunting party succeeds in killing a mature mother boar, and sometimes a hunter dies in the attempt, and the others have to flee. The latter happens most often when they rush in to kill a young boar and are surprised by the mother. When they cautiously encircle a mother boar, they generally succeed in killing it, but not always.

With this in mind, imagine that our signaling-simulator engages in a sort of branching simulation process: she first tokens "there is a mother boar hiding" and then "we attack the baby straightaway." This simulation leads to images of (or other simulations of) death to tribe members and a failed hunt. This makes our signaling-simulator feel bad. Now she tokens "there is a mother boar" and "we slowly encircle the area with spears out," and the simulation ends with a happy and successful killing of two boars and a feeling of great joy. Next she runs two other simulations: one involving "there is no mother boar" and "we slowly encircle." Result: successful hunt with a bit of wasted time, but an underlying sense of fear. And "there

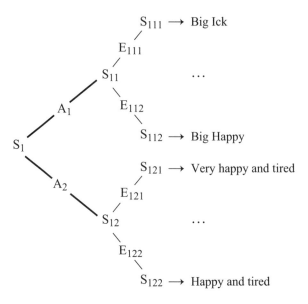

$S_{111} \longrightarrow$ Big Ick

$S_{112} \longrightarrow$ Big Happy

$S_{121} \longrightarrow$ Very happy and tired

$S_{122} \longrightarrow$ Happy and tired

Figure 5.3

is no mother boar" and "we rush in and kill the young one." Result: successful hunt with no wasted time—Big Happy.

Finally, let's add one more step of complication. What the signaling-simulator has done so far is to simulate a simple branching process that we, from the outside, can represent as in figure 5.3.

At the first branch, A_1 represents the decision to rush in against the young boar, and A_2 the decision to slowly encircle it. E_{111} and E_{121} represent the result that there is a mother boar hiding, while E_{112} and E_{122} represent the result that there is not.

Suppose now that a signaling-simulator is what we call a "pragmatic optimizer," someone wired in such a way as to allow such a process of reasoning to guide its behavior in a pragmatically rational manner. He will be deterred by the Big Ick in the topmost branch—Big Ick, recall, being the negative reaction to a simulation of someone in the group being killed by a boar. The net result of our four simulations and behaviorally guiding reactions, for a pragmatic optimizer, is that he will be motivated to actualize, A_2.

5 Taking Stock

Let us compare our pragmatic optimizing signaling-simulator to fully conceptual reasoners. There is a nontrivial structural similarity to practical reason in that our creatures go through a series of states, the functionality of which is aptly represented *from the outside* by seeing them as considering hypothetical possibilities and engaging in counterfactual and evidential reasoning, leading to a practical conclusion such as "If I were to rush out, then if there is a mother boar nearby, I would likely be killed; whereas if there is no mother boar, I will succeed. On the other hand, if we were to all slowly surround the area, then we would likely succeed in our hunt whether or not there is a mother boar around. The risk of death is to be avoided, so we should surround the area." We have stipulated that our hunters go through no such reasoning and that all sorts of complex inferential ability that would go with the literal truth of such ascriptions is beyond our critters. But their behavior tracks this reasoning in this limited context.

Recognizable discrete states may be in play—those brought about by uptake of the various signaling moves—but these states fall short of fully conceptual belief in two ways. First, the state itself is highly contextual. Nothing we have posited in pragmatic optimizing signaling-simulators implies that they would be in the same dispositional A-state after giving uptake to "boar by the oak" on a hunt as they would be after hearing this signal while gathering herbs or cooking. Or, indeed, that looking at a boar would produce the internal representation, if any, in the two contexts. And the social correlate of this point is that the significance—the appropriate response to uptake—of signal production will differ from context to context. Second, and related, no aspect of their inner life or outer production stands as potential fodder for more-complex contexts. Unless one is able to say "there is a boar, but there is no squirrel" or "there is a boar and he is lying down" or "if he stands up on two legs and talks to me, he isn't a boar after all," and so on, then whatever states are functionally in play are not propositional attitudes—beliefs that there is a boar by the oak. This is to say that no element is separable from the local context that functions within a broader semantic system. Rather, the behavior-guiding and environment-prompted internal states of pragmatically optimizing signaling-simulators are aspects of the complex totality of engagement with the hunt-world that we on the outside can distinguish by applying our own conceptual skills to the interpretation of theirs.

Since no distinct compositional states with context-independent significance are at play, we can see pragmatic optimizers only as what, following

Haugeland, we might call *"derivative* reasoners" as opposed to *"original* reasoners."[10] With this distinction in mind, let's call what they engage in when
they carry out action-guiding simulations "pseudoreasoning." Aspects of
this sort of pseudoreasoning nonetheless square quite naturally with the
phenomenology of genuine reasoning in the midst of skillful engagement.
The temporal possibility trees of conceptual creatures have indefinitely
many branches. One with access to a language *could* simulate the hypothetical possibility of taking a nap, of singing a love song to the boar, or of
setting one's beard on fire. But it is precisely part of being skillfully engaged
in a hunt that such agential possibilities do not figure into one's reasoning
at all, any more than do environmental possibilities such as that a giraffe is
folded up behind the tree in a little bamboo box, or that the mother boar is
made of water. It is not that we have, or need, a reason to ignore such considerations—to suppose that we do would generate a regress that would
obviate the entire point of thinking of understanding as fundamentally a
matter of skillful engagement. For skillfully engaged reasoners, that is, the
space of possibilities that are serious candidates for the capacity of active
reason is given by the underlying skill, rather than by the specifically
reasoning faculty.[11]

Equally dependent on nonlinguistic skill are the inferences one draws
from the skillfully obtained starting hypotheses. The fact that I rush the
boar and the mother is nearby does not *entail* that I will be killed. It is my
skillful understanding of situations like this that leads me to foresee a bad
outcome, but I'm quite aware that there are possible circumstances in
which this doesn't happen. The point of both ways that our reasoning is
limited within the space of mere possibility is that while the pseudoreasoning signaling-simulators have available only a scattered collection of signaling moves and a narrow range of simulational developments from them,
that collection of scattered points remains significant within the skillful
economy of more mature conceptual descendants because these points are
the ones that present themselves as serious elements in the process of active
reasoning.

That is all to say that reasoning in the context of ongoing comportment
ineliminably involves what Sellars called "material inferences," specifically
nonmonotonic material inferential proprieties, proprieties based not
on logical form but on the totality of our background understanding.
Now, it is a common assumption in the philosophy of language and

logic—everyone from Quine and Davidson to Brandom assumes this—that this propriety-determining background is itself to be understood as a system of auxiliary beliefs.[12] But to take this route is to come very close to giving up entirely on the centrality of skillful engagement to human understanding. If we acknowledge, as I think we must, that even the most mundane and everyday human social practice is shot through with signaling, simulating, and various forms of nonmonotonic, contextually significant practical inference, then we will see the question of the relevant background as arising in any context of skillful human engagement. Thus it is not a question that can be seen as purely linguistic, as a matter of theoretical reason.

The idea of basing material inferential propriety on background belief is that the difference between, say, a situation in which it is reasonable to infer from "we attack the young boar" and "there is a mother boar waiting nearby" to "we will be killed" and a situation in which it is not, lies in the presence or absence of a background premise (e.g., the boar is already injured). But adding this premise does not turn the inference into an entailment (maybe the injury is superficial). The point is that no matter how many premises we add, the same issue arises until we terminate at an entailment, an inference that holds in all circumstances. But to suppose that the propriety of inference is always a matter of entailments from a set of premises, some of which are implicit, is just the assumption of what Haugeland calls "Good Old-Fashioned Artificial Intelligence"; it is implicitly to accept that engaged coping has nothing at all to do with reasoning, and to treat reasoning as a strange kind of exception to an otherwise general theory of skillful actions.[13]

Far better, I suggest, to say that while we are doing much else besides, we—in a fully conceptual mode—are engaging in the environment in the same way that pragmatic optimizing signaling-simulators are, that even in the context of explicit reasoning and propositional attitudes, there is understanding that is ontologically grounded in skillful engagement with a practical context. The guidance that skillful engagement gives to reasoning is mediated not by further background beliefs but by the skillful integration of speech acts into the practical context of acts in general, via simulation, signaling, and pragmatic optimization.

6   From Pseudoreasoning to Reasoning, Signaling Moves to Claims

The advantage of the account I am developing is that it allows us to see how material inference can be a continuous part of a social practice, how it can facilitate collective action and improve the utility of practice in the face of skillful imperfection. All of this can be understood in the same basic terms that are familiar to neo-Heideggerians. But the disadvantage, really the flip side of these very features, is that we seem to have no way to understand any of this as involving genuine propositional content. Pragmatic optimizing signaling-simulators merely coordinate locally appropriate behavior, and this is not what goes on when one, for example, makes a claim about the world—as is made clear by the fact that the truth is sometimes less useful than a well-chosen fiction. Our methodological consistency, that is, seems to leave us stuck with contextually embedded Heideggerian pseudoreasoning. The structures of use that we have focused on—and out of which we are attempting to build a picture of reasoning—are essentially tied to local contexts of practical activity in precisely the way that propositional content cannot be.[14]

The key to resolving this issue, I suggest, lies in recognizing that while there is something to the idea that meaning tracks inferential role, it is not accurate in the most obvious interpretation of this phrase, namely, that two utterances have the same meaning only if they are used (or properly used) to make similar inferences. This, as examples like the difference between genuinely linguistic versions of our hunters and gatherer/cooks makes clear, is just false. Although the pseudo-inferential function, and indeed proper function, of "boar by the oak," in the context of gathering among the cooks, is utterly different from its function for the hunters, if the hunters were to say (i.e., if they were to produce something that was the actual speech act) "there is a boar behind the oak," and the cooks, from a vantage point up on the hill, were to deny this, they would have disagreed, de jure. But to disagree is to engage with *the same content.* If I assert P and you deny something very close to P, we have not necessarily disagreed. So how can there be such a thing as factual disagreement or incompatibility—how can there be an act of denying an asserted content—if the skillful use of language is essentially contextual? Doesn't semantic incompatibility dissolve into nothing more than clashes in usefulness? I find it handy to perform a particular verbal act; you find it not so. Perhaps you even find my noisemaking to be a problem leading to conflict. But such disputes are in no sense semantic, in no way involve logical denials of contents.

Let us return to the practice of argument. Pragmatic optimizing signaling-simulators hold one another to norms. One context in which this might arise is when conflict occurs among signaling creatures. One produces a signal, and the other a different signal that pushes toward a different action—that is, a pair of signals are produced that are practically incompatible in the sense that others cannot perform the acts appropriate to each. If this occurs among pragmatic optimizing signaling-simulators, we can easily imagine the practice gaining another dimension of complexity so as to more closely resemble a practice of argument, of defending claims for reasons. So far, our pseudo-inferential behavior has shared functional structure with practical inference, but it can also share similarly thin structure with theoretical inference. α signals "there is a boar behind the tree." β signals "the boar has run away." These lead us on two practically incompatible courses: to surrounding the tree, and to continuing to track. But it is easy to imagine there coming to be a practice in which β offers two signals, "the boar is behind the tree" and "I throw a rock at the tree." This, for our simulating creatures, will lead to an anticipation of the boar running out from cover. Now imagine that in the midst of our hunters engaging in such simulation, β actually throws a rock, and nothing happens. Then, if our creatures are wired in the right way, this will function as a theoretical refutation of the signaling move "there is a boar behind the tree." Again, this demonstration is utterly contextual, tied closely and essentially to our skillful understanding of the world of hunting; but aside from all the reasons why none of this is genuinely conceptual, it is perfectly good pseudo-theoretical refutation.

A key point here is that the addition of new normative constraint on our signaling practice has something of the telos of classical rule utilitarianism. When introducing new norms, there is no call to defend them on the basis of previously existent normative regularity. It is not as if the reason why a more complex practice of signaling-signaling moves (the pseudoversion of Sellarsian language-language moves) is useful is that there was already implicit in previous practice some entitlement for that move. Rather, we are genuinely adding new normative material in instituting such a propriety, and the utility of it is that the new system with the new constraints lets us accomplish our social goals more readily than did the previous simpler system. Quasi-inferential proprieties are instituted, not found, and then survive or not on the basis of the overall utility of the resulting system. But for all of that, the utility in question is contextually local.

But now let us return to the possibility of dispute between the hunters and the gatherer/cooks. Suppose that the hunters employ their social simulating, signaling, and pragmatic optimizing capabilities to reach a situation in which they are entitled to produce their version of the signal "there is a boar behind the sacred oak." Suppose that at the same time, the gatherer/cooks, up on the hill with a clear view of the hunters, license an utterance that functions contextually like our own "there is nothing behind the sacred oak." Have the gatherers *challenged* the hunters or simply done something sonically similar, but practically different, in a distinct context?

The key insight is that this is not a matter to be determined by looking for evidence—even social normative evidence—but a matter to be normatively instituted by their intra-sub-practical holdings of one another. Suppose that the gatherers, seeing the hunters cautiously circling, yell out their signal to them. One way this situation could develop is that the hunters could come to habitually ignore such interventions. After all, the cooks are not expert at hunting, and they do not know what is useful for the hunt. Even after the tree is circled and no boar found, the hunters might not take themselves to be vulnerable to normative correction by gatherers. They may, as it were, practically instantiate the attitude "better safe than sorry," taking the signaling move "there is a boar behind the tree" to simply be a demand for caution, and hence not something that turns out to have been incorrect *in any sense* if they surround an empty tree. To do so is precisely not to take the utterance to be making a claim about an objective world.

On the other hand, the hunters might treat the utterance by the gatherers as *a challenge*, as an utterance that is de jure incompatible with their own in the sense that it is socially impossible for both to be entitled. To take it in this way does not require any particular similarity in local inferential roles. That the cooks use different skills to arrive at their judgment, or draw different consequences from it, does not preclude our taking their utterance to be a challenge to our own. (Switching to a maturely conceptual, and indeed highly institutionally sophisticated issue, consider the claim "Jones is the person who assaulted Smith" as uttered by a police detective on the scene, on the one hand, and by a crime lab technician examining DNA evidence, on the other. Though the evidence the police detective uses to justify or falsify this claim bears little to nothing in common with the evidence used by the lab technician—interviewing witnesses and analyzing DNA are, after all, about as practically divergent as two practices can get—their judgments have the same content. If the lab determines that it was not Jones, the detective cannot treat this as a locally reasonable move in the world of science and still go on taking himself to be justified in his judgment of guilt.)[15] Whether such cross-practically instituted incompatibility relations will survive will depend on how useful it is to base our practices on the facts responsiveness to which are thus instituted. Even someone with no hunting skills might nonetheless be *right* that there is no boar behind the tree, and a skillful hunting practice that takes such an observation as binding on one's hunting might well do better overall than one that ignores the ignorant cooks. My claim is that a necessary condition on there being such a status of *right*—correct, true, getting the world as it is—is that we institute cross-contextual incompatibility relations.

There are always many dimensions of normative appraisal. Of course, in a mature practice, we can hold that there is a sense in which we were not wrong—not imprudent, not unjustified, not reckless—in having said that there was a boar, even if no boar shows up. But my point here is that there must also be a sense in which we turn out *wrong*, and that this wrongness, and our entitlement to criticize one another on such grounds, must be capable of arising from a normative conflict between performances made in very different contexts of practical coping. To engage with a common content requires being embedded within a system of cross-practical rules that determine when two utterances cannot be jointly entitled and how we must proceed to revise the disputes so instituted.

To be a person—a language user, an intentional being, a truth claimer—is to operate in a system of mutually interdependent subpractices, held together by incompatibility norms. One must be part of a social system such that it is a general norm that certain performances in one context cannot be jointly entitled alongside performances in other contexts, regardless of the differing practical goals and internal norms of the two contexts.

A clarification is essential here. I am not purporting to analyze some commonsense notion of sameness of content. Rather, I am interested in specifying a necessary condition on the pragmatic institution of an epistemically and ontologically important function of judgments of content sameness. One essential structural aspect of what it is for a performance to count as a truth claim, an assertion about how the independent world is, is that it shows this sort of normative context independence. To be a truth claim requires being judged not merely by standards of local contextual usefulness but by survival across all argumentative contexts, come what may, including contexts in which we receptively hold one another to engagements with the world.[16] Truth is not truth-for-this-project, or truth-for-us, but truth simpliciter, and whether entitlement to some performance is context neutral in this way is something to be normatively instituted, not discovered in other aspects of use. The aspect of the intuitive idea of utterances in two contexts meaning the same thing that I am interested in here is captured by cross-contextual de jure coentitlement. That is something that can be socially instituted in any community with multiple practical contexts and the social skills necessary for operating in and across them. If the norms of a group of critters, locally different though they are, give rise to normative pronouncements that stand in practical incompatibility relations across contexts, we have performances that are on their way to functioning as content bearers in the sense that matters ontologically and epistemologically. If those contents also begin to differentiate in terms of the types of normative functional roles typologized in Kukla and Lance 2009, then these performances will count as having a genuinely pragmatic function.[17] If we can evaluate both in terms of the local contextual propriety of the performance, and the standing that is beholden to the de jure coentitlement norms that contexts share, we have the beginnings of a force-content distinction. Indulging in a bit of sloganeering—the virtue of which is another of Haugeland's insights, of course—the norm of truth, of getting the world right, is a matter of joint extradition treaties rather than similarity in local policing.

Though there is far more to a practice being genuinely linguistic—importantly, the way that our receptive involvement with the world functions in any such practice[18]—a number of distinctively conceptual elements begin to fall in place once this sort of cross-sub-practical differentiation and dependency is on board. Let's call a social practice that is divided into subpractices among which there is this kind of normative dependence a "structurally articulated" practice. For a practice to be structurally articulated, we need to begin thinking of the creatures engaged in it as having distinct states corresponding to their commitment to repeatable contents. Recall that in the case of a single local context, there was no reason to treat our creatures as having a "believing that P" state. We could simply treat them as being in a total state of engagement with this local context. Though we could use our own conceptual resources to account for their behavioral regularities in linguistic terms, there was no practical upshot to claiming that this creature in this circumstance had the same belief as this one in this other, or to separating off an element of their holistic engagement that was the belief-that-P state.

Things are different when practices are structurally articulated, when entitlements are instituted as standing in cross-sub-practical incompatibility relations. This performance by α must be taken by the normative economy of the entire community as *the same as* this one by β in precisely the sense that a vindication (refutation) of the one counts as a vindication (refutation) of the other. Similarly, a sotto voce signal by α in the course of a simulational argument must count as the same state as such a signal in β because of their mutual evaluative dependence. Explicit assertions and inner states, that is, can be individuated by de jure coentitlement. Thus it is the practice of intercontextual scorekeeping, rather than any complexity that can occur in any given context, that first amounts to the assignment of distinctive repeatable content-like statuses. An agent in a given subpractice, who is in this way beholden to the norms of another, we call "generically embedded" in the structurally articulated practice of the entire community. Though he may exclusively hunt and never gather or cook, the status of his signaling performances within a hunt can be affected by actions within the context of gathering.

Though I cannot go into any detail here, it is precisely such practices of coevaluation that are the crucial necessary step toward compositionality, as well. Once we have de jure reidentifiable performances that can be assigned discrete normative statuses, then they function as a *normative locus of significance* across different contexts. That is to say, the principle that this utterance and that must have the same entitlement status induces an utterance type—an equivalence class under de jure coentitlement— that is available for reidentification in multiple pragmatic contexts, some of which will be defined by the existence of other signaling performances, performances we might as well now call speech acts. The normative locus that is asserted and challenged by different people in different contexts is thereby available as a recognizable subperformance in more complex linguistic performances, and norms can arise that evaluate the normative status of the entire performance as a function of various aspects of the normative status of the parts.

Sketchy as all of this is, it suggests something surprising about the conditions of the possibility of a practice that purports to disclose objective reality. There cannot so much as be a practical distinction between performances judged by local success and performances that make truth claims, unless there are distinct contexts of practice, Heideggerian subworlds, that stand in this sort of relationship of normative beholdenness. Only when critters are generically embedded in structurally articulated practices can they count as fully conceptual.

Notes

1. See Kukla and Lance 2014 for an attempt to cash out this metaphor.

2. Haugeland's work contains another tension. In papers like "Truth and Rule-Following," his concern is objectivity, and this involves highlighting the way that our highly theoretical practices can be refuted by what happens in the world (and by the way those theoretical practices allow events in the world to have such significance.) In other places—most notably "Mind Embodied and Embedded"—he is concerned, essentially, to explicate Heideggerian comportment, focusing not on the normative significance of objects in refuting theories but on our thoroughgoing involvement with them in competent behavior. This paper tries to bring these sides of Haugeland's thought together as much as it attempts to highlight the sociality of all such human practice.

3. Early mild skepticism emerged in chapter 2 of Lance and Hawthorne 1999, but I now think that insofar as a systematic account of language is possible, it is via a systematic pragmatics that sees semantic facts as arising here and there, for particular pragmatic purposes rather than via a systematic semantics. And I doubt that there can, in principle, be such a thing as a complete and systematic pragmatics. Indeed, I would like to argue one day that it is part of the essence of language that this is impossible.

4. Readers who are familiar with Haugeland's work will recognize that my paper not only draws substantially on many of his ideas but is something of a formal homage to "Truth and Rule-Following," in which Haugeland similarly tries to build

up a notion of objective purport via adding successive layers to a simple form of practice.

5. I take these to be questions for the cognitive sciences and think that it is just a mistake to think that the phenomenological points Heidegger is making about the difference between the mode of understanding one has when engaged with the ready to hand, and when explicitly theorizing, have direct implications for questions about the way that either is engineered, or with, for example, the question of whether there is information processing going on at some level in the former case. Of course, I offer no argument for this view here.

6. Cf. Kukla and Lance 2009 for a definition and vindication of this terminology and the distinctions it codifies.

7. In the terms of that paper, our hunter-gatherers are at best super-squirrels, not claim makers.

8. The idea of a primitive simulation mode like this plays an explanatory role in two previous papers. In Lance 1996, I use it to explain quantification in inferentialist terms. In Lance and White 2007, Heath White and I use the idea—especially the difference between the sort of simulation initiated by signaling correlates of environmental changes and those instituted by signaling correlates of actions—to clarify a range of notions like freedom, agency, and responsibility, as well as to provide a radically different account of the indicative-counterfactual distinction. Haugeland's (1987) discussion of the use of mental imagery in imagined scenarios intersects with this discussion in a number of ways that I do not pursue here.

9. My cognitive science informant tells me that something like this sort of simulation does seem to be what is going on when we reason through what would happen if x were done in a situation. We tend to engage in imaginative playing through of scenes, and this involves an activation of the same parts of the brain that are activated in the course of a similar actual performance. But I make no claims in this regard, leaving these details up to the experts.

10. Cf. the distinction between derivative and original intentionality in, e.g., "Semantic Engines" (Haugeland 1981).

11. Again, compare the discussion in Haugeland 1987.

12. In Brandom, this assumption shows up as the thesis of the autonomy of declarative practice. Since, in his view, one can understand the inferentially articulated practice of committing oneself to, and attributing assertions independently of, any other human social practice, it follows that the proprieties inherent in this system must be constituted by the relations among declaratives themselves. Of course, this is in tension with Brandom's frequent, and to my mind correct, insistence that nonmonotonic inference is ineliminable. He is not committed to the idea that we can reduce nonmonotonic inferential propriety to monotonic. He is, I claim, committed

to the idea that the propriety of a given nonmonotonic inference depends in principle on nothing other than the totality of scorekeeping commitments and entitlements across the linguistic community.

13. For what it's worth, I suspect that something like this line of thought drives Dreyfus's continued resistance to the idea that conceptuality is involved in all human understanding.

14. Again, pragmatically optimizing signaling-simulators are essentially the super-squirrels in Kukla and Lance 2014.

15. See Lance 2015 for a discussion of just such cases and the way that intercontextual de jure coentitlement institutes objective norms. The sense of content instituted here is a special case of that more general notion.

16. For the details, see Kukla and Lance 2014.

17. In Kukla and Lance 2009, we took care not to make claims about content or meaning. I do not take myself to be going back on that in any important way. I am neither using content to explain pragmatic roles nor suggesting any reduction of content to pragmatic facts. Rather, I merely claim that the cross-contextual norms that institute incompatibility relations among performances—the norms that allow for a distinction between force and content—must be in place before anything counts as a genuine speech act.

18. I see both Kukla and Lance 2014 and Lance 2015 as companion pieces to the current discussion.

References

Brandom, R. 1994. *Making It Explicit: Reasoning, Representing, and Discursive Commitment.* Cambridge, MA: Harvard University Press.

Haugeland, J. 1981. *Mind Design.* Cambridge, MA: MIT Press.

Haugeland, J. 1987. An overview of the frame problem. In *The Robot's Dilemma: The Frame Problem in Artificial Intelligence*, ed. Z. W. Pylyshyn, 77–93. Norwood, NJ: Ablex.

Haugeland, J. 1998. *Having Thought: Essays in the Metaphysics of Mind.* Cambridge, MA: Harvard University Press. (Abbreviated as *HT*.)

Haugeland, J. 2013. *Dasein Disclosed.* Cambridge, MA: Harvard University Press. (Abbreviated as *DD*.)

Kukla, R., and M. Lance. 2009. *'Yo!' and 'Lo!' The Pragmatic Topography of the Space of Reasons.* Cambridge, MA: Harvard University Press.

Kukla, R., and M. Lance. 2014. Intersubjectivity and receptive experience. *Southern Journal of Philosophy* 52 (1): 22–42.

Lance, M. 1996. Quantification, substitution, and conceptual content. *Noûs* 30 (4): 481–507.

Lance, M. 2015. Life is not a box-score. In *Meaning without Representation: Essays on Truth, Expression, Normativity, and Naturalism*, ed. S. Gross, N. Tebben, and M. Williams, 279–305. Oxford: Oxford University Press.

Lance, M., and J. Hawthorne. 1999. *The Grammar of Meaning*. Cambridge: Cambridge University Press.

Lance, M., and H. White. 2007. Stereoscopic vision: Reasons, causes, and two spaces of material inference. *Philosophers' Imprint* 7 (4): 1–21.

6 Being Minded

Danielle Macbeth

To constitute is to bring into being. This formulation, while strictly defensible, is deliberately provocative. For it seems to force a choice between two hopeless ways of understanding the idea that ordinary objective phenomena are constituted, one incredible, the other self-defeating. That provocation sets a stage on which the merits of a more careful formulation—constituting is *letting be*—can come into view.
—John Haugeland, "Truth and Rule-Following" (*HT*, 325)

In "Mind Embodied and Embedded," Haugeland develops and defends a conception of our mindedness that in its explicit opposition to the Cartesian divide of mind and body (or matter or world) is reminiscent of the ancient Aristotelian conception. He holds, in particular, that

the meaningful is not in our mind or brain, but is essentially worldly. The meaningful is not a model—that is, it's not representational—but is instead objects embedded in their context of references. And we do not store the meaningful inside of ourselves, but rather live and are at home in it. (*HT*, 231)

Such a view is defended also by, for instance, McDowell (1994) and Thompson (2008), and both emphasize the Aristotelian resonances of the conception. It appears, then, that we are faced with a choice between the ancient Aristotelian conception of our mindedness and the modern Cartesian conception (or more usual today, some suitably naturalized version of that latter conception). And as critics are happy to point out, if we are faced with such a choice, then it must be the modern conception that we choose. My aim is to show that this inference rests on a mistake: the conception of our mindedness that Haugeland defends is not the premodern conception but instead one that is essentially and recognizably "postmodern" in incorporating insights of *both* the premodern Aristotelian conception and that of early modernity.[1]

It may seem that there is no room to move here, that the philosopher is no more in a position to criticize the Cartesian conception than she is positioned to argue that the sun circles the earth. This concern is evident in, for example, Huw Price's review of McDowell's *Mind and World*:

Physiology ... teaches us that Kant was right: what we get from our sensory apparatus depends on quite contingent features of our physical construction, as well as on the nature of the external world. Arguably, the same is true of our entire conceptual apparatus, but certainly it is true of experience. This product of the sideways-on scientific perspective is not a kind of comatose version of transcendentalism, but a plausible first-order theory about the way in which our brains are linked to the environment. Nor is it a kind of philosophical opening bid, which we can abandon on the grounds that it causes problems elsewhere in philosophy. To all intents and purposes it is a fact of modern life, within the constraints of which philosophy must operate. (Price 1997, 174)

According to Price, it is not philosophy but empirical science that tells us that the meaningful is not essentially worldly, that the meaningful does indeed reside in our minds or brains, that it is representational. And Haugeland is sensitive to this concern: it is from the perspective of science, specifically, "from principles of intelligibility drawn from systems theory" (*HT*, 5), that Haugeland argues that his way of slicing up the mind–body–world pie holds out greater promise of insight than the Cartesian alternative. It is not clear, however, how exactly this is to work, how we are to recover meaning in the world without falling back into premodern superstition. What has *changed* to render the idea of worldly meaning so much as intelligible in the context of modernity?[2]

In the ancient Aristotelian conception, the meaningfulness of the world is a given. The things we find in it, plants and animals, the sun, moon, and stars, mountains and rivers, all have their natures and characteristic powers and behaviors, and also sensory properties. And we, conveniently enough, have the sense organs requisite for sensing those properties, the capacity to take in, for instance, the redness of ripe McIntosh apples, the warmth of the sun, and the drumming of the rain. And because things have natures that are expressed in their characteristic behaviors, we can learn by observing those behaviors what it is to be, say, an animal of a certain sort, or fire, or a heavenly body. Two features of this worldview are especially salient for our purposes here: the understanding of a living being as an instance of a form of life with characteristic powers and behaviors, and the understanding of inanimate nature on the model of animate nature.

The paradigm of being for Aristotle is a living organism, a cat, say, engaged in the perceptual and motor activities that are characteristic of cats. These activities are expressive of its nature and are intelligible as the activities they are only in light of the form of life as a whole. What the animal *does* makes sense only on the basis of an antecedent understanding of what it *is*, an instance of some particular life-form. An animal is in this regard strikingly different from a mere mechanism or other inanimate system, which can be understood wholly in terms of what it does. Of course, Aristotle had no notion of a mechanism in our sense. But because we have this notion, and furthermore have a tendency to think of living organisms in reductive, mechanistic terms, in terms of what they do (for example, reproduce themselves), it will help briefly to consider Thompson's motivation for distinguishing (formally) between what living things do and the characteristic behaviors of nonliving things, in particular his discussion of the respects in which both respond to stimuli.

The warming of an asphalt roadbed and *the train of photosynthetic events in a green leaf* are both of them, in some sense, the effect of sunlight. And *the thawing of icy ponds* and *the opening of maple buds* are each occasioned by rising spring temperatures. It is natural, though, to think that the two vegetative phenomena belong together as instances of a special type of causal relation, or a causal relation with special conditions, distinct from any exhibited in asphalt or water. ... On the other hand, though, the effect of *the hydrogen bomb* on a rose, and on a roadbed, will be pretty much the same—at least if they are both at ground zero. I mean not only that the effects will be similar, but also that the type of causality will be the same. It is in a more restricted range of cases that we seem to see a difference, if the affected individual is an organism. (Thompson 2008, 40)

Intuitively, the difference Thompson points to in this passage should be clear.[3] Our understanding of the process *sunlight warming an asphalt roadbed*, on the one hand, and our understanding of the process *sunlight triggering a train of photosynthetic events in a green leaf*, on the other, are very different. In the first case, we have a merely causal connection, one that is fully explicable by appeal to certain molecular structures (in asphalt) and the effects of sunlight on such structures. Thompson's interest lies in the fact that if we try to conceive the second process as another instance of such a causal connection, something essential goes missing, something that can be highlighted by the question, what happens next?

In the case of a photosynthesizing leaf, there is a natural answer to the question of what happens next, namely, whatever then goes on in the course of a properly functioning, that is to say, living, thriving, plant leaf. There is no comparable answer in the case in which the connection is merely causal.

> In a description of photosynthesis, for example, we read of one chemical process—one process-in-the-sense-of-chemistry, one "reaction"—followed by another, and then another. Having read along a bit with mounting enthusiasm, we can ask: "And what happens next?" If we are stuck with chemical and physical categories, the only answer will be: "Well, it depends on whether an H-bomb goes off, or the temperature plummets toward absolute zero, or it all falls into a vat of sulphuric acid ..." That a certain enzyme will appear and split the latest chemical product into two is just one among many possibilities. Physics and chemistry, adequately developed, can tell you what happens in any of these circumstances—in *any* circumstance—but it seems that they cannot attach any sense to a question "What happens next?" *sans phrase.* ... It is not just that "the rose and maple are subjects of processes of their own": they are also subjects of a special type or category of process—"biological" processes, if you like, or "life-processes." (Thompson 2008, 41–42)

Such processes, Thompson argues, are distinguished not by their content but by their form: they exhibit a different sort of unity from that of merely physical, that is, nonbiological, processes.

In the case of a living organism, in this Aristotelian conception, the question "what happens next?" (*sans phrase*) makes sense because living organisms are instances of characteristic forms of life. What happens next is what happens in the life not of this or that individual but of that species of organism: "When springtime comes, and the snow begins to melt, the female bobcat gives birth to two to four cubs" (Thompson 2008, 63). The bobcat is a species or life-form, something about which just such a judgment can be made. Furthermore, such an answer to the question "what happens next?" can be correct even if the answer is in fact false of most instances of the kind: it is perfectly correct to say that the mayfly breeds shortly before dying although most die long before breeding. A judgment such as that the mayfly breeds shortly before dying "may be true though individuals falling under both the subject and predicate concepts are as rare as one likes, statistically speaking" (Thompson 2008, 68). To understand something as alive is in this way to see it as an instance of a biological kind and so as having a distinctive form of life, one that takes the form of a story, a narrative with a characteristic beginning (what happens first), middle (what happens next), and end (what happens last), whether or not all or even most instances of the kind in fact realize such a life.

What an animal does is expressive of its nature, its form of life; it is an actualization of something that is in the animal's power to do.[4] And this notion of a power, like that of a life-form, is something we heirs of modernity tend to ignore, or forget, or simply know nothing about at all. As Rödl explicates it, a power contrasts not only with the dispositions of mere physical objects such as stones and pieces of iron but also, and more tellingly, both with a habit and with what, corresponding to a power, an infinite being might be said to have. First, a power is unlike a habit in having what Rödl (2007, 141) calls a "normative measure": an act of a power, unlike an act of habit, brings that act under a standard. An animal (of the hunting sort) does not, for instance, have merely the habit of hunting for food, though it may habitually do so in one location rather than another. Instead it has the power of hunting, a capacity to pursue prey in characteristic ways, and on any given occasion is successful or not. The end for the sake of which the animal hunts is to be fed; and this end, what it is for, provides the measure. Being a good hunter is having the power to bring about the relevant end. But in some cases the animal, even one that is a highly skilled hunter, will fail, perhaps because the prey runs too quickly to be caught. And this is true of powers generally; they can be exercised—as the powers they are and to the ends for which they are powers—only in favorable circumstances. This second aspect of powers, their fallibility, the fact that they can be thwarted when the circumstances are not propitious, is furthermore, as Rödl (2007, 153) emphasizes, not a *limitation* of the power but instead a "logical or metaphysical fact." Powers are by their natures as powers inherently fallible; the successful exercise of a power requires propitious circumstances. An infinite or eternal being does not have powers for precisely this reason. Such a being engages in activity (we can suppose) but without thereby exercising a power. The act of such a being is "pure act; the contrast of power and act does not apply to it" (Rödl 2007, 153; see also Beere 2009, 296). Only living, finite beings, beings that are instances of life-forms, have powers.

The things that are to be found in inanimate nature, rocks and stones, rivers, fires, and winds, for example, likewise have their natures and (fallible) powers in Aristotle's account, in particular the power of self-movement. It is in the nature of, for example, fire to "go up," to seek the highest point, and in the nature of earth to "go down," to seek the lowest point, and in both cases unpropitious circumstances can result in the failure of the stuff to achieve its end. In this view, then, living things such as cats and cacti are natural in precisely the same sense that water, stones, and stars are natural. All are equally and *similarly* a part of the natural world, as, of course, are we rational animals.

With the rise of modern science, this conception of the workings of inanimate nature came to seem utterly naive, a childishly anthropomorphizing conception of nature that self-indulgently projects distinctively human meaningfulness onto what is now understood to be properly objective, mathematically describable fact. A stone is not trying to get anywhere when it sinks to the bottom of a pond; it has no nature in the Aristotelian sense, and no power to act. Its behavior is instead to be explained by appeal to empirically discovered forces such as gravity, together with exceptionless (Newtonian) laws of motion. And as Haugeland emphasizes, this (derivational-nomological) form of explanation constituted "a *radical* advance in *ways* of understanding … a totally new way of talking about what happens, and a new way of rendering it intelligible; mathematical relationships and operations defined on universal measurable magnitudes became the illuminating considerations, rather than the goals and strivings of earth, air, fire, and water" (*HT*, 38).[5] Goals and strivings, and more generally the sort of meaningfulness that is found in lives shaped by such things, have no place in nature as it is now to be conceived. Not reasons but only causes are to be found in nature.

From the perspective achieved in early modernity, the ancient Aristotelian conception was grounded in a confusion of what belongs properly to the mind and what is characteristic of mere bodies, physical stuffs. But not only was a new distinction drawn; there was also a *reversal* in what is understood in terms of what. Whereas the ancients had understood inanimate nature on the model of animate nature, for the early moderns, animate nature, now focused in particular on rationality, *our* mindedness, came to be modeled on the newfound understanding of nature in terms of exceptionless physical laws. To be, in the modern view, is to be governed by (exceptionless) law, laws of physics on the side of nature and laws of freedom on the side of reason. To act freely is to act for a good reason (the better the reason, the freer the action, as Descartes notes in the Fourth Meditation); and as Kant was the first explicitly to emphasize, what constitutes the goodness of a reason is its bindingness on *all* rational beings, whatever their form of life. The notion of a form of life, and with it the concept of a power as a fallible capacity of a living thing to do things toward some end, no longer has any explanatory role to play and simply falls away. *Everything* is to be explained instead by appeal to laws, either laws of nature or laws of reason and freedom.

This Cartesian conception of the world and the place and nature of mind in it is very familiar to us and is usually assumed to be so obviously true as to be, as Haugeland notes, "almost invisible" (*HT*, 208), the unquestioned and unquestionable backdrop for all reflection and inquiry, whether in philosophy or in the empirical sciences. On the side of philosophy, Haugeland singles out "accounts based on interpretation and the 'principle of charity,' such as those of Donald Davidson and Daniel Dennett," and "social practice accounts, such as those of Richard Rorty and Robert Brandom" (*HT*, 207). We focus here on Brandom's project in *Making It Explicit.*

Brandom's starting point is that it is not the world but only our *responses* to its impacts that are or can be meaningful and, in particular, normative: "The natural world does not come with commitments and entitlements in it. ... They are creatures of the *attitudes* of taking, treating, or responding to someone in practice *as* committed or entitled" (1994, xiv). "Apart from such practical acknowledgement ... performances have natural properties, but not normative proprieties" (63). On this basis, Brandom aims to make explicit what something would have to be able to do to count as saying, to specify under what circumstances a response to something is not merely causal but instead an instance of speaking, of deploying concepts, of judging. Brandom explains:

Classification by the exercise of regular differential responsive dispositions may be a necessary condition of concept use, but it is clearly not a sufficient one. Such classification may underlie the use of concepts, but it cannot by itself constitute discursiveness. The chunk of iron is not conceiving its world as wet when it responds by rusting. Why not? What else must be added to responsive classification to get to an activity recognizable as the application of concepts? What else must an organism be able to do, what else must be true of it, for performances that it is differentially disposed to produce responsively to count as applications of concepts to the stimuli that evoke those responses? (1994, 87)

As Brandom sets things up, to see and so to say, for instance, "that's yellow" when presented with a ripe Chiquita banana is similar to a chunk of iron's rusting in the rain insofar as both responses involve a reliable differential responsive disposition, a merely causal process. But they are, of course, also importantly different. Here Brandom follows Kant in taking it that we have two options: either we think of the sequence of events *presentation of banana followed by an utterance of "that's yellow"* as behavior in accordance with a rule or norm, that is, as causally necessitated, or we think of it as an action according to one's conception of a rule or norm, that is, as rationally necessitated. The first sort of necessity is clearly not sufficient for an adequate understanding of the response as a piece of language use; in speaking, one is not merely causally necessitated to make the noises one makes. So we must choose the second: to say "that's yellow" when presented with a ripe Chiquita banana is to act according to one's conception of a norm or rule. But this will not do either, because such an action presupposes rather than explains one's capacity to use language with meaning and understanding. To follow an explicitly formulated rule is already to be able to assign it the requisite meaning, already to have the conception in question, on pain of a vicious regress. (This, Brandom suggests, is precisely the point of Wittgenstein's regress argument in the *Investigations*.) Brandom's thought is to stop the regress of interpretations by appeal to norms implicit in practice:

Norms explicit as rules presuppose norms implicit in practice because a rule specifying how something is correctly done (how a word ought to be used, how a piano ought to be tuned) must be applied to particular circumstances, and applying a rule in particular circumstances is itself essentially something that can be done correctly or incorrectly. A rule, principle, or command has normative significance for performances only in the context of practices determining how it is correctly applied. (1994, 20)

To understand the essential difference between a person's saying "that's yellow" in the presence of something yellow and, say, a parrot's (trained) squawk of "yellow" in the same circumstance, according to Brandom's account, we need to consider the norms implicit in practice that for the person but not the parrot underlie her commitment to explicit linguistic rules such as that enjoining us to call yellow things yellow.

Brandom's project in *Making It Explicit* is to start with norms implicit in practices that do not presuppose conceptual content, and so truth, and then build up to such content. Any such account faces three major challenges. First, because Brandom assumes that nature and norms are essentially opposed, the normative attitudes that are the fundamental building blocks of the account are constitutively nonnatural; the norm-laden responses that are the unexplained explainers of the account cannot arise within the causal realm. Like Descartes's mind-stuff, these normative attitudes are just *there* alongside physical, nonnormative nature. But can we countenance something so mysterious as nonnatural attitudes? The theory demands these unexplained explainers, and we do have normative attitudes, but the dualism, like Descartes's original dualism, strains credulity.

Brandom aims to avoid more blatantly Cartesian projects by starting not with full-blown intentionality, that is, with rational beings answerable to the norm of truth, but more modestly with creatures capable of taking up attitudes that institute, at least in the first instance, social (rather than objective) proprieties of correctness and incorrectness. His task, the project of the book as a whole, is to build up to the sort of intentionality that is characteristic of rational beings answerable to the norm of truth. Although we do not start with them, the aim is to end up with something that is evidently an account of rational beings. Thus, in chapter 8, we are provided with what Brandom describes as his "objectivity proofs" to show that the notion of truth is distinguished in his account from the notion of assertibility. Brandom claims in particular that those proofs

define a robust sense in which the facts as construed in this work are independent of what anyone or everyone is committed to. The *claim-making* practices described here are accordingly properly understood as making possible genuine *fact-stating* discourse, for they incorporate practices of assessing claims and inferences according to their *objective* correctness—a kind of correctness that answers to how things actually are, rather than how they are *taken* to be, by anyone (including oneself) or everyone. (1994, 606–607)

In fact, as many readers have recognized, Brandom does not, and cannot, prove what is needed, that what is objective is what is the case anyway, however we take things to be, *and* that in judging we are answerable to what is objectively the case. As Haugeland puts the problem:

What the proofs show is that there is no legal move, in Brandom's system, from "Everyone believes *p*" (or: "I believe *p*") to "*p*." But they don't show anything at all about what *could* legitimate "*p*" instead; in particular, they don't begin to show how any of the moves could be *claim*-makings or *fact*-statings. (*HT*, 358n14)

On Brandom's account, judgment is not answerable to things as they are, nor can it be, given his conception of the way the normative is contrasted with the causal. If what is "outside" is merely causal, if it is normatively inert, then there is simply no possibility of its rationally constraining judgment.[6] And then, as Haugeland notes, there is no reason to call it *judgment* at all.

The final difficulty concerns the conception of reason and freedom that Brandom finds in Kant.[7] We have seen that on Kant's account, to be rational is to act according to one's conception of a law (as contrasted with a physical object, which merely behaves in accordance with a law). But what binds one to this rule? If I cannot help but act according to the rule as I conceive it, then I am not free, not really acting rationally at all. And if I choose so to act, then on pain of a vicious regress I must, at bottom, act with no reason at all. But if I act with no reason, then I do not act rationally. Given the way things have been set up, these are the only possibilities. Either the law is imposed from without, or it is chosen from within, and neither option leaves any room for rational freedom.

Modernity is founded on an in-principle separation of meaning from nature. But it also effects an inversion relative to the ancient mode of understanding. Instead of understanding the inanimate on the model of the animate, as Aristotle had, now we take our hard-won understanding of the inanimate as governed by exceptionless physical law to provide the model for the animate or, more exactly now, the rational. Intelligence is to be understood not by appeal to perceptual and motor skills and capacities, and to understanding and insight, but by appeal to the rule-governed manipulation of symbols.[8] The model that is often invoked here is, of course, mathematics and mathematical reasoning, but recent work suggests that *even mathematical reasoning* cannot be accounted for by appeal to the rule-governed manipulation of signs, that it too relies on learned skills, on insight and understanding.[9] But if so, then the modern Cartesian conception suffers from the same structural flaw as the ancient conception, the flaw of trying inappropriately to model one domain of being on another. Perhaps neither inanimate being nor animate/rational being can provide the model for the other. Perhaps there is room for *both*, both a largely Aristotelian conception of animate being in terms of forms of life and characteristic (fallible) powers, including the power of reason, and a modern conception of inanimate being in terms of exceptionless laws, or whatever is the successor of that conception in contemporary fundamental physics.[10]

Aristotle conceives animate beings and inanimate stuffs as equally and similarly a part of the natural world. We have since learned that this is not so, that living things come into being only through a very long process of evolution by natural selection. Independent of such a process, there is only inanimate stuff. And this process, Gibson has urged, realizes a new sort of significance not only for the animal but also for the world in which the animal arises:

The words *animal* and *environment* make an inseparable pair. Each term implies the other. No animal could exist without an environment surrounding it. Equally, although not so obvious, an environment implies an animal (or at least an organism) to be surrounded. This means that the surface of the earth, millions of years before life developed on it, was not an environment, properly speaking. (Gibson 1979, 8; quoted in *HT*, 221)

On a Gibsonian account, the form of life that the animal instantiates articulates *both* the "inside," the living, perceiving, feeling animal, *and* the "outside," its environment; and neither is fully intelligible without the other. Water comes to have the significance of hydrating the animal; a certain formation of rocks acquires the significance of a safe haven, a den; fire is now dangerous. The notion of a form of life in this way mediates between the animal and its environment.[11] That the animal has the sense organs it has and that things in its environment have the perceptible properties they have are not now to be conceived merely as given, as they are in the ancient account. They are instead explained by the processes of evolution. And once we have this picture, we can immediately recognize that what an animal perceives are not "physically simple properties of light, like color and brightness, but rather visual features of the environment that matter to it," that is, what Gibson calls affordances (*HT*, 222). Although the animal's *eyes* are responsive to physically simple properties of light, the *animal* is not so responsive. What the animal is responsive to are things that matter to it as the sort of animal it is, opportunities for and hazards to its biological existence.

According to the Gibsonian account, evolution by natural selection realizes a new order of intelligibility in nature (affordances), a new mode of being (animate), and new characteristic behaviors that are not reducible to the behaviors exhibited by mere physical stuffs (exercises of powers). This is not yet an account of *rational* animals, but even so, it would seem to have put us on the right track insofar as it enables us to understand how there might be animals *within a meaningful world*. And yet the Cartesian is unmoved, and he is unmoved because if one is a Cartesian, one will *not* understand biological evolution as Gibson does. One will instead understand it as a kind of as-if design of complex biological mechanisms. Haugeland finds the Gibsonian account compelling. The Cartesian does not. Why? Again: what has *changed* to render the idea of worldly meaning so much as intelligible in the context of modernity?

On the conception of mind and world that Haugeland urges in *Having Thought*, "the constituted objective world and the free constituting subject are intelligible only as two sides of one coin" (*HT*, 6). That is, they form a kind of unity, and Haugeland's imagery of sides of one coin suggests that this unity is an *essential* unity. This, we need to see, is not quite right; it is not what Haugeland intends, though it is a good place to start.

An essential unity is a whole that is prior to its parts in the sense that one cannot understand the parts except through the whole, *as* parts of the whole. A human body, Aristotle argues, is an essential unity insofar as its parts, hands, say, are what they are *only* in relation to the whole; to separate the parts from the whole is to destroy them. (A hand that has been severed from the body is a hand in name only; because it can no longer function as a hand, it is not really a hand.) Similarly, for the ancient Greeks, mind and world (that is, the individual human being and the cosmos) form an essential unity. Because the cosmos is like a living being, to understand ourselves adequately is to understand ourselves in relation to the whole.[12] It is just this that the Cartesian denies; for the Cartesian, mind and world are only accidentally related.[13] Indeed, the project of modern science is precisely to understand the world *independent* of how it is in relation to us. The aim is to discover how things *are* as contrasted with how they appear to perceivers like us, with our sort of sensibility, our sense organs and brains. Because Haugeland's imagery of sides of a coin seems to suggest that mind and world form an essential unity, we are again faced with our either/or: either the ancient conception (an essential unity of mind and world) or the Cartesian conception (a merely accidental unity). And if so, then again it would seem to be the Cartesian conception we must choose.

In an essential unity, the whole is conceptually prior to the parts. In an accidental unity, the parts are instead conceptually prior to the whole. That neither is adequate for the case of mind and world is indicated by the sorts of problems each raises for our capacity for knowledge. First, if mind and world form an essential unity, then it is impossible to understand how we can, even if only occasionally, fail to be in cognitive contact with reality. Because the unity is conceived as an essential one, a mind separated from the world (in, for example, perceptual illusion or false belief) would thereby cease to be a mind. As Burnyeat explains:

The problem which typifies ancient philosophical enquiry in a way that the external world problem has come to typify philosophical enquiry in modern times is quite the opposite. It is the problem of understanding how thought can be of nothing or what is not, how our minds can be exercised on falsehoods, fictions, and illusions. The characteristic worry, from Parmenides onwards, is not how the mind can be in touch with anything at all, but how it can fail to be. (1982, 33)

The modern problem, Burnyeat indicates, is different. If the unity of mind and world is only accidental, as it is in the Cartesian view, then, as Kant already saw, it is impossible to understand how we can, even if only occasionally, *achieve* cognitive contact with it. Knowing (as Gettier-type arguments highlight) requires an *internal* relationship between the knower and the known. But if mind and world form only an accidental unity, such an internal relation would seem to be impossible to achieve. Neither appeal to an essential unity nor appeal to an accidental unity enables us to understand the relation of mind and world. Is there, then, some *other* sort of unity that we might appeal to instead? Could we appeal, for instance, to a unity in which *neither* the whole *nor* the part is conceptually prior? Does it even make sense to suppose a whole in which there are independently intelligible parts but that is nonetheless not *reducible* to those parts? The answer, perhaps surprisingly, is that it does.

Consider a Euclidean figure, say, an equilateral triangle. Clearly such a figure has parts, for instance, sides, three of them, all of which are equal in length. But the triangle is not reducible to its parts in relation. A triangle is not merely an accidental unity, the chance (spatial) arrangement of parts. It is *one* thing, a geometrical figure of a certain sort. But it also is not an essential unity as a human body is (if Aristotle is right). The parts *are* intelligible independent of the whole in which they figure. The side of a triangle is, for instance, a straight line, and straight lines are fully intelligible independent of any reference to triangles. Perhaps one could object that nonetheless sides of triangles are not intelligible independent of any reference to triangles. And this is true: one cannot have the concept of a side of a triangle independently of having the concept of a triangle. We know that the unity of an equilateral triangle is not accidental because the straight line *is* a side of the triangle, and sides of triangles are not independently intelligible. But nor can the unity of the triangle be essential because the side *is* a straight line and straight lines *are* independently intelligible. The unity such a Euclidean figure exhibits is neither accidental nor essential. It is, as an essential unity is, a *real* whole, but it also has, as an accidental unity has, *real* parts. Following Grosholz (2007, chap. 2), I will refer to this sort of unity as an *intelligible* unity. An intelligible unity such as an equilateral triangle is a real whole of real parts, and because it is, it poses a problem for understanding. And in the case of an equilateral triangle, at least in the context of Euclidean diagrammatic practice, to solve the problem the triangle poses is to find a means of constructing such a figure, to learn how to construct it out of canonical elements, points, lines, and circles.

In an intelligible unity, the whole has real parts but is not reducible to its parts. And as just indicated, this is possible because the parts regarded one way are independently intelligible and regarded another way are not independently intelligible. Regarded as a straight line, the side is independently intelligible. Regarded as a side, the straight line is not independently intelligible. And both are needed fully to comprehend the triangle as the Euclidean figure it is. We do not merely *take* the straight line to be a side, for it would be equally correct to say that we take the side to be a straight line.[14] Nor do we *make* the straight line to be a side, say, through our act of construction—as, for instance, one makes a stone to be a keystone by putting it in a certain place, namely, at the summit of an arch.[15] The straight line is not merely serving or functioning as a side as the stone serves or functions as a keystone. It *is* a side. That is, we have an identity: the side is identical to, one and the same as, the straight line. And we need to understand this identity as Frege has taught us to understand identities generally. To say that the line is a side, or alternatively that the side is a line, is to say that they are one and the same thing (*Bedeutung*), though cognized through different senses (*Sinn*). To understand triangles in the context of Euclidean diagrammatic practice is to grasp *both* senses, and to grasp both senses is to understand the triangle in its characteristic intelligible unity.[16]

A constituted unity in Haugeland's sense is not, I think, an essential unity, as the sides of a coin are, but instead an intelligible unity, as a

Euclidean triangle is. Nevertheless it is one that is distinctive insofar as it emerges only as the fruit of some process. Unlike triangles, constituted unities are not timelessly what they are but only emerge *in time*. The processes of biological evolution by natural selection, for example, realize the constituted unities that are animals and their environments.[17] Although we can conceive the animal and its environment in ways that are independently intelligible—the animal as a collection of various chemical processes that are also observable in a test tube, and the environment as mere physical stuff—we can (and must if we are to understand the animal as the animal it is) also regard them differently, the animal as alive, as an instance of a particular form of life, and the fire (say) as dangerous. What ancient Greek philosophers had taken to be an essential unity, and the early moderns as only an accidental unity, is instead a constituted unity, a whole of parts, animal and environment, that are (suitably regarded) independently intelligible, but to which that unity cannot be reduced because, differently regarded, the parts are what they are only in relation to the whole. The fire *is* dangerous (in the context of an animal's life). But it is equally true that the dangerous thing *is* fire, that is, something that is intelligible as what it is wholly independently of any appeal to any living being. It is one thing thought of in two radically different ways.[18]

In much the same way, we rational animals and the world as the locus of the truth of our judgments together form a constituted unity. Through some sort of sociocultural process of development that need not concern us here, we and the world (as world, as that to which our judgments are answerable) emerge out of animals in their environments in much the way that animals and their environments emerge out of merely inanimate stuff through the processes of evolution by natural selection.[19] Knower and known are co-constituted in an *intelligible* unity, that is, in a whole of intelligible parts that is nonetheless not reducible to its parts. And this is furthermore the conception we *need* if we are to understand the notion of a power, whether the power of an animal to perceive things or our own power to know how things are, both of which are (as any power is) inherently fallible. Were it the case that living things, whether rational or not, and their surroundings together formed an essential unity, then, as already indicated, errors, and more generally the *fallibility* of powers, would be wholly inexplicable. Were it the case that living things, rational or not, and their surroundings together formed only an accidental unity, veridical perception and knowledge would be impossible, as there could not in that case be any internal or necessary relation between the perceiver/knower and the perceived/known. In this case, it would be the *efficacy* of powers, the fact that they are *powers* to achieve certain ends, albeit fallibly, that would be wholly inexplicable. We need both the irreducible unity of the whole subject/object nexus to understand the successful exercise of powers and also the independent intelligibility of the parts to account for the failures of powers in unpropitious circumstances. And again it is Frege, with his distinction of sense and significance or meaning, who shows us how we *can* have both.

I have argued that the notion of a constituted unity that Haugeland needs is different both from that of an essential unity and from that of an accidental unity. But as may already be evident, these are not merely three possibilities among which one might choose. It is not an *accident* that it was the ancient Greeks who understood the relationship of subject and world in terms of the notion of an essential unity, that early modern philosophy following Descartes took mind and world instead to form an accidental unity, and that only very recently have we seen philosophers such as Haugeland defend (if not in quite these terms) yet a third sort of unity. The relationship between these possibilities is Hegelian and dialectical: we begin with an essential unity, in immediacy, then through a reversal take ourselves to be in an ineluctably mediated relationship to a reality to which we are only accidentally related, and only at the third stage, through an understanding of mind and world as a constituted unity, reach equilibrium in a recognizably mediated, and thus stable, immediacy. But if that is right, then we need self-consciously to have in view not only the processes of evolution by natural selection that realize animate beings and their environments, and whatever the sociocultural processes that realize rational beings in the world, but *also* our own intellectual history through which we finally achieve the conceptions of things that we need to render them intelligible. In particular, we need to see that although the early modern conception of mind and world in terms of exceptionless laws of nature and freedom was an advance, it was not (as it was taken to be) the last word, the truth of the matter, but only a stage along the way, something that had to be taken up and gone through but eventually left behind. The revolutionary advances that realized early modern mathematics, science, and philosophy need to be followed by *further revolutionary advances*—and Haugeland points the way.[20]

We saw that from the Cartesian perspective, premodern thinkers naively and mistakenly projected onto nature, the merely physical world, what in fact belongs only to our experience of nature. Inanimate stuffs do not have their own characteristic natures and powers of self-movement. And similarly (so it is thought) they do not have sensory properties but only cause in us characteristic qualitative experiences. We learned in this way to separate the meaningful from the worldly, from nature. The meaningful is what is "inside," and nature, now conceived as merely causal, is "outside." But in fact the "outside" is *not* unintelligible to us as a result. Instead it manifests a radically new sort of intelligibility; we come to know it in an essentially new way, namely, as governed by exceptionless physical laws. This is, however, *incoherent* given the Cartesian framework—as Kant already saw. If nature *really is merely causal*, as *contrasted* with what is normative, then its impacts on us can have no normative significance, and we cannot in our judgments be answerable to how things actually are. To resolve the difficulty, as already indicated, we need to draw a *further* distinction, one that the early moderns do not draw. Instead of identifying the realm of freedom (reasons, meaning) with the inner and the realm of nature (causes, exceptionless physical laws) with what is outside, we need to recognize that the two distinctions are orthogonal to each other. Meaning is not simply a given of the physical world, as Aristotle thought. Nor is it something distinctively mental, as the Cartesian thinks. Instead it is the medium (the Fregean sense) through which both the animal and its environment are articulated. And the notion of an exceptionless physical law similarly does not belong to the "outside" realm of nature merely as a given but is instead (a constitutive aspect of) yet another medium through which both the animal, that is, we rational animals as we have learned to understand ourselves, and our environment, as we have come to conceive it, come to be articulated in the course of our intellectual history. We made just this distinction—between the two distinctions, inner and outer, and meaningful and lawful—when we suggested, following Gibson, that an animal's form of life articulates not only what is "inner," that is, the animal with its characteristic abilities, but also what is "outer," things in its environment. And it is made again when we realize that, in the course of our intellectual investigations, we develop new forms of mathematical practice, such as that inaugurated by Descartes, that provide the medium for a radically new understanding not only of nature but also of ourselves as rational beings. The immediacy of our cognitive involvements in the world, first and foremost, through our powers of perception that enable us to grasp things in their natures, is in this way *ineluctably* mediated, by our biological, sociocultural, and intellectual form of life as it has been realized through our biological, sociocultural, and intellectual history of growth and transformation.

The Cartesian conception of an inner, meaningful realm set over against the outer realm of merely physical, causally efficacious nature that we find, for instance, in Price and Brandom is overcome when we see that we must not conflate, as the early moderns do, the inner–outer distinction with that of meaning as it contrasts with what is merely causal. Not only can what is merely physical, merely causal, acquire meaning and significance through the course of biological and sociocultural evolution (the signpost itself, for instance, telling one the way to go), but even being "merely causal" is a way of being meaningful, a form that our understanding of the world can take. And that is just to say that the world as it is the object of our scientific investigations, as much as it is the stage on which we live out our everyday lives, is constitutively a *constituted* one in Haugeland's technical sense (see *HT*, 325–327). We do not *make* the world to be thus and so, as Price seems to suggest we do; nor do we merely *take* it to be thus and so, as in Brandom's account. Instead various processes of growth and transformation give us the eyes to see and by the same token the world a face by which to be seen, whether in our everyday lives or in our ongoing scientific investigations.[21]

Here, then, is our answer to the question of what has changed to render the idea of worldly meaning intelligible: we have learned to make a new distinction that the moderns do not make. As from a modern, Cartesian perspective, the ancient understanding of a meaningful world for a subject conflates the meaningful inner realm with a merely causal outer reality, so from the perspective we have achieved here, that modern distinction conflates two very different distinctions, that of inner and outer with that of the justificatory and the merely causal. We can experience the rich, meaningful world of everyday life, and we can think the austere world of the scientist. They are one and the same reality, only thought very differently.

Our relationship to the world is, in either case, a constituted one. Together with the world, we form an intelligible unity.

Although the conception of mindedness as embodied and embedded that Haugeland defends is often criticized as being one that we have known since the seventeenth century to be mistaken, it is in fact this criticism that is mistaken. The conception does incorporate fundamental Aristotelian notions, foremost among them that of a form of life together with that of a power to act and be acted on, but it does so while explicitly acknowledging the profound differences between animate and inanimate nature that were first recognized by early modern thinkers. And it corrects the early modern error of modeling animate or, more exactly, rational nature on that of inanimate nature and thereby enables us to recognize that understanding, even mathematical understanding, "pertains not primarily to symbols or rules for manipulating them, but to the world and living in it" (*HT*, 39).[22] The progression, we have seen, is recognizably Hegelian: from the immediacy of an Aristotelian understanding of our being in nature, through an inversion and ineluctably Cartesian mediation, to an essentially Hegelian mediated immediacy that incorporates the insights of both earlier stages. And Hegel is for Haugeland an important intellectual forebear. What Haugeland does not embrace is Hegel's essentially historicist dialectical method. This, I think, is not only a strategic but a philosophical mistake. We, with the self-understanding we have achieved, are essentially a late fruit of more than two and one-half millennia of rational reflection. And because we are, any fully adequate philosophical understanding of our being in the world must be historicist and dialectical. It must take the form of a philosophical narrative of our becoming the beings we have become. As in the case of organic beings more generally, as Aristotle already saw, we will not understand ourselves, the beings that we are and have become, until and unless we see ourselves grow from the beginning.

Acknowledgments

My thanks to Joe Rouse and Zed Adams for very helpful comments on an earlier draft.

Notes

1. My interest in this topic was first developed while studying Heidegger's work with John nearly thirty years ago. Heidegger's diagnosis of the ancient mistake of erroneous objectivizing followed by the early modern mistake of erroneously subjectivizing seemed to me deeply plausible. The problem was to understand what a third alternative might be. I wrote my dissertation on the topic under John's direction and have only recently come to what seems to me a fully satisfying solution. The ideas sketched here are developed in detail in Macbeth 2014.

2. My concern is thus with a broad and fundamental issue that arises not only in the context of John's work but for anyone wishing to recover a more Aristotelian understanding of the world and our being in it. That John further distinguishes between rational animals in general and subjects of authentic intentionality in particular is important, but it is not immediately to the purposes of this discussion because the form of unity of subject and world that is my concern is in either case the same—and indeed, as will become clear, something to be found already in the case of plants and nonrational animals.

3. I do think that in a sense the distinction Thompson is after should be intuitively clear. Nonetheless it is likely to seem not at all clear to one deeply enmeshed in the modern Cartesian conception. One way to approach my concerns here is through the idea that with the rise of modernity, we lost sight of something that should really be intuitively clear, and need now to learn how to bring it back into view.

4. This is, of course, not true of everything an animal in some sense does. Though an animal can eat what is poisonous to it, can fall down or fall ill, or be devoured by a predator, it is not in its nature to do, or have done to it, these things.

5. Haugeland further argues that this ability we have to make sense of radically new ways of thinking about things constitutes a significant hurdle for cognitivism, for the view that "intelligent behavior can be explained (only) by appeal to internal 'cognitive processes'—that is, rational thought in a broad sense" (*HT*, 9).

6. To think that something merely causal can nonetheless serve as a reason, have normative significance, is to fall into what Sellars calls the Myth of the Given.

7. Rödl (2007, 114–120) argues that this conception is not Kant's. The question whether it is or is not is not one that can be taken up here.

8. As Haugeland notes, cognitive psychology takes as its paradigm the workings of a computer, automatic data processing. This paradigm, he suggests (1998, 43), is an "imposter paradigm" in essentially the way that the earlier behaviorist paradigm, founded on the notion of a conditioned response, was.

9. See, e.g., Rav 1999. That mathematical logic, which understands reasoning to be the essentially mechanical manipulation of signs, is of no use in understanding

actual mathematical practice is a central tenet in much recent work in the philosophy of mathematical practice.

10. Once having recovered the notion of reason as a power of a certain sort of living animal, we can also overcome the difficulty raised earlier in my discussion of Brandom's views regarding the bindingness of rules on reason. See again Rödl 2007, 114–120.

11. We thus take meaningfulness to supervene not merely on the animal, or its part, but on the whole of the (physical) world. See Haugeland's "Weak Supervenience," in *HT*.

12. As Taylor (1975, 6) explains: "The subject is defined in relation to a cosmic order. … The view of the subject that came down from the dominant tradition of the ancients … was that man came most fully to himself when he was in touch with a cosmic order, and in touch with it in the way most suitable to it as an order of ideas, that is, by reason. This is plainly the heritage of Plato; order in the human soul is inseparable from rational vision of the order of being. For Aristotle contemplation of this order is the highest activity of man. The same basic notion is present in the neo-Platonist vision which through Augustine becomes foundational for much medieval thought."

13. Taylor (1975, 7) continues: "To dispense with the notion of meaningful order was to re-define the self. The situation is now reversed: full self-possession requires that we free ourselves from the projections of meanings onto things, that we be able to draw back from the world and concentrate purely on our own processes of observation and thought about things. … Self-presence is now to be aware of what we are and what we are doing in abstraction from the world we observe and judge."

14. Such a taking would correspond to a Brandomian view according to which it is our responses to things that ground their significance, insofar as they have any.

15. Such a making would be like Price's understanding of our brain activity, of the way it gives rise to experience.

16. I develop this conception in much more detail in Macbeth 2014, chap. 2.

17. Again, the processes of biological evolution by natural selection do not thereby realize the unities that are rational animals and the world, let alone that of an authentic subject in the world who is, Haugeland argues, responsible in a highly distinctive way for the constitution of her world. The unity that is the fruit of processes of biological evolution is only the first and most well-understood species of the genus.

18. Notice that in the case of the triangle, it is one and the same order of intelligibility that reveals the whole and the parts, in particular, the side as an independently intelligible straight line and the straight line as a dependently intelligible side. In the case of a living thing, we must appeal to two different orders of intelligibility to

see all that we need to see, both the everyday order that reveals the unity of a living being, and the essentially modern natural scientific order through which the independently intelligible parts are discernible. The problem of understanding our being in the world has proved so intractable for just this reason: only very late do the resources needed to resolve it become available.

19. In Macbeth 1994, based on my dissertation written under John's direction, I provide an extended analogy aimed at clarifying to some extent both how such sociocultural evolution might work and the kinds of significance that can be realized through such a process. See also Macbeth 2014.

20. This is particularly evident in his "Two Dogmas of Rationalism" (this vol.), in which he calls for a postrationalist epistemology. Because rationalism is, as John notes, an essentially modern phenomenon, made possible by the rise of modern science, to call for a postrationalist epistemology is to call for a *third* phase in our most fundamental understanding.

21. Haugeland (*HT*, 325–326) describes this as *letting be* and contrasts it both with the "incredible" idea that we *create* (make) the world to be thus and so and with the "self-defeating" idea that we *count* (take) it as thus and so.

22. As I argue in Macbeth 2011, even mathematical understanding often constitutively involves perceptual and motor skills.

References

Beere, J. 2009. *Doing and Being: An Interpretation of Aristotle's Metaphysics Theta*. Oxford: Oxford University Press.

Brandom, R. 1994. *Making It Explicit: Reasoning, Representing, and Discursive Commitment*. Cambridge, MA: Harvard University Press.

Burnyeat, M. 1982. Idealism and Greek philosophy: What Descartes saw and Berkeley missed. In *Idealism Past and Present*, ed. G. Vesey, 19–50. Cambridge: Cambridge University Press.

Gibson, J. J. 1979. *The Ecological Approach to Visual Perception*. Boston: Houghton Mifflin.

Grosholz, E. R. 2007. *Representation and Productive Ambiguity in Mathematics and the Sciences*. Oxford: Oxford University Press.

Haugeland, J. 1998. *Having Thought: Essays in the Metaphysics of Mind*. Cambridge, MA: Harvard University Press. (Abbreviated as *HT*.)

Macbeth, D. 1994. The coin of the intentional realm. *Journal for the Theory of Social Behaviour* 24:143–166.

Macbeth, D. 2011. Seeing how it goes: Paper-and-pencil reasoning in mathematical practice. *Philosophia Mathematica* 20 (1): 58–85.

Macbeth, D. 2014. *Realizing Reason: A Narrative of Truth and Knowing*. Oxford: Oxford University Press.

McDowell, J. 1994. *Mind and World*. Cambridge, MA: Harvard University Press.

Price, H. H. 1997. Mind and world. *Philosophical Books* 38:167–177.

Rav, Y. 1999. Why do we prove theorems? *Philosophia Mathematica* 7:5–41.

Rödl, S. 2007. *Self-Consciousness*. Cambridge, MA: Harvard University Press.

Taylor, C. 1975. *Hegel*. Cambridge: Cambridge University Press.

Thompson, M. 2008. *Thought and Action: Elementary Structures of Practice and Practical Thought*. Cambridge, MA: Harvard University Press.

III Intentionality

7 Truth, Objectivity, and Emotional Caring: Filling In the Gaps of Haugeland's Existentialist Ontology

Bennett W. Helm

Existential commitment … is no sort of obligation but something more like a dedicated or even a devoted way of living: a determination to maintain and carry on. It is not a communal status at all but a resilient and resolute first-personal *stance*. … [I]t is a *way*, a *style*, a *mode* of playing, working, or living—a way that relies and is prepared to insist on that which is constitutive of its own possibility, the conditions of its intelligibility.

—John Haugeland, "Truth and Rule-Following" (*HT*, 341)

My dissertation, which John directed, was on the nature of animal thought. I argued that we need to distinguish between mere goal-directedness, of a sort illuminated by Dennett's intentional systems theory, and genuine desire. Genuine desires involve our finding their objects worth pursuing; consequently mere intentional systems, which exhibit merely a kind of informationally mediated goal-directedness, therefore fall short of robust agency. I presented an account of the sort of worth that is possible for animals—an account of what it is for an animal to *care* about something— in terms of projectible, rational patterns of emotions. By appealing to the emotions, I was able to articulate a distinctive kind of rationality that was simply not in view in Dennett's appeal to instrumental and epistemic rationality, thereby enriching our understanding of the mind.

At my dissertation defense, John asked only one question, which I remember clearly: "Okay," he said, "so you've shown that desire is richer than goal-directedness in terms of an appeal to caring and the emotions. What about belief? How is belief richer than a mere informational state?" It was clear to me then that John was thinking of Heidegger and the place of care in Dasein's disclosedness of the world, a topic that was central to John's own thought at the time. In part the intuition behind this question was that just as genuine desire differs from mere goal-directedness by being a part of our caring about what is good, so too belief differs from a mere informational state by being a part of our caring about truth. Needless to say, I had little to say in response about how exactly the notions of robust belief, truth, and caring are connected.

Nonetheless, John's question has haunted me ever since, even as my own research took a turn toward moral psychology and apparently away from John's central concerns. Thus I went on to distinguish persons from animals in part in terms of our capacity to value things as a part of the sort of life worth our living, and subsequently embedded this within a broader account of our capacity to love both ourselves and others, thereby arguing that persons are essentially social animals. More recently I have focused on the nature of responsibility, respect, and dignity, as I set my sights on laying the groundwork for an account of metaethics. Throughout, I have tried to articulate an account of the sort of practical rationality central to our understanding of what it is to be a person, an account in which emotions play a central role. Yet we persons are not simply practical creatures; we are theoretical creatures as well. And so I find myself continually being brought back to John's question and ultimately to his incisive work on objectivity and truth and their relation to caring. I now think that the account of emotions and of several varieties of caring that I have been developing is directly relevant to this work of John's and can help us fill in some of the missing pieces in his account of ontical and ontological responsibility for truth. After all, if in perceiving and understanding the world we are doing something for which we can be responsible, then perhaps the practical and theoretical sides of rationality and of persons are not as distinct as they might initially have seemed. This is the project I aim to sketch here.

1   Haugeland's Beholdenness Theory of Truth

In a remarkable series of papers, Haugeland lays out what is both a striking interpretation of Heidegger and a compelling account of objectivity and truth (Haugeland 1994; the essays on truth in both *HT* and *DD*).[1] Not being a Heidegger scholar, I will leave interpretive questions to others; I focus here on the account of objective truth, which I attribute to Haugeland rather than Heidegger.

Haugeland argues that the objectivity of entities in the world can be made intelligible through what he calls the *beholdenness theory of truth* (*HT*, 346–348). The basic idea is that something is objective—is an entity—just in case it is independent of, and criterial for, our perceptions of it; thus an objective phenomenon must be *accessible* (potentially something to which we can be responsive, perhaps with the aid of some equipment), *authoritative* (that in terms of which our perceptions are understood to be correct or incorrect), and *autonomous* (independent of both particular perceptions and group consensus) (Haugeland *HT*, 325). Objects that are autonomous and both accessible to and authoritative for our perceptions, Haugeland claims, are that to which our perceptions are *beholden* in a way that makes intelligible both the objectivity of the objects as something we might come genuinely to discover and the potential truth of our perceptions (*HT*, 348). The task, then, is to flesh this idea out.

Following Heidegger, one of Haugeland's central claims is that the discovery of objects—and so their being accessible, authoritative, and autonomous—is possible only together with a disclosure of their being. The basic idea is this:

The *being* of entities is that in terms of which they are *intelligible as entities*. The qualifier "as entities" (as I am using it) is short for this: with regard to the fact *that* they are (at all) and with regard to *what* they are. Understanding an entity *as an entity*—and there is no other way of *understanding* it—means understanding it in its *that*-it-is and its *what*-it-is. Disclosing the being of entities amounts to letting them become accessible in this two-fold intelligibility—that is, as phenomena that are *understood*. When taken with sufficient generality, a pretty good colloquial paraphrase for "disclosing the being of" is *making sense of*. (*DD*, 191)

For example, the rules of chess, including not just the rules governing how pieces can be moved but also rules governing the perceptions and actions of players, are what make chess phenomena (from pieces like rooks to moves like castling to statuses like being in check) intelligible as such; the being of particular chess phenomena, therefore, is their place within such a system of constitutive rules or norms. Likewise, the norms of physics, including not just the laws of nature but also the norms governing the perceptions and actions of physicists, are what make physical phenomena intelligible as physical; the being of particular physical phenomena, therefore, is their place within such a system of constitutive norms. The puzzle, of course, is to see how we can have a role in disclosing the being of some domain of phenomena and thereby constituting those phenomena as the phenomena they are while still being able to make sense of those phenomena as accessible, authoritative, and autonomous—as objective.

Haugeland's solution to this puzzle appeals to a distinction between ontical and ontological understanding.[2] *Ontical understanding* is our ability to identify and cope appropriately with phenomena as the phenomena they *in fact* are, in light of the relevant background norms of a particular domain. Thus in chess our ontical understanding allows us to recognize and respond to knight forks, undefended bishops, opportunities for en passant, and so on. By contrast, *ontological understanding* is our mastery of what is possible or impossible for phenomena—of their intelligibility—again in light of those constitutive norms, a mastery in which it matters to us that the phenomena we discover not be impossible. This makes intelligible what Haugeland calls the *excluded zone*: phenomena we might purport to recognize that nonetheless are ruled out as impossible, such as rooks moving diagonally (or neutrinos moving faster than the speed of light). Moreover, our ontical understanding of chess pieces (or neutrinos) depends on our ontological understanding of the total set of norms constituting the domain of chess (or physics), norms that in part govern us as "players" in that domain—as chess players (or physicists). So an ontical understanding of entities depends on an ontological (and ontical) understanding of ourselves as "players" and so of what is possible or impossible for *us* (what we can and cannot do) and not simply for entities. Once again, such an ontological understanding involves its mattering to each of us that we not act in violation of the norms—in ways that are unacceptable; to do so, Haugeland claims, would be *irresponsible* (*DD*, 204).

This provides a preliminary sense in which entities are accessible and authoritative, for entities will be *accessible* insofar as the norms constituting them as such are interwoven with norms governing our responsiveness to those very entities. And entities are *authoritative* insofar as our responsibility to ourselves is, because of the excluded zone, also a responsibility to the entities we purport to discover. After all, a chess player who is indifferent to a rook apparently moving diagonally or a physicist who is indifferent to a neutrino apparently moving faster than the speed of light is being irresponsible, for these apparent phenomena belong to the excluded zone. So responsibility for oneself as a "player" is also responsibility for the truth of the phenomena one discovers; or, to put the same point another way, in being responsible for truth, we are thereby bound to entities as authoritative over our perceptions.

Of course, this is not yet good enough for a robust notion of objectivity, for what is missing so far is the autonomy of these entities: not merely their independence of particular perceptions of ours (that's what the authority already discussed provides) but also their independence of mere consensus: the mere fact that the overwhelming majority of physicists think nothing can move faster than the speed of light doesn't make it so. Unless we can make sense of that independence of mere consensus, the norms constituting the being of entities (and ourselves as players) may seem to be arbitrary because of the possibility that they fabricate a fictional world that is disconnected from reality. What is needed for the autonomy of entities is that particular norms, and indeed the whole system of norms itself, be answerable to the entities themselves. Haugeland's primary example of this is in science: when a scientist finds an apparent phenomenon that seems to violate a law of nature, such as neutrinos traveling faster than the speed of light, her first response to such a finding in the excluded zone is to reject it and so search for an alternative explanation for why things appeared that way. Yet in the face of apparent phenomena in the excluded zone that cannot be explained away, she must confront the possibility of giving up (and revising) the laws or other norms of science, potentially in revolutionary ways,[3] thereby changing her ontological understanding so as to resolve the apparent impossibility of the manifest phenomena and thereby make their discovery possible. (The same need to confront the possibility of revising the norms is present in other domains as well: as Haugeland says, chess[4] would not be a playable game if the pieces moved around on their own accord and not under control of the players; and baseball would be unplayable if the pitcher had to make the ball "hang" for a moment over home plate before proceeding on to the catcher. The inertness of chess pieces and the behavior of baseballs are part of the "empirical content" of chess and baseball to which the norms must be answerable in order for the game to be playable.)

I said that a scientist must confront the possibility of giving up the norms of a domain; indeed, she has a responsibility to do so.

"Refusing to accept" intransigent impossibilities has a double meaning. One way of refusing to accept is bullheadedly refusing even to see—blinding oneself. *Existentially*, that kind of refusal—running away and hiding—is *irresponsible*. (*DD*, 216)

The responsibility here flows from her ontological understanding not only of the "game" of science but also of herself as a "player" of that game—as a scientist—and so as bound by its norms through a commitment she has both to herself as a player and to the game itself. As such, what matters from the perspective of ontological understanding is not simply that discovered phenomena not be impossible but also that she can coherently fulfill her responsibility as a player. The sort of responsibility at issue here is for the playability—the viability—of the "game" of science itself, what Haugeland calls *ontological truth* (*DD*, 217). Thus, we might say, in contrast to *ontical truth*, which is that which determines success in a particular exercise of our ontical understanding (of a particular identification of and response to an object, a particular discovery of how things in fact are), *ontological truth* is that which determines success in a particular exercise of our ontological understanding (of a particular way of making intelligible what phenomena are possible or impossible, what the being of these phenomena is as the phenomena they are). The commitment to and resulting responsibility for ontological truth, Haugeland says, are *existential* (*HT*, 341–342; *DD*, 206–207, 216–218).

It should be clear that in undertaking this existential commitment to ontological truth, we are *beholden* to the very entities whose discovery that commitment makes possible, for it is in the face of the apparent discovery of impossible phenomena whose appearance cannot be explained away that we confront ontological falsity: the failure of the norms to constitute a "playable game." Indeed, this explains the sense in which entities are *autonomous*: they are "independent" of those norms "in the concrete and inescapable sense that they are *out of control*" (*DD*, 218). Because such norms are beholden to autonomous entities, the norms of a "playable game"—of a viable ontical and ontological understanding—are not simply arbitrary; rather, they are a kind of *achievement*, and the entities whose discovery they make possible are *objective* (*HT*, 279, 292–293, 298, 331, 353). This is what Haugeland calls the *beholdenness theory of truth* (*HT*, 346–348).

2   Reexamining Existential Commitment

Haugeland understands the objectivity of entities in terms of their being criterial not only for correct recognition (their authority) but also for the constitutive norms of the domain (their autonomy); hence objective phenomena are discoverable, and their intelligibility can be an achievement. It is not clear to me, however, that Haugeland has yet captured another aspect of our ordinary notion of objectivity, namely, the *publicness* of objective phenomena: the idea that the objective world is one that we all (and not just particular individuals) share in common. I shall bring out the worry here in two ways.

First, there is at least conceptual space for entities that are accessible, autonomous authorities that are nonetheless relative to particular individuals. Personal values—the sort of values that enter into a particular person's understanding of the sort of life worth living—are in effect accessible, autonomous authorities that are relative in this way. As I have argued elsewhere (Helm 2001), personal values are constituted by a certain sort of projectible, rational pattern in a particular person's emotions and evaluative judgments. In this account, for values to be accessible and authoritative is for them to be that to which particular emotions are responsive, and for them to be autonomous is for the rational norms governing that pattern (as articulated in part by evaluative concepts) to be answerable to those values themselves. Nonetheless, insofar as such values are personal, they are relative to the individual in that what is and ought to be valuable in my life need not be the same as what is and ought to be valuable in yours. Indeed, the difference here is not one we can understand simply in terms of manifest differences between you and me, such as our talents or tastes, for insofar as one's personal values define the kind of life worth one's living, one's *identity* as the particular person one is, we think it is in part up to the individual to define what one's values shall be, through an exercise of the individual's capacity for autonomy. Whether or not this account is successful, the possibility of such an account ought to make us worry about whether an understanding of objectivity in terms of accessible, autonomous authorities is sufficient.

Second, in part in light of the analogy Haugeland makes to games, we might worry about the possibility of "forking the game." Imagine that Alice and Bob have happily been playing chess (perhaps even esoteric chess; see n. 4) until Alice makes a particular move to which Bob objects, saying that bishops can't move like that. Alice replies that the bishop had turned into a knight two moves back when it was surrounded by a certain configuration of pieces, in much the same way that pawns can turn into other pieces once they reach the last row. Both Alice and Bob are stubborn, and both have their followers, resulting in a group of people who decide to play chess Alice's way, and another group Bob's way. In such a case, it might seem, there is no clear sense in which one of them is right and the other wrong: the game has forked in two. Indeed, this is what often happens with games like Crazy Eights or Monopoly, in which there are many variations or "house rules" delineating what are in effect different (albeit related) games. What has gone wrong from the perspective of public objectivity is that there is no pressure for Alice and Bob (and their followers) to play the same game: there seems to be no requirement that they agree on the same constitutive norms in the case of chess. Yet this seems false of science: science at least purports to be publicly objective in that the failure to agree on the entities or background theory of a particular science is something that scientists ought to overcome so as to achieve a single ontical and ontological understanding of the relevant domain of entities. Insofar as Haugeland's account of objectivity applies equally well to chess as to science, it is not clear that he has a way of making sense of such publicness, of such shared normativity, in domains like science in which it matters.

This last claim becomes especially worrisome when we look carefully at how Haugeland articulates the sort of existential commitment and existential responsibility at the root of his account of the autonomy of entities. As partially quoted in the epigraph, Haugeland claims that an existential commitment to ontological truth is

a dedicated or even a devoted way of living: a determination to maintain and carry on. It is not a communal status at all but a resilient and resolute first-personal *stance*. ... [I]t is a *way*, a *style*, a *mode* of playing, working, or living—a way that relies and is prepared to insist on that which is constitutive of its own possibility, the conditions of its intelligibility. ... The governing or normative "authority" of an existential commitment comes from nowhere other than itself, and it is brought to bear in no way other than by its own exercise—that is, by self-discipline and resolute persistence. A committed individual holds him- or herself to the commitment by living in a resilient, determined way. (*HT*, 341)

Note that for Haugeland, it is *individuals* that undertake and hold themselves to existential commitments through *self*-discipline.[5] This emphasis on individuals is even more emphatic in remarks he makes about responsibility. In defining responsibility in general, Haugeland says, "a responsiveness that finds what is ruled out in the responding entity's *own* actions to be unacceptable *to that entity itself* is *responsibility*" (*DD*, 204). And he says that existential responsibility (or "conscience") in particular is "a more originary *self*-responsibility—one that cannot be public but can only be taken over by an individual. Conscience, understood existentially, calls upon Dasein in each case to take over and own *this* responsibility" (*DD*, 209).[6]

What Haugeland means here is not entirely clear. At a minimum, it seems he must mean that a particular person's *taking* responsibility for how she understands the world is conceptually prior to others *holding* her responsible. It is precisely this claim that I think is a mistake: being able to take responsibility, I claim, is not intelligible apart from one's being a part of a community in which others can hold one responsible, so that self-responsibility is not "more originary."[7]

If existential commitment and existential responsibility are fundamentally individual matters, then it might seem that there is no pressure for different "players" to play the same "game." It would not be *we scientists* (or chess players or musicians, etc.) who jointly bear responsibility for upholding a common way of life that discloses a domain of entities. With what right, then, can you criticize me for too quickly changing my understanding of the laws of nature so as to resolve apparently anomalous experimental results? Even if your existential commitment is to an understanding of norms according to which my acceptance of the anomaly is rash and irresponsible, if existential commitment and responsibility are individual, then you have no reason to think I am or ought to be committed in the same way and so no reason to think that either I or you are failing in our individual existential responsibilities. My refusal to accept your criticism is, in effect, a matter of my forking the "game" of science and thereby responding to different entities than you are, so that we in a sense inhabit different "science worlds," in much the same way we could fork the game of chess and so inhabit different chess worlds. That is not the sort of shared, public objectivity we need for science or that Haugeland is trying to make intelligible.

Of course, we might assume that what makes intelligible such public objectivity is the world itself: given that there is a single world to which we scientists are all individually committed and beholden, we can expect there to be a convergence in our individual ontological understandings of that common world. However, it is not clear in advance that there is a single best way to understand the world and so that our individual understandings will converge; that this is so is part of the existential commitment we scientists undertake as scientists. As the example of personal values shows, not all phenomena need be publicly accessible in the sense of being an aspect of the world that everyone ought to recognize on pain of criticism from others: not everyone need share my personal value of doing philosophy, and those who also happen to value doing philosophy need not change their value (or criticize me or be subject to criticism themselves) when I do. What is needed for existential commitment to public objectivity is that we be *answerable* to each other: I cannot escape your criticism simply by claiming to inhabit a different "science world" than you do. Rather, for your criticism to be genuine, the conflict between us must be one that can be resolved by appeal to a single shared authority as that to which we are beholden together. Consequently, the existential commitment to a public, objective, scientific world must be one *we* undertake *jointly*, so that we are responsible not only to that shared world but also to each other. And this means that my existential commitment and responsibility to a shared, public world require not merely that I be able to take responsibility for my understanding but also that you be able to hold me responsible for it. Existential commitment and responsibility cannot therefore be individual matters, contrary to Haugeland's claim, on pain of abandoning the sort of objectivity we thought the world has.

This all raises the question of how we should make sense of existential commitment and responsibility in such a way as to make intelligible how they can be *ours jointly*. Haugeland never clearly addresses what such commitment and responsibility are, even in the individual case, let alone the joint case. My aim in the remainder of this paper is to explain this. In doing so, I follow Haugeland's (and Heidegger's) lead in thinking of existential commitment as involving a distinctive kind of caring. In section 3, briefly and without argument, I outline my background account of caring in terms of the emotions. There are many different types of caring, including valuing, loving, and respecting, each distinguished by a distinct class of emotions. Section 3 gives an account of a basic type of caring that we share with the animals, and I proceed in section 4 to lay out an account of respect in terms of a distinctive rational pattern of reactive attitudes—emotions like resentment, gratitude, indignation, approbation, and trust. In part (and partly following Strawson 1962 and Darwall 2006), I claim that we respect others as members of particular communities of respect, and it is within such communities of respect that we hold each other responsible and so can take responsibility for the joint commitments that define that community. In section 5, I argue that when public objectivity is in play, our existential commitments can best be understood in terms of our forming a community of respect committed to truth, thereby filling in the gaps in Haugeland's account of objectivity.

3 Caring and the Emotions

In general, to care about something is to have a concern for its well-being, a concern in which one finds it to be worthy of both one's attention and action.[8] As I have long argued (see, e.g., Helm 2001), caring is constituted by rational patterns of emotions. To understand this, it is first necessary to say something about the emotions and their objects.

Emotions in general involve implicit evaluations, with each type of emotion having its own characteristic evaluation—its own *formal object*; the object one evaluates in having a particular emotion is that emotion's *target*. For example, I might be afraid that the kids playing baseball in the street will damage my car, or I might be angry at you for stealing my car. In these cases, the kids and you are the targets of my fear and anger, and in having these emotions, I am evaluating the kids as dangerous (the formal object of fear) or you as offensive (the formal object of anger). One question these evaluations raise is why they are appropriate. Here the answer cannot simply be that the kids have the potential to damage my car, for they also have the potential to damage the piece of cardboard they are using for home plate, and yet that fact wouldn't normally inspire my fear. The difference is that I just don't care about that piece of cardboard, whereas I do care about my car: it is only because of the relationship between their playing and something I care about (my car) that my emotional evaluation of them as dangerous makes sense. We can formalize this idea by understanding emotions to have a third object in addition to a target and a formal object: an emotion's *focus* is the background object the subject cares about whose relation to the target makes intelligible the evaluation of the target in terms of the formal object. So both my fear of the kids and my anger at you have my car as their focus.

This notion of an emotion's focus is important for understanding the way emotions are rationally connected to each other, for the sense in which each emotion "involves" an evaluation should be understood in terms of a commitment to the worth—to the *import*—of the focus of that emotion and thereby of its target. This means that in having one emotion and so being committed to the import of its focus, one is thereby committed to having other emotions with the same focus in the appropriate circumstances: committed in the sense that, other things being equal, one rationally ought to have such emotions. For example, there would be something rationally odd about my being afraid of the kids and yet not also being relieved were my car to escape unscathed; this indicates that in feeling particular emotions, we undertake what I have called *transitional commitments* from forward-looking emotions (like hope and fear) to corresponding backward-looking emotions (like relief and disappointment). Similarly, there would be something rationally odd about my being relieved that my car avoided the danger if, in the relevant counterfactual situation in which my car's windshield is broken, I would not also feel saddened or angry; this indicates that in feeling particular emotions we undertake *tonal commitments* between positive emotions (like relief and satisfaction) and negative emotions (like sadness and anger). Moreover, these rational connections among emotions apply even when the emotions do not share a common target: my fear of the kids is rationally connected to my fear of the impending hailstorm and my anger at you for stealing my car. (How do you suppose I would—*should*—feel were I to discover that sometime during the night a large tree branch fell on my car?)

Because a condition of the intelligibility of one's having any particular mental capacity is that one is by and large rational in the exercises of that capacity (see, e.g., Davidson 1980), one's emotions must in general come in projectible patterns with a common focus. Other things being equal, isolated emotions not falling within such patterns manifest emotional irrationality because they thereby fail in their commitments to other emotions; conversely, once a pattern of emotions is established, one generally ought to have other emotions with the relevant focus when these are otherwise appropriate. To exhibit such a pattern of emotions is therefore to be disposed to attend to the focus of that pattern and to act on its behalf; moreover, the rationality of the pattern is such that one ought so to attend and act. Consequently, I claim, because caring about something (its having import to you) is a matter of finding it worthy of your attention and action, we can see that what it is to *care* about something (what it is for that to have *import* to you) just is for it to be the focus of such a projectible, rational pattern of emotions.[9] Particular emotions, then, can be assessed for warrant depending in part on whether they fit into such a pattern of emotions with a common focus—on whether they are properly responsive to what has import to one.

In my discussion so far, "import" is intended to be a generic term for the worth something has to one. Clearly there are many distinct ways in which something can have import. In earlier work, I have distinguished caring from valuing: whereas caring is a kind of evaluative attitude we share with at least some higher animals like dogs and cats, *valuing* is deeper in that it involves finding something worthwhile as a part of an overall life worth living. Consequently, the emotions constituting values must be similarly "deep" in their characteristic evaluations. These emotions, such as pride, shame, and anxiety, I call *person-focused emotions* to reflect their engagement with the quality of life of particular persons. To value something, then, is for it to be the focus of a projectible, rational pattern of person-focused emotions. In general, distinctive kinds of import are constituted by distinctive classes of emotions, classes defined by the way in which such emotions form rational patterns with common focuses (Helm 2010). As I have recently begun to argue (Helm 2011, 2012, 2014a, 2015) and summarize here in section 4, we can make intelligible a further evaluative attitude, respect, in terms of another distinctive class of emotions, namely, the reactive attitudes. This will prove fundamental to understanding Haugeland's notion of an existential commitment.

4   Respect and the Reactive Attitudes

According to Strawson, the "participant reactive attitudes are essentially natural human reactions to the good or ill will or indifference of [people toward each other], as displayed in *their* attitudes and actions" (1962, 195).[10] Strawson distinguishes three types of reactive attitudes: the personal reactive attitudes (such as resentment and gratitude), the vicarious reactive attitudes (such as indignation and approbation), and the self-reactive attitudes (such as guilt and self-approbation). As I shall suggest, that there are these three types of reactive attitudes is fundamentally important.

Some controversy has arisen about exactly which emotions we should classify as reactive attitudes. Strawson himself was rather liberal, including such responses as esteem, indifference, contempt, love, loss of security, goodwill, affection, and forgiveness (1962, 191). Others, such as Jay Wallace (1994), are much more restrictive, limiting the reactive attitudes to just a few negative emotions: resentment, indignation, and guilt. As indicated at the end of section 3, I think this controversy can be settled, and so the class of reactive attitudes can be delineated, by looking to the kind of rational patterns the reactive attitudes form with a common focus, thereby constituting a distinctive kind of import. Such rational patterns will involve various rational commitments among the reactive attitudes.

Consider first a tonal commitment: if I resent you for harming me in some way, then I ought also feel gratitude toward you in other circumstances were you notably to benefit me. If this is right, then we should not, contra Wallace, restrict the reactive attitudes to simply negative emotions. Yet what is interesting about the reactive attitudes is that the relevant rational commitments are not simply *intra*personal; they are *inter*personal as well: one person's reactive attitudes are rationally tied to those of others, *addressing* them and *calling on* them for a response (see, e.g., Darwall 2006; Watson 2008). Thus if you resent me for harming you in some way, then other things being equal, I ought to feel guilty, and others ought to feel disapprobation or indignation. The rational connections here are among the personal reactive attitudes of the "victim," the self-reactive attitudes of the "perpetrator," and the vicarious reactive attitudes of witnesses. In each case, there would be something rationally amiss about your feeling resentment and yet my not feeling guilt or their not feeling disapprobation: either you are responding in a case in which I have not really wronged you, or I am failing to take responsibility for what I did, or others are failing to hold me responsible. In each case, the failure is a failure to respond as we ought.

What about transitional commitments? If the reactive attitudes form the sort of rational patterns I have claimed are constitutive of caring, then we ought to find transitional commitments between forward-looking and backward-looking reactive attitudes. Yet here it might seem that my attempt to understand the class of reactive attitudes in terms of the rational patterns they form constituting a distinctive kind of caring breaks down: what is distinctive of the reactive attitudes as *reactive*, it might seem, is that they are responses to what has already happened—that they are essentially backward-looking emotions. In reply, we do find transitional commitments between trust and distrust and the reactive attitudes: other things being equal, if you uphold my trust, I ought to feel gratitude (and you ought to feel self-approbation), whereas if you betray my trust, I ought to feel resentment (and you ought to feel guilty).[11] This indicates that we should construe the reactivity of reactive attitudes not as being to what is done to us or others but rather as being to persons as proper objects of our address—as having a certain *standing* or *status* as victims, perpetrators, or witnesses.[12] After all, inanimate objects and nonhuman agents are not appropriate targets of reactive attitudes precisely because they do not have the requisite standing; and the difference between reactive attitudes like resentment and their nonreactive analogues like anger seems to involve the perpetrator's failing properly to acknowledge or respond to that standing. If this is right, then trust and distrust (as distinct from their nonreactive analogues like reliance) would seem to be reactive attitudes insofar as they exhibit precisely this sort of reactivity to the standing others (or we ourselves) have.

The upshot is that the reactive attitudes are a distinctive class of emotions defined by the distinctively interpersonal rational patterns that they form. This leads to two questions: (a) what is the common focus of such patterns to the import of which we are committed in feeling the reactive attitudes, and (b) what is the distinctive form of caring that these rational patterns constitute? I shall consider these questions in turn.

The focus of a rational pattern of emotions is the common background object, concern for which makes intelligible the evaluations implicit in these emotions and explains the rational interconnections among them. What can play this explanatory role in the case of the reactive attitudes? We might think that the victim is the focus, for it is the harm or benefit done to the victim that makes intelligible the blameworthiness or praiseworthiness of the perpetrator, who is the target of the reactive attitudes. So having a background concern for the victim as having a certain kind of standing— a claim on the perpetrator's regard—might seem to explain the rational interconnections among the victim's resentment, the perpetrator's guilt, and witnesses' disapprobation.

Promising as it is, this understanding of the victim as the focus of reactive attitudes fails because it does not explain some of the rational interconnections among the reactive attitudes. In particular, first, it does not explain the way in which the victim's current resentment is rationally tied to the victim's later guilt were he similarly to harm the perpetrator. As Darwall (2006) notes, the reactive attitudes generally involve a kind of symmetry of the demand for each to recognize the other's standing and consequent claim to one's own regard. This symmetry is evident in the victim's feeling of resentment, which involves not only the demand that the perpetrator recognize the victim's standing but also the victim's recognition of the perpetrator's standing—and potential authority to make similar demands of the victim. Second, this understanding of the focus does not explain the way in which the victim's resentment is rationally tied to other reactive attitudes toward witnesses. As I indicated, the victim's resentment calls on witnesses to respond with disapprobation or indignation as a way of recognizing the victim's (and, we can now see, the perpetrator's) standing. In thus calling on the witnesses to respond, the victim's resentment likewise involves a recognition of the standing the witnesses have—a standing not merely for them to hold the perpetrator responsible in this case, and to hold others responsible in other cases, but also for themselves to be held responsible for their own failures to respond. In thus calling on witnesses to respond, the victim's resentment of the perpetrator is rationally tied to the victim's later resentment of witnesses who fail to respond with disapprobation (or the victim's later gratitude toward witnesses who prove supportive in holding the perpetrator to account). It is this rational connection between the victim's initial resentment of the perpetrator and the victim's later resentment (or gratitude) toward witnesses that is not explained by understanding the victim to be the focus of these reactive attitudes.

So what background concern can explain these rational interconnections among reactive attitudes? The common element is the standing that the victim, perpetrator, and witnesses all have as responsible to certain norms—as engaged in certain activities or practices or way of life defined by these norms—and so as having the authority to hold each other responsible. Of course, depending on what norms are at stake, not just anyone will have this standing as responsible to the norms and as having authority to hold others responsible. The norms of my family or club for how to greet one another (with hugs or a secret handshake), how to conduct meetings concerning family or club business and thus when and how someone can be recognized to speak (by passing the "talking stick" or being called on by the Grand Poobah)—these norms are binding only on members of the relevant community, not to outsiders, and it is only members who have the relevant authority to hold other members responsible to these norms. Thus, in general, the relevant standing and authority belong only to members of the relevant community, and our recognition of the standing of others is a recognition of them as fellow members of the community. Indeed, our concern for the standing and authority of one member of the community is tied to our concern for the standing and authority of other community members, as well as to our concern for the norms to which we hold each other as fellow members. This suggests that our concern for each member and each norm is a *part* of a more general concern for the community itself. It is, I submit, this concern for the community that explains all these rational interconnections among reactive attitudes and so is their common focus.

The claim that our concern for fellow members and for norms is a "part" of our concern for the community needs further explanation. Consider instrumental concerns. If I care about an end, then, other things being equal, I ought to care about the relevant means to that end. Applying the basic account of caring presented in section 3, we might think this implies that I ought to experience a pattern of emotions focused on the means but would ignore the way in which my concern for the means is not independent of my caring about the end and, indeed, is connected to my concern for other means that are also relevant to achieving the end. The relevant pattern of emotions clustered around each means is itself a *subpattern* of the pattern of emotions focused on the end; in this sense, I care about the means only as a part of my caring about the end. To mark this connection, I shall understand, for example, my disappointment at failing to achieve a means to the end as *subfocused* on the means but *focused* on the end.[13]

Likewise, one's concern for the standing of a fellow member is connected to one's concern for the standing of other fellow members and one's concern for the norms, none of which is independent of one's more general concern for the community itself. That is, the pattern of reactive attitudes clustered around each particular member or norm is a subpattern of the more general pattern focused on the community itself. The caring commitment each member of the community normally has to its members (including herself) and to its norms presupposes a caring commitment to the community itself and so is intelligible only as a part of the commitment to the community. Again, we can mark this connection by understanding, for example, the victim's resentment, which targets the perpetrator, to be subfocused on particular members (as victim, perpetrator, and witnesses) and norms and focused on the community itself.[14]

Given this account of the focus and subfocus, what can we say about the form of caring such patterns of reactive attitudes constitute? As I have suggested, to feel particular reactive attitudes is to hold others responsible (or yourself to take responsibility) as a part of recognizing the import of their (or your) standing as a member of the community. Indeed, such standing just is one's standing as a responsible agent accountable to others in the community, and to have a concern for such standing just is to *respect* the person. Such respect is what Darwall calls *recognition respect*, a matter of recognizing the standing others have, as opposed to what he calls *appraisal respect*, which is "an assessment of someone's conduct or character" in light of the norms of the community—an assessment that itself is a particular reactive attitude.[15] Consequently we might call the type of concern members have for the community itself *reverence*, so that they respect each other as a part of revering the community and its way of life. Such communities, therefore, I shall call *communities of respect*.

Two facets of this account need to be brought out. First, I have understood the patterns of reactive attitudes to be essentially interpersonal: my gratitude for you is rationally connected to your self-approbation and others' (appraisal) respect. Such interpersonal structure to these patterns implies that the underlying commitment to the import of the community and its members and norms is not the commitment of the members individually but rather their jointly: it is *we* who respect each other as a part of *our* reverence for the community as a community of respect. Consequently, *my* respecting other members is something I do—indeed, *ought* to do—only *as one of us*. Second, the standing one has as a member of the community and to which each ought to respond with the reactive attitudes in the appropriate circumstances is the standing as a responsible agent, not only able to take responsibility for what one does but also possessed of the authority to hold others responsible to the norms of the community. Insofar as the community as a community of respect, in which each has this standing the others ought to recognize, is itself constituted

in substantial part by these interpersonal rational patterns of reactive attitudes, we can see that our being responsible agents at all depends on our
having such standing within a community of respect and so is not intelligible apart from others' standing to hold us responsible: neither taking
responsibility nor holding responsible is "more originary" than the other,
contrary to Haugeland's assumption (see n. 7). Responsibility, therefore, is
essentially social.[16]

5   Reverence for Ontological Truth

We are now in a position to return to the questions raised at the end of
section 2 concerning how we ought to understand Haugeland's notion of
existential commitment in such a way as to make sense of the shared,
public nature of objective truth. Recall that for Haugeland, objects are
accessible, autonomous authorities—are *objective*—only insofar as they are
what our thought and understanding are beholden to as criteria of correctness that are at least partly independent of both particular thoughts
and group consensus. Such independence is possible, Haugeland thinks,
only if we can make sense of our ontological understanding—our understanding of what is possible and impossible for entities—as itself beholden
to entities in virtue of our existential commitment to, and hence responsibility for, ontological truth. The trouble is, I argued, that because he
understands existential commitment (and the consequent responsibility)
to be fundamentally an individual matter, Haugeland does not seem to
be able to make sense of the *publicness* of objective phenomena: their
being a part of a world to which we all are beholden, and thus a world
in this sense shared in common by all. How, then, can we make sense of
such commitment and responsibility in such a way as to make sense of
publicness?

    As I suggested, public objectivity requires that the requisite existential
commitment be something we undertake jointly. It should now be apparent how the account of communities of respect that I have been sketching
can make sense of this: for public objectivity to be possible, we must form
a community of respect in which we jointly revere ontical and ontological
truth (perhaps among other things). Such joint reverence just is our existential commitment to ontical and ontological truth in which we each
have standing to criticize the other and thereby to hold each (including

oneself) responsible to the phenomena as a common authority for their "moves" in this "game" of "giving and asking for reasons" (Brandom 1994) for how things objectively are—of "justifying and being able to justify what one says" (Sellars 1963, sec. 36; see also Brandom 1994). This standing to criticize and so to hold each other responsible means that disagreements among us will in general be genuine and not merely verbal, grounded in a shared ontological understanding of the world. Because we are each answerable to the others from within a community that jointly reveres ontological truth, we cannot simply escape criticism by "forking" our ontology; rather, we are jointly beholden to a shared, public world. This needs further explanation.

In a game like basketball, a player might commit a foul, in a sense "breaking the rules," and so be penalized for it—turning the ball over to the other team, for example. Such sanctions are ways of holding players accountable to the rules, and yet it is important to note two things about them. First, they are not sanctions imposed from outside the game itself as a way of ensuring that players remain in the game; rather, they are themselves part of the rules, and other things being equal, in committing fouls, whether intentionally or not, players are playing the game of basketball just as much as when they are not committing fouls. Second, such sanctions are intelligible as sanctions—as ways of holding players *accountable*—and not merely as an arbitrary harm sometimes imposed on some players only because they are embedded within the relevant patterns of reactive attitudes. Although we do not feel resentment or disapprobation for players who commit fouls—indeed, sometimes we may (appraisal) respect a player for her ability to commit smart fouls—we do feel these reactive attitudes when players temporarily fail in their reverence for the game itself in part by failing in their respect for other players. Thus a player who refuses to accept the referee's call thereby manifests a lack of respect for the referee and her standing as such in a way that, we might think, merits the resentment or disapprobation of other players. (Of course, the referee herself might abuse her standing and authority in intentionally calling the game unfairly, and so she, rather than the player who refuses to accept her call, might be the proper object of resentment and disapprobation.) It is in feeling these reactive attitudes that we hold players (including the referee) accountable to the norms defining the practice of basketball and so call on each properly to respect the others as players.

The same is true in the "game" of giving and asking for reasons—the game defined by our joint reverence for objective, public truth. Claims I make about what is true are not simply claims about how things are for me; that would not be a "move" in this "game". Rather, such claims are about how things are within the public world we share, and so are claims about how others should see the world, too. Of course, such a claim can be based on false or inadequate evidence or on bad reasoning, and so I might thereby fail to uphold the rules—the norms—of this game in a way that is something like the commission of a foul in basketball, and someone might call me to account for such a failure by offering a criticism or asking for a justification. Such criticism or demand for justification is intelligible as a way of holding someone accountable only because it involves one's having the standing to criticize, a standing that others in general ought to respect. Once again, although we do not feel resentment or disapprobation toward those who make mistaken assertions, we do when they fail properly to revere objective, public truth. Thus, other things being equal, to fail properly to acknowledge and respond to criticism or a demand for justification is to fail properly to respect the other's standing to criticize in virtue of which each is answerable to the others and, thereby, to the truth that we jointly revere; it is such notable failures to revere truth (or notable successes in revering truth) that merit the reactive attitudes. For example, we might feel resentment or disapprobation toward a psychologist who videotapes subjects in the experimental condition and then intentionally scores (and puts pressure on his graduate students to score) the videos in such a way as to confirm his hypothesis, for he thereby flouts the norms of science and betrays the trust others have placed in him; or someone might feel gratitude toward an adviser who trusted and supported her in pursuit of an unpopular research program that eventually bears fruit.

Of course, other things are not always equal. Those who criticize me may be prejudiced or in other ways blind to the truth, or they may have ulterior motives leading them to distort the truth, or they may lack the necessary expertise, so that their criticisms utterly miss the mark. In such cases, attending to and responding to criticism may be otiose. Yet I must still have reasons to think this, reasons grounded not just in my understanding of others as responsible to a shared, objective world but also in my understanding of *how* particular others are responsible: I must appraise their responsibility, at least within a particular domain of truth, and so have reason, for example, to trust or distrust them. Moreover, I must similarly have such an understanding of myself, as having this standing as a responsible epistemic agent, both beholden to the shared world and accountable to others. Yet such appraisals of oneself and others are themselves criticizable; reasons for them can be questioned or demanded, and I can be praised or blamed for how I handle the criticism and reasons of others. Consequently, in having a caring commitment to objective, shared truth, I must care about and be responsible (and so beholden to the world) for not only the judgments and understanding I have of the world but also how I respond to others in taking or failing to take their criticisms seriously. In thus blaming others for epistemic irresponsibility, we call on them to accept such blame or account for themselves, thereby taking responsibility for objective truth.

Consider how this plays out not simply in the case of our ontical understanding (as has featured in my examples thus far) but also in the case of ontological understanding. Assume that after doing a single experiment, a firmly established physicist calls a press conference to celebrate his discovery of a new elementary particle or fundamental force and so a change in our ontological understanding of the world. Here it may seem that the physicist is being irresponsible: attempting to change the game, to reject his "ontological heritage" (DD, 215), in the face of evidence that is too scanty, so that he is at best being sloppy. In the face of such criticism, it is not an option for the physicist simply to fork the ontology and thereby escape criticism by claiming to inhabit a different "world." To do so would be to fail properly to respect the standing others have to criticize him, demand justification, and so hold him accountable, in part through the reactive attitudes. Consequently, as a member of a community of respect that reveres objective, public truth, in choosing to reject his ontological heritage, the physicist is making not simply a personal choice, as when he finds the life worth pursuing for him to involve science rather than music or to involve his having children rather than not. Rather, in so choosing, he is, more or less responsibly, taking a stand on what is right for others, thereby leaving himself open to precisely this sort of criticism as a part of the community of respect that reveres ontological truth.

All of this implies that in being committed to objective, public truth, each individual must have a certain understanding of herself as responsible to a shared world, with shared standards regulating her ontological understanding, and so as accountable and answerable to others, as well as herself. In short, making sense of the autonomous authority of publicly objective phenomena requires understanding Haugeland's notion of an existential commitment in terms of communities of respect in which our responsibility to the world essentially involves our accountability to each other via the reactive attitudes.

It should be apparent that a large objection is looming. I have claimed that one cannot escape criticism by forking the ontology because one's membership in a community of respect makes one answerable to others who can hold one responsible to the objective world. However, why cannot someone fork the ontology and thus escape this sort of accountability to others simply by opting out of the group? An initial reply might go as follows: membership in the group, in the first instance, is determined by the actual patterns of reactive attitudes. So someone is a member of this group just in case we generally hold him responsible to the norms of the group: you are accountable to us just in case we actually hold you to account and so accord you the standing as a member. Yet this initial reply is inadequate: surely the group can be wrong about its membership: prejudice of various sorts can prevent us (white men, say) from recognizing that others (women, various minorities) really are members of this community of respect and treating them accordingly, but such prejudicial treatment is wrong whether we recognize it or not. So the question, of who should be treated as a member of the community of respect that reveres objective truth, so far seems to have no answer. Even worse, there seems to be no compelling reason to think that there will be a single such community: someone who wanted to fork our ontology (rather than convince the rest of us that we misunderstand the world) may accept my claim that objective truth and so existential responsibility can be made intelligible only in terms of communities of respect and so form a distinct community with its own ontology, claiming that objective truth, while public, is nonetheless relative to a community. This may be an advance on Haugeland's appeal to individual existential commitment, but the resulting account of objective truth seems to fall short of what we were looking for.

I can provide no convincing answer to this objection here. For all I have said so far, it may be that we have to bite the bullet and accept that objective, public truth is relative to a community. Nonetheless what I have said does point us in a promising direction for giving a more complete response. Among the truths—both ontical and ontological—that a community of respect that reveres objective, public truth is fundamentally concerned with will be truths about itself, including not only the ontical truth of who in fact are its members (i.e., those whom we ought to respect and hold responsible to the standards of truth) but also the ontological truths both of what it is to be an objective, public truth and of what it is to be a member of this epistemic community. Notice that these two ontological questions, of what it is for something to be an objective, public truth and of what it is to be a member of this epistemic community, are plausibly connected: the public for whom something is publicly true just are the members of the epistemic community. If so, then forking the ontology is something that can only be done wholesale: someone cannot agree with us and so be a member of our epistemic community when it comes to, say, the truths involved in our ordinary practical lives and yet claim to live in a different science world from us in which things are true for him that are not true for us (or vice versa). To be a member of our epistemic community in practical contexts is to revere objective, public truth as one of us and thus to be both answerable to us for claims about what is true and responsible to the shared, public world; such answerability and responsibility hold for all truths on pain of equivocation on the notion of "truth." Objective, public truth then would turn out to be relative to an entire worldview, a complete way of life, and perhaps this is enough to make such relativity not so hard a bullet to bite.

We can go further. It is plausible to assume that what it is to be a member of an epistemic community just is to be a person, so that to be a member of any epistemic community is to be a member of all, which seems to suggest (but does not on its own show) that there is only a single epistemic community. Indeed, the plausibility of this claim is bolstered by the kind of account I have given of an epistemic community insofar as it reveals the essential interconnectedness of theoretical and practical rationality: theoretical rationality, embedded as it is within the norms of an (epistemic) community of respect, is grounded in the very notions of standing, authority, respect, and responsibility that make us autonomous agents.

Consequently, the ontical and ontological understanding the community must have of itself as a part of its reverence for objective, public truth must include the ontical understanding of who in fact is a person and the ontological understanding of what it is to be a person as the public that ought to revere objective, public truth. As usual, such ontical and ontological understandings are beholden to ontical and ontological truth and so are open to both potential dispute and potential resolution. That is, understanding the nature of objective, public truth itself is possible only simultaneously with understanding what it is to be a person; indeed (dare I say it?), this sounds like what Heidegger is after with the existential analytic of Dasein.

6 Conclusion

I began with the question John asked at my dissertation defense: how is belief richer than a mere informational state? Now, more than twenty years later, I think I have the beginnings of an answer. As Searle makes clear, to believe something in the rich sense distinctive of persons is not merely a matter of "holding something true":

> Think hard for one minute about what would be necessary to establish that that hunk of metal on the wall over there had real beliefs, beliefs with direction of fit, propositional content, and conditions of satisfaction; beliefs that had the possibility of being strong beliefs or weak beliefs; nervous, anxious, or secure beliefs; dogmatic, rational, or superstitious beliefs; blind faiths or hesitant cogitations; any kind of beliefs. The thermostat is not a candidate. (Searle 1980, 420)

That beliefs can be strong, nervous, dogmatic, hesitant, and so on, and that these possibilities for belief are essential to the rich sense of belief characteristic of us persons, make sense given a background of caring that is simultaneously a matter of caring about (revering) objective, public truth and, as a part of such reverence, caring about (respecting) oneself and others as responsible epistemic agents. In short, to *believe* something is to hold it true in such a way that one cares about its truth, as a part of revering objective truth from within a community of respect. Moreover, the capacity for belief in this rich sense is intelligible only as a part of a more general account of persons as fundamentally caring, social creatures. I can think of no better tribute to John than to continue to pursue this line of thought.

Acknowledgments

Portions of this paper have been presented at the Collective Intentionality VII conference in Basel, at the Thumos Seminar at the University of Geneva, at a workshop with the Applied Phenomenology Research Group at Durham University, and at a joint Carleton–St. Olaf College colloquium, all of which led to many helpful discussions. Thanks go especially to Hans Bernhard Schmid, Kevin Mulligan, Fabrice Teroni, Julien Deonna, and Matthew Ratcliffe. Thanks also to Kathryn Kutz and Daniel Kaplan, with whom I had many profitable conversations, supported by Franklin & Marshall College's Hackman Scholars Program. An early ancestor of this paper was coauthored with Dan and presented at the Psycho-Ontology Conference in Jerusalem in 2011. Finally, thanks to Zed Adams and Jacob Browning for detailed, helpful comments on the penultimate draft.

Notes

1. In *HT*, especially those in the section on Truth: "Objective Perception," "Pattern and Being," "Understanding: Dennett and Searle," and especially "Truth and Rule-Following"; in *DD*, especially "Truth and Finitude" and "Letting Be."

2. Haugeland (*DD*, 194–201), following Heidegger, distinguishes understanding (*Verstehen*), telling (*Rede*), and "findingness" (*Befindlichkeit*), all of which come in ontical and ontological varieties. I am essentially compressing this distinction into a single notion of *understanding* for simplicity of exposition. As will perhaps become clear later, I think Haugeland fails properly to make sense of the place "findingness" has in disclosing entities as objective, but I do not have space to discuss this idea in any detail.

3. Haugeland's model here, of course, is Kuhn 1986—the gospel according to "Saint Thomas A-Kuhn-is," as we sometimes teased him.

4. Especially "esoteric chess": chess played in a medium in which it is initially very difficult for the uninitiated to be able to identify the pieces and positions, let alone manipulate those pieces to make a move. The initiated, after considerable training, are able to do this effortlessly, much as experienced musicians have the ability easily to recognize certain patterns of sound as particular chords or chord progressions and can manipulate the sounds so as to continue those patterns. For details, see *HT*, 327–329.

5. It should be acknowledged that Haugeland describes existential commitment as essentially first-personal, saying that "the first person doesn't mean particularly the

first-person *singular*" (*HT*, 339). Yet nowhere does he spell this out, and (as I point out in the text just below) in later work he is more explicit in understanding existential commitment and responsibility as individual matters.

6. Note that for Haugeland, "'Dasein *in each case*' means each individual person, whether or not one accepts my controversial suggestion that Dasein as such is not individual or personal" (*DD*, 202; see also Haugeland 2004).

7. Although I have something to say about this claim in section 4, I have argued for it in more detail in Helm 2012, 2015.

8. This section borrows heavily from Helm 2012.

9. This is, of course, an oversimplification, for desires and evaluative judgments can also be a part of the relevant patterns constituting import; indeed, that this is so is partially constitutive of them as desires and evaluative judgments (see Helm 2001).

10. The passage from which this quote was taken was intended to describe only the personal reactive attitudes; I have generalized it somewhat to include the vicarious and self-reactive attitudes, as well.

11. This is a point made by Holton (1994). However, Holton does not explicitly infer that trust itself is a reactive attitude. Helm 2014b provides more complete arguments for the idea that trust is a reactive attitude.

12. Darwall (2006) provides an extended discussion of reactive attitudes as responsive to the standing others have as persons.

13. For further details of this account, see Helm 2001, esp. chap. 4; and, for its application to making sense of how in love we value what our beloved values as a part of loving him, Helm 2010, esp. chaps. 4–5.

14. Implicit in this account is the idea that we can belong to many different, potentially overlapping communities. This runs counter to the presumption found in the literature that the reactive attitudes involve a single, "human" community—the community of all persons. To the contrary, I think we can make some headway in understanding how the community of all persons can be a *moral* community by first examining the more general category of communities of respect.

15. Darwall 2006, 122. Darwall does not understand this distinction in terms of my distinction between particular emotions and the evaluative attitudes constituted by patterns of emotions (but he should: failing to do so leads him to question whether this distinction can ultimately be maintained [124n9]).

16. Of course, I have not really argued for this here. For some detail, see Helm 2012, 2015.

References

Brandom, R. 1994. *Making It Explicit: Reasoning, Representing, and Discursive Commitment*. Cambridge, MA: Harvard University Press.

Darwall, S. L. 2006. *The Second-Person Standpoint: Morality, Respect, and Accountability*. Cambridge, MA: Harvard University Press.

Davidson, D. 1980. *Essays on Actions and Events*. New York: Clarendon Press.

Haugeland, J. 1994. Remarks on machines and rule-following. In *Philosophy and the Cognitive Sciences*, ed. Roberto Casati, Barry Smith, and Graham White, 127–138. Vienna: Hölder-Pichler-Tempsky.

Haugeland, J. 1998. *Having Thought: Essays in the Metaphysics of Mind*. Cambridge, MA: Harvard University Press. (Abbreviated as *HT*.)

Haugeland, J. 2004. Closing the last loophole: Joining forces with Vincent Descombes. *Inquiry* 47:254–266.

Haugeland, J. 2013. *Dasein Disclosed: John Haugeland's Heidegger*. Ed. J. Rouse. Cambridge, MA: Harvard University Press. (Abbreviated as *DD*.)

Helm, B. W. 2001. *Emotional Reason: Deliberation, Motivation, and the Nature of Value*. Cambridge: Cambridge University Press.

Helm, B. W. 2010. *Love, Friendship, and the Self: Intimacy, Identification, and the Social Nature of Persons*. Oxford: Oxford University Press.

Helm, B. W. 2011. Responsibility and dignity: Strawsonian themes. In *Morality and the Emotions*, ed. C. Bagnoli, 217–234. Oxford: Oxford University Press.

Helm, B. W. 2012. Accountability and some social dimensions of human agency. *Philosophical Issues* 22 (1): 217–232.

Helm, B. W. 2014a. Emotional communities of respect. In *Collective Emotions*, ed. C. von Sheve and M. Salmela, 47–60. Oxford: Oxford University Press.

Helm, B. W. 2014b. Trust as a reactive attitude. In *Oxford Studies in Agency and Responsibility: "Freedom and Resentment" at Fifty*, ed. D. Shoemaker and N. Tognazzini, 187–215. Oxford: Oxford University Press.

Helm, B. W. 2015. Rationality, authority, and bindingness: An account of communal norms. In *Oxford Studies in Agency and Responsibility*, ed. D. Shoemaker, 189–212. Oxford: Oxford University Press.

Holton, R. 1994. Deciding to trust, coming to believe. *Australasian Journal of Philosophy* 72 (1): 63–76.

Kuhn, T. S. 1986. *The Structure of Scientific Revolutions*. Chicago: University of Chicago Press.

Searle, J. R. 1980. Minds, brains, and programs. *Behavioral and Brain Sciences* 3:417–424.

Sellars, W. 1963. Empiricism and the philosophy of mind. In *Science, Perception, and Reality*, ed. W. Sellars, 253–329. London: Routledge & Kegan Paul.

Strawson, P. F. 1962. Freedom and resentment. *Proceedings of the British Academy* 48:187–211.

Wallace, R. J. 1994. *Responsibility and the Moral Sentiments*. Cambridge, MA: Harvard University Press.

Watson, G. 2008. Responsibility and the limits of evil: Variations on a Strawsonian theme. In *Free Will and Reactive Attitudes: Perspectives on P. F. Strawson's "Freedom and Resentment,"* ed. M. McKenna and P. Russell, 115–141. Burlington, VT: Ashgate.

8   Constancy Mechanisms and the Normativity of Perception

Zed Adams and Chauncey Maher

To recognize something is to respond to it in a way that distinguishes it from other things; to recognize is to tell apart. But differential response cannot be the whole story, for two deeply related reasons. First, what is recognized is always some determinate item, feature, or characteristic of the confronted situation, whereas a given response can equally well be taken as a response to any of several distinct things. Second, recognition, unlike response, is a normative notion: it is possible to *misrecognize* something, to get it *wrong*, whereas a response is just whatever response it is to whatever is there. These are related because: only insofar as something determinate is supposed to be recognized, can there be an issue of recognizing it rightly or wrongly; and it is only as that which determines rightness or wrongness that the object of recognition is determinate.

—John Haugeland, "Objective Perception" (*HT*, 272)

1   The Problem of Perceptual Representation

Imagine that John is scouring the Arizona desert for a scorpion, to encase it in resin for the aiguillette on his bolo tie. As he looks under a small rock, he sees something that looks like a scorpion, and he swings his net down on it. But it is not a scorpion. It is, in fact, a plastic scorpion left there by a mischievous graduate student. In frustration, John sits down on a large rock, thinking he might just give up wearing bolo ties altogether. As he does so, he suddenly feels a sharp pain in his lower regions, realizing that he has found what he was looking for, albeit unintentionally.

Consider John's visual experience as he looks under the small rock. It makes sense to assess it as accurate or inaccurate, and it turns out that it is inaccurate. Contrast this with John's pain experience as he sits on the large rock. There are many things we could say about this experience, but it would not make sense to assess it as accurate or inaccurate.

What marks the difference between these two sorts of experience? Why can some be assessed as accurate or inaccurate, whereas others cannot? More generally, what does it take for a sensory experience to represent things as being a certain way, such that it can be assessed in terms of accuracy and inaccuracy? Call this *the problem of perceptual representation*.

In *Origins of Objectivity*, Tyler Burge attempts to solve this problem. He proposes that a sensory experience is representational if and only if it is an exercise of a perceptual constancy mechanism. As he puts it, "Perceptual constancies are necessary as well as sufficient for perceptual objectification and perceptual representation" (Burge 2010, 413).

In this essay, we draw on John Haugeland's work to argue that Burge is wrong to think that exercises of perceptual constancy mechanisms suffice for perceptual representation. Although Haugeland did not live to read or respond to Burge's *Origins of Objectivity*, we think that Haugeland's work contains resources that can be developed into a critique of the very foundation of Burge's approach. Specifically, we identify two related problems for Burge. First, if (what Burge calls) *mere* sensory responses are not representational, then neither are exercises of constancy mechanisms, since the differences between them do not suffice to imply that one is representational and the other is not. Second, taken by themselves, exercises of constancy mechanisms are only *derivatively* representational, so merely understanding how they work is not sufficient for understanding what is required for something, in itself, to be representational (and thereby provide a full solution to the problem of perceptual representation).

In section 2, we offer a concise summary of Burge's account of perceptual representation. (It is worth noting that Burge's own articulation of this account spans 656 pages.) In section 3, we draw on Haugeland's work to identify and spell out the problems for Burge's account just mentioned. In section 4, we conclude by spelling out why we think the failure of Burge's account nevertheless makes an excellent starting point for solving the problem of perceptual representation.

2   Burge on Perceptual Representation

In this section, we summarize Burge's account of perceptual representation. We begin by canvassing his criticisms of two familiar theories of perceptual representation, before presenting his own positive proposal.

2.1.1 Information-Theoretic Theories of Perceptual Representation

Information-theoretic theories (hereafter *IT theories*) are based on the idea that sensory states can correlate with or "indicate" states of the environment.[1] There are different versions of this sort of theory, which vary mainly according to how correlation or indication is understood.[2] An influential version of the theory emphasizes causation: X indicates Y when X is caused by Y.[3] Why think this sort of relationship suffices for perceptual representation? Consider an example from outside the realm of sensation. The fossil of a megalosaurus's femur unearthed in 1676 at Oxfordshire was caused to be the way it is by a megalosaurus's femur in that same place 166 million years ago. A number of this fossil's features causally depend on the femur's features: most obviously, the fossil's size and shape causally depend on the femur's size and shape. Thus the size and shape of the fossil indicate—or "carry information about"— the size and shape of the femur. For this reason, one might think of the fossil as a representation of the femur.

An IT theory of perceptual representation holds that indication suffices for representation. Schematically, a sensory state S of an organism O represents a state of the environment E if S indicates E.

Following many others, Burge alleges that the problem with IT theories is that they do not allow for the possibility of error; they do not allow for *mis*indication.[4] A sensory state either indicates an actual state of affairs or does not indicate anything at all. It is impossible for it to make a mistaken indication. Consider the fossil. Although it can indicate many things about the megalosaurus's femur, it cannot mistakenly indicate anything about this femur—or anything else, for that matter. To misindicate would be to indicate incorrectly, for example, to indicate that the femur was larger or shaped differently than it actually was. However, a property of the fossil can indicate only whatever it was that actually caused the fossil to have that property. Consequently, misindication is impossible. For this reason, IT theories cannot explain the possibility of misrepresentation, which is a necessary part of explaining the possibility of representation.[5]

Of course, *we* can misunderstand what the fossil indicates and thereby come to have false beliefs about the megalosaurus. For instance, we might mistakenly take part of the rock surrounding the fossil to be part of the fossil and thereby come to believe that the megalosaurus's femur was larger than it actually was. So there is a sense in which fossils can misrepresent: namely, we can take them to misrepresent something. But this misrepresentation is our mistake. In general, the only way in which a fossil can misrepresent is in the derivative sense that something that already has representations can take the fossil to indicate something that it does not. But fossils themselves are not capable of misrepresenting anything.

2.1.2 Teleosemantic Theories of Perceptual Representation *Teleosemantic theories* (hereafter *TS theories*) attempt to improve on IT theories by making room for misrepresentation. They appeal to the idea of a biologically evolved function. Roughly, misrepresentation is a form of malfunction. If X has the function of indicating Y, then if X malfunctions, it misindicates Y.

What is it for something to have a biologically evolved function?[6] According to the most prominent versions of TS theories, the function of a trait is what that trait has been selected for. A trait gets selected for something when possession of that trait is part of an explanation as to why organisms with it have survived and reproduced more than organisms without that trait. And what a trait is selected for—hence what its function is—is whatever that trait has done to make organisms with that trait "fitter" than organisms without such a trait.[7]

Take a typical example, the human heart. Roughly speaking, hearts have the biological function of pumping blood because they were selected for pumping blood. That means that ancestors of humans with hearts that pumped blood were fitter than ancestors of humans with hearts that did not pump blood; ancestors with hearts that pumped blood reproduced more successfully than ancestors without such hearts. In this sense, pumping blood is what hearts *should* do. The force of this "should" is differential fitness. In a given environment, all else being equal, ancestors who had hearts that did not pump blood were less adaptive than ancestors who had hearts that do. This implies that if a heart does not pump blood, it is malfunctioning.

TS theories of perceptual representation apply this idea of a function to sensory states. The basic idea is that a perceptual representation is a sensory state that has the function of *indicating* some state of the environment.[8]

What is it for a sensory state S of an organism O to have the function of indicating some state of the environment E? S has been selected for indicating E. That means that past Os that have had Ss that indicate Es have been fitter than Os that have had Ss that do not indicate Es; Os that have had Ss that indicate Es have survived longer or reproduced more successfully than Os that have lacked such Ss. Thus, indicating E is what S should do. As with the human heart, the force of this "should" is differential adaptiveness. In a given environment, all else being equal, Os that have Ss that do not indicate Es are less fit than Os that have Ss that do indicate Es. This implies that an occurrence of S is a malfunction if it occurs when no E occurs, or if it is not caused by an E.

For Burge, the key problem with TS theories is that proper and improper functioning is independent of representational success or failure. That is, although functions make room for a sort of error, that sort of error is not the same as representational error or inaccuracy. A sensory state can fulfill its function without being accurate, and it can fail to fulfill its function without being inaccurate. After all, fast but inaccurate sensory systems are often adaptive, just as slow but accurate systems are often maladaptive.[9] Burge claims that no account of perceptual representation in terms of practicality or usefulness is adequate.[10] Yet TS theories are just that sort of theory. They attempt to account for perceptual representation in terms of practical success. For that reason, Burge thinks TS theories fail to account for perceptual representation.[11]

Burge does not hold that TS theories are completely off track. While they fail to state a sufficient condition for perceptual representation, they succeed in stating a necessary condition for it. The necessary condition is that, like all sensory states, genuine perceptual representations have a biologically evolved function of indicating a state of the environment.[12] But only some such states are genuinely representational. So at this point, the key question is: What else must be true of a sensory state for it to qualify as a genuine perceptual representation? What distinguishes genuine perceptual representations from mere sensory responses?

2.2 Veridicality Conditions and Constancy Mechanisms

Burge proposes that perceptual representations differ from mere sensory responses in virtue of having what he calls *veridicality conditions*, conditions under which the state is accurate or true. He thinks that sensory states manage to have such conditions if they are exercises of what he calls *constancy mechanisms*. In short, if a sensory state is the exercise of a constancy mechanism, then it is a perceptual representation. As Burge puts the point, "Perceptual constancies are the key to understanding the nature of perception" (2010, 349); "Their presence in a sensory system is necessary and sufficient for the system's being a perceptual system" (413).

Constancy is a pervasive feature of our perceptual engagement with the world. For instance, if you walk around a square table, viewing it from different angles and distances, the input to your visual system from the light reflected off the table varies considerably, producing a variety of different retinal images.[13] On their own, each of these individual images could equally have been produced by a differently shaped table, and the sequence of changing images could equally well have been produced by a table that is itself changing shape. Yet you are nevertheless able to perceive a *single* table with an *unchanging* shape; you are able to perceive *one constant* shape, not merely a sequence of *many different* shapes. This is a case of shape constancy, which is one of several visual constancies, including size, distance, and color constancy. Constancy is also exhibited by nonvisual perceptual systems. For instance, as you walk closer to or farther from a speaker emitting music at a set volume, you are able to have a constant or nonvarying perception of the music's volume. Your perception of the volume of the music does not change across the changing conditions in which you hear it; you are able to perceive it as remaining the same volume, not increasing or decreasing. In general, constancy "is nothing more or less than a stability in perceptual response across a range of varying perceptual conditions" (Cohen 2015, 624).[14]

Given that the input to our perceptual system at any given moment underdetermines what caused that input, how do we manage to have a determinate perception of one aspect of this input rather than any of the other aspects? Following Burge, we will call this problem the *underdetermination problem*.[15] Constancy requires solving the underdetermination problem.

The underdetermination problem is not easy to solve. As noted, even a simple, regularly shaped object (such as a square table) can produce a tremendous variety of retinal images under different viewing conditions. Thus it is tempting to assume that the ability to solve the underdetermination problem must involve a higher-level cognitive capacity for reasoning: a conscious, voluntary, and general ability to think thoughts like SQUARE TABLES SEEN FROM CHAIRS LOOK LIKE TRAPEZOIDS. Although tempting, this assumption is at odds with the dominant view in vision science for the last century and a half.[16] The dominant view holds that solving the underdetermination problem does not require such higher-level cognitive processes. Instead the underdetermination problem is solved by constancy mechanisms that are unconscious, automatic, and domain specific.[17]

A constancy mechanism is a capacity to generate a single type of response to varying proximal input in such a way that this response tracks or correlates with some single type of distal cause. Constancy mechanisms solve the underdetermination problem by relying on what Burge calls *formation principles* that transform proximal input into responses to distal causes.[18] Here is how Burge describes these transformations:

These laws or law-like processes serve to privilege certain among the possible environmental causes over others. The net effect of the privileging is to make the underdetermining proximal stimulation trigger a perceptual state that represents the distal cause to be, in most cases, exactly one of the many possible distal causes that are compatible with (but not determined by) the given proximal stimulation. (2010, 92)

Formation principles transform proximal input into responses to distal stimuli by privileging some possible causes over others. As a way of explaining how they work, Burge compares formation principles to *filters* that eliminate a host of possible distal stimuli.[19] Obviously, randomly eliminating possible distal stimuli will not work. Formation principles work—they solve the underdetermination problem—by disregarding unlikely distal stimuli. More exactly, formation principles exploit the fact that because we tend to be in certain specific kinds of environments and tend to move around in them in certain specific kinds of ways, many possible distal stimuli are extremely unlikely to be causes and can simply be discounted.[20] For example, in our world, there are not very many table-like objects that constantly change shape. Formation principles solve the underdetermination problem by simply disregarding unlikely possibilities.

Relying on formation principles, constancy mechanisms solve the underdetermination problem without any conscious work or reasoning by perceivers. In Burge's words:

Formation principles describe and explain laws instantiated in transformations in the system. They are not *applied* in reasoning or cognition, even "implicit" reasoning or cognition, within the system. Thinking of them as applied by the system hyper-intellectualizes the system. (2010, 97)

Formation principles are not rules that we perceivers entertain and follow but simply lawlike regularities that are exhibited by constancy mechanisms. They do not involve any sort of general ability to reason about how one-and-the-same object will look or sound across varying conditions. Rather, they are specific to individual types of constancies (e.g., shape, color, volume, etc.), automatic, and instantiated independently of any general reasoning ability. In a word, constancy mechanisms are modular.[21]

Burge appeals to constancy mechanisms to explain perceptual representation, and specifically to explain how they have veridicality conditions. Why does Burge think that exercises of constancy mechanisms have veridicality conditions? At bottom, it is because he thinks they distinguish between *perceiver-independent* facts or properties and *perceiver-dependent* facts or properties; according to him, there is a sense in which they distinguish appearance from reality.[22] Consider again the visual experience of the shape of the table. For a shape constancy mechanism to solve the underdetermination problem in that case, allowing us to have a perception of a square table rather than a trapezoidal one, the mechanism must be able to distinguish the real, unchanging, perceiver-independent shape of the table from its changing, perceiver-dependent appearances across different viewing conditions.

Burge is not alone in thinking that genuine perceptual representation requires being able to distinguish reality from appearance. For instance, two of Burge's primary opponents, P. F. Strawson and Gareth Evans, also think that perception requires it. As Evans puts this point, genuine perceptual representation involves being able to distinguish between "that of which there is an experience (part of the world) and ... the experience of it (an event in the subject's biography)" (1996, 277).[23] As Burge sees it, however, making such a distinction does not require general, higher-order, or conscious thinking. Instead Burge contends that constancy mechanisms do this when they distinguish distal causes from proximal sensory activity. Burge calls this ability *objectification*:

Objectification lies in marking off states that are *as of* specific system-independent elements in the environment from states idiosyncratic or local to the perceiver. ... Objectivity is the product of separating what occurs on an individual's sensory surfaces from the significance of those stimulations for specific attributes and particulars in the broader environment. In this way, perception is the product of objectification. (2010, 400)

Because an exercise of a constancy mechanism "objectifies"—because it distinguishes the distal world from proximal effects on the organism—it purports to be of or about some distal stimulus, such as an object's shape. Because it does that, it can be accurate or inaccurate; it can get the distal stimulus right or wrong. Thus, according to Burge, it has veridicality conditions. By contrast, mere sensory systems do not objectify; they cannot distinguish perceiver-independent, distal causes from perceiver-dependent, proximal sensory input. Thus they do not have veridicality conditions.

2.3 Burge's Account of Misrepresentation

If exercises of constancy mechanisms have veridicality conditions, then it must be possible for them to be accurate as well as inaccurate. But what is it for an exercise of a constancy mechanism to be inaccurate?

Here is a short answer: misrepresentation occurs when a constancy mechanism fails to solve the underdetermination problem because it is exercised in abnormal conditions.

Here is a longer answer: A constancy mechanism has the function of solving the underdetermination problem, of tracking some single distal cause across changes in sensory inputs. It relies on formation principles to do this. Formation principles determine these distal causes by privileging certain possible causes over a host of possible causes. In particular, they privilege the statistically likely, or normal, distal causes of sensory inputs. However, since the actual cause of a type of sensory input is not always the normal cause of that type of input, formation principles are fallible.[24] Thus a constancy mechanism can respond as though the cause of the input is something normal when in fact the cause is not something normal. In other words, Burge holds that a constancy mechanism can identify something as the distal cause of a sensory input that is not the cause of that input. Since the function of the constancy mechanism is to determine the distal cause of an input to the sensory system, in these cases, the constancy mechanism has erred; it has gone wrong. For that reason, Burge contends that it is possible for constancy mechanisms to misrepresent.[25]

In sum, Burge holds that perceptual misrepresentation occurs under the following three conditions: there is a constancy mechanism that has the function of responding to a type of distal cause across changes in sensory input; that constancy mechanism is exercised in a particular instance; in this particular instance, the conditions under which it is exercised are abnormal, leading it to be exercised in response to something other than the type of distal cause that it has the function of responding to, thereby leading it to fail to solve the underdetermination problem.

2.4 Summary

To see what is distinctive about Burge's view, compare it with the views he rejects. Burge holds that IT theories do not make room for the possibility of error; as such, they do not make room for the possibility of misrepresentation. He admits that TS theories make room for the possibility of error, but he alleges that this sort of error is not the same as misrepresentation. Failing to fulfill a function is failing to do what's useful; and that need not have anything to do with accuracy or inaccuracy. Burge nevertheless holds that TS theories provide a necessary condition for perceptual representations: like all sensory states, genuine perceptual representations have the function of indicating a state of the environment (2010, 317). It is only when such sensory states have the more specific task of solving the underdetermination problem that they count as representations, states with veridicality conditions.[26] In the next section, we argue against this aspect of Burge's account.

3 Constancy Is Not Enough for Perceptual Representation

Burge's account has two problems. First, if mere sensory responses are not representational, then neither are exercises of constancy mechanisms, since the differences between them do not suffice to imply that one is representational and the other is not. Second, taken by themselves, exercises of constancy mechanisms are only derivatively representational, so merely understanding how they work is not sufficient for understanding what is required for something, in itself, to be representational.

3.1 Proximity Is Irrelevant

The gist of our first criticism is that mere sensory experiences and exercises of constancy mechanisms stand or fall together, and since Burge gives good reasons for holding that mere sensory experiences are not representational, he should not hold that exercises of constancy mechanisms are either.

For Burge, exercises of constancy mechanisms and mere sensory responses are similar in that they both have indicator functions (that is, the function of indicating something). Having a function brings with it the possibility of malfunction. Accordingly, a first key similarity between exercises of constancy mechanisms and mere sensory responses is that they both allow for the possibility of misindication. Consider, for example, a paradigmatic example of a mere sensory response: the shiver response.[27] Warm-blooded animals shiver in response to hypothermia. Shivering has the function of indicating that the organism's core body temperature has dropped below the level required for normal metabolic activity. It is possible, therefore, for shivering to misindicate a drop in core body temperature (if it is triggered in the absence of such a drop). Given this, it would be a mistake to think that exercises of constancy mechanisms differ from mere sensory responses on the grounds that exercises of constancy mechanisms are uniquely capable of misindication. This is something they both share.

There is a second, subtler similarity between exercises of constancy mechanisms and mere sensory responses. The possibility of misindication requires a difference between (a) being in some state or exhibiting some response, and (b) correctly or incorrectly being in that state or exhibiting that response. For example, in the case of the shiver response, whether the organism is shivering must be independent of whether the shivering is correctly or incorrectly indicating a drop in core body temperature. The fact that a warm-blooded animal is shivering does not by itself imply that its core body temperature has dropped. So it would be a mistake to think that this distinction between (a) and (b) holds uniquely for exercises of constancy mechanisms. The distinction holds equally well for mere sensory responses.

Let us return now to why Burge supposes that mere sensory responses differ from exercises of constancy mechanisms. Burge thinks that exercises of constancy mechanisms are distinctive because they distinguish between appearance and reality; they "objectify." By this, he means that exercises of constancy mechanisms distinguish proximal effects on the organism from the distal causes of those effects. For instance, an exercise of a shape constancy mechanism indicates (and has the function to indicate) the shape of a distal object, not the shape of the retinal image caused by the object. Mere sensory responses do not do that; they only have the function to indicate proximate stimuli. For instance, the shiver response has the function to indicate a proximal state of the organism, not a distal state in the organism's environment. So, for Burge, the key difference between exercises of constancy mechanisms and mere sensory responses is that exercises of constancy mechanisms have the function of indicating distal stimuli, whereas mere sensory states have the function of indicating proximal stimuli.

Burge then contends that this difference is decisive for the issue of perceptual representation. Because exercises of constancy mechanisms objectify and thereby indicate distal stimuli, not proximal stimuli, they have accuracy conditions and are thereby representational. We think Burge's reasoning here is mistaken. Our key question for Burge is: why think a difference in *proximity* is what matters for having accuracy conditions?

Burge's argument appears to proceed as follows:

1. If something has the function to indicate distal stimuli, as opposed to proximal stimuli, then it is able to distinguish between perceiver-independent and perceiver-dependent states of affairs.

2. Being able to distinguish between perceiver-independent and perceiver-dependent states of affairs is akin to distinguishing between reality and appearance.

3. Being able to distinguish reality from appearance suffices for having accuracy conditions.

4. Exercises of constancy mechanisms have the function to indicate a distal stimulus.

5. Thus they have accuracy conditions.

While this argument is valid, its second premise is dubious. What reason do we have for thinking that something that has the function of indicating a distal stimulus distinguishes between appearance and reality? Burge might contend that something that has the function of indicating a distal stimulus (e.g., the shape of a distal object) rather than a proximal stimulus (e.g., the shape of a retinal image) is able to distinguish distal things from proximal things. That might be true, but it does not follow that such a thing distinguishes appearance from reality (or has the function to do so). Distinguishing between proximal and distal things is not the same as—and does not suffice for—distinguishing appearance from reality. It merely involves being able to distinguish two, equally real, things: proximal stimuli and distal stimuli. Furthermore, the fact that the organism's or system's response to the distal stimuli is mediated by the proximal stimuli does not make either sort of stimuli any less real. In short, it is a mistake to think that the shape of a retinal image is merely apparent; the shape of the retinal image is just as real as the shape of the object that causes it.

Underlying the idea that exercises of constancy mechanisms distinguish between appearance and reality is a further mistake. It requires the proximal stimulus (e.g., the retinal image) to be an appearance. But for something to be an appearance, it must be a representation; it must appear to represent things as being a certain way.[28] However, Burge rightly does not think that proximal stimuli (such as retinal images) are themselves representational; he does not think that they have accuracy conditions. But then he cannot allow that proximal stimuli are appearances. So the idea that exercises of constancy mechanisms distinguish between appearance and reality is deeply problematic.

Why else might one think that the proximity of a stimulus matters? One might think it matters because exercises of constancy mechanisms are mediated responses, whereas mere sensory responses are immediate responses. That is, one might contend that an exercise of a constancy mechanism is representational because there is an intermediate state between an exercise of a constancy mechanism and the state of affairs that it has the function of indicating. For instance, consider once again an exercise of a shape constancy mechanism. One might contend that what matters is that a retinal image mediates the connection between an exercise of a constancy mechanism and the shape that the exercise has the function of indicating. But why think the presence of an intermediary matters? As we have just noted, the intermediary is not, in any sense, representational. Even Burge himself does not think that it is.

One might think instead that the crucial difference is that exercises of constancy mechanisms have the function to indicate something despite *variation* in an intermediary. For instance, an exercise of a shape constancy mechanism has the function to indicate the shape of a distal object despite variation in the retinal images caused by the object. Again, why think such variation matters? Would exercises of a shape constancy mechanism cease to be representational if the proximal retinal image did not vary? That seems implausible. Although not decisive, it helps to think about cases such as spheres, whose retinal images do not vary across changes in line of sight. Would that suffice to show that perceptions of spheres are not representational? Are perceptions of cubes representational, while perceptions of spheres not? In short, it seems implausible to think that whether or not the intermediate state varies is what makes the difference between representational sensory responses and nonrepresentational sensory responses.

Summing up, Burge says there is a difference in kind between two sorts of sensory responses: one sort is representational; the other is not. One sort is representational because it has the function of indicating a distal stimulus; the other is not representational because it has only the function of indicating a proximal stimulus. We have claimed that this difference does not suffice for the first sort of sensory response to be representational. The proximity of a stimulus seems to be irrelevant to whether or not a response to that stimulus is representational. We acknowledge that there are further differences between exercises of constancy mechanisms and mere sensory responses. But these differences do not seem to suffice for exercises of constancy mechanisms to be representational. Thus no good reason has been given for thinking that merely being an exercise of a constancy mechanism suffices for having accuracy conditions. (Perhaps such exercises do have accuracy conditions when they are suitably involved in larger systems.)

3.2 Constancy Mechanisms Are Only Derivatively Representational

The gist of our second criticism is that exercises of perceptual constancy mechanisms are only derivatively representational, whereas Burge's proposal needs them to be *non*derivatively representational. To spell out this criticism, we introduce a pair of distinctions. The first is Haugeland's important and influential distinction between original and derivative intentionality. The second is our own distinction between original and derivative beholdenness to norms of accuracy. In brief, we argue that perceptual constancy mechanisms are only derivatively representational *because* they are only derivatively beholden to norms of accuracy.

The first distinction is Haugeland's distinction between original and derivative intentionality. "Intentionality" is just another word for being about something, in precisely the way that representations are about something. Here is how Haugeland introduces this distinction:

Intentionality … is not all created equal. At least some outward symbols (for instance, a secret signal that you and I explicitly agree on) have their intentionality only derivatively—that is, by inheriting it from something else that has that same content already (such as the stipulation in our agreement). And, indeed, the latter might also have its content only derivatively, from something else again; but, obviously, that cannot go on forever.

Derivative intentionality, like an image in a photocopy, must derive eventually from something that is not similarly derivative; that is, at least some intentionality must be original (nonderivative). And clearly then, this original intentionality is the primary metaphysical problem; for the possibility of delegating content, once there is some to delegate, is surely less puzzling than how there can be any in the first place. (*HT*, 129)

In short, Haugeland's distinction is between intentionality that is derived from something else, and intentionality that is not so derived—or, as he calls it, *original*.

We take for granted that in aiming to solve the problem of perceptual representation, Burge aims to show how acts of perception have original intentionality, not merely derivative intentionality. The problem of perceptual representation is a difficult problem precisely because it requires making sense of the original intentionality of acts of perception. (Derivative intentionality is easy to explain.) Our second criticism of Burge is based on the claim that exercises of constancy mechanisms do not have original intentionality, but at most derivative intentionality. Our argument for this claim has two steps: first, original intentionality requires being beholden to norms of accuracy in a certain way; second, constancy mechanisms are not beholden to norms of accuracy in that way.

To understand the first step of our argument, think about the phenomenon of being beholden to norms more generally. Consider two cases. In the first case, Nat says, "The window should be shut." In this case, the window is being held to a norm. It should be shut. But the window, in itself, is not capable of modifying its own behavior in response to a failure to abide by this norm. Of course, something else can modify the window's behavior (say, if Wyeth shuts it), but the window alone cannot do it. As such, the window is only derivatively beholden to this norm. In the second case, Nat quietly says to himself, "I should shut my mouth." Here Nat is being held to a norm. And it is he who is holding himself to it, in thinking and saying it. (Whether he complies is another matter.) In this case, he is nonderivatively or originally beholden to a norm.

In general terms, a system S is nonderivatively or originally beholden to a norm N if and only if S can modify its own behavior in response to failures to abide by N. For a system to be nonderivatively or originally beholden to a norm of *accuracy*, therefore, it must be possible for that system to modify its behavior when it is inaccurate. Accordingly, we think that original intentionality—hence the sort of intentionality that Burge thinks perceptual representations have—requires original beholdenness to norms of accuracy.

To understand the second step of our argument, consider what constancy mechanisms can and *cannot* do. Burge is right to be impressed by constancy mechanisms. He is also right to think that earlier discussions of perceptual representation—by both philosophers and psychologists—failed to see what is impressive about constancy mechanisms, because they did not fully appreciate how constancy mechanisms involve sensory states that have the function of responding to distal stimuli across changes in sensory input. Peter Schulte provides a useful summary of how our understanding of constancy mechanisms has shifted in this respect:

In the early days of neurological research on frogs, scientists thought that certain retinal ganglion cells, which respond strongly to the movement of small, dark spots on the retina, were almost solely responsible for generating the frog's feeding response. ... These neurons were colloquially called "bug detectors."

But it soon became clear that matters were not that simple. Further research showed that frogs do not respond to all small, dark spots moving on their retina. If these spots are caused by large objects that are far apart, frogs usually show no feeding response. And conversely, they do respond to large, dark spots if these spots are caused by small objects that are very close. In other words, frogs respond to the real size of objects, not to the size of retinal stimuli.[29] (Schulte 2012, 489)

This summary is doubly useful. First, it is useful because it perfectly captures what Burge thinks is significant about constancy mechanisms for understanding perceptual representation. Second, it is useful because, like Burge, it glaringly overlooks a crucial feature of constancy mechanisms. That feature is underconstancy.[30]

Underconstancy is simply the fact that well-functioning constancy mechanisms do not always track the same distal stimuli across changes in proximal stimuli.[31] Size constancy mechanisms, for instance, often systematically under- and overestimate the sizes of objects, in predictable ways. That is, built into the way in which constancy mechanisms track distal stimuli is the fact that they often end up tracking a range of distal stimuli, owing to the effect of differences in proximal stimuli on their responses. This means that it is false to say, as Schulte does, that size-distance constancy mechanisms "respond to the *real* size of objects, not to the size of retinal stimuli" (2012, 489; our italics). It is false because size-distance constancy mechanisms only respond to the real size of objects *some* of the time. At other times, their responses are affected by the size of retinal stimuli. As Mazviita Chirimuuta puts this point in general terms, "Constancy is not perfect" (2015, 55). In sum, Burge overemphasizes the degree to which constancy mechanisms exhibit constancy, thereby overlooking the effect that differences in proximal stimuli sometimes have on the exercise of constancy mechanisms.[32]

Underconstancy exposes a problem for Burge's account because it reveals how constancy mechanisms are not beholden to norms of accuracy *in the right sort of way* to produce genuine perceptual representations. What is the right sort of way? Burge himself does not raise this question, because he does not distinguish two important ways in which representations can be beholden to norms of accuracy. This is a fatal oversight for Burge's theory, because any talk of representations (perceptual or otherwise) being beholden to norms of accuracy is ambiguous.

Since constancy mechanisms are unable to respond to failures of constancy, they are unable to correct themselves in response to errors. They simply do not have any ability to modify their responses in response to underconstancy. For a sensory state to be a genuine perceptual representation, however, we think it would have to be *part* of a system that is non-derivatively beholden to a norm of accuracy. Constancy mechanisms *by themselves* are not such a system.[33]

To prevent misunderstanding, two qualifications are in order. First, underconstancy in itself is not the problem. It is how constancy mechanisms *respond* (or do not respond) to underconstancy that is the problem.

We are not claiming that representational systems must be perfect, or that they must always be accurate. Far from it: we agree with Burge that genuine representational systems must make room for the possibility of misrepresentation. Rather, we are saying that such systems must be able to respond to errors for such errors to count as inaccuracies.

Second, it is not that constancy mechanisms do not, in fact, respond to underconstancy that is the problem. The problem is their *inability* to respond to underconstancy. We are not saying that representational systems must always respond to inaccuracies. We agree with Burge that representational error is in principle independent of practical consequences (see sec. 2.1.2). So we agree with Burge that a system can misrepresent how things are without that misrepresentation leading to any actual consequences whatsoever. Furthermore, it would be absurd to think that for a representational system to be beholden to a norm of accuracy, it must always correct its misrepresentations, for two reasons. First, that a system is in error need not entail that it recognizes that it is in error. Second, it would imply that uncorrected misrepresentations are not representations. Speaking generally, the fact that a system is beholden to a norm does not imply that it must actually correct failures to abide by that norm. Again, our claim is only that a system must be *able* to respond to errors for such errors to count as inaccuracies.

Summing up, our second criticism of Burge is that exercises of constancy mechanisms do not suffice for perceptual representation, for they do not have original intentionality but are only derivatively beholden to norms of accuracy. Constancy mechanisms are not able to correct themselves in response to their own failures to be accurate.

4   Conclusion

In this essay, we have summarized Tyler Burge's recent proposal for solving the problem of perceptual representation and drawn upon John Haugeland's work to offer a pair of criticisms of Burge's proposal. We will close by spelling out why we think that the failure of Burge's account is nevertheless an excellent starting point for solving the problem of perceptual representation.

Exercises of constancy mechanisms, in themselves, do not have veridicality conditions. Something more is needed. But what?

Much of Burge's rhetoric in *Origins of Objectivity* makes it seem as though the only alternative to his view of perceptual representation is a hyperintellectualized view, according to which genuine perceptual representation requires that a perceiver be able to represent, to herself, the conditions under which her perceptions would be accurate or inaccurate.[34] This is, self-evidently, an individualistic and intellectualized view of perceptual representation, one that would most likely imply that only humans are capable of genuine perceptual representation. But, following Haugeland, we think that it is a mistake to assume that the only way in which exercises of constancy mechanisms might be supplemented (to produce genuine perceptual representations) is for a perceiver to be able to represent to herself the veridicality conditions of her perceptions. There are alternatives that do not have this manifestly individualist and intellectualist form.

For an example of nonindividualist, nonintellectualist, nonhuman behavior that makes a good starting point for thinking about perceptual representation, consider the case of vervet monkeys. Vervet monkeys are social animals, living in troops, who use a system of alarm calls to alert the troop to different types of threats, thereby eliciting in the troop different types of evasion responses.[35] It is widely accepted that vervets are genetically predisposed to make such calls as well as respond to them. But some evidence seems to indicate that adult vervets train their infants to make such calls only in response to actual predators, by punishing incorrect calls and rewarding correct calls.[36] If this evidence is correct, then it suggests that vervet monkeys *might* have visual experiences that are nonderivatively beholden to norms of accuracy. As such, these experiences *might* be genuine perceptual representations. Whether this is, in fact, true of vervet monkeys is an open question that deserves to be examined at length (in another essay). But the fact that it might be true shows that Burge's rhetoric relies on a false dichotomy. There are not only two options, Burge's view or a hyperintellectualized view.

The failure of Burge's account is not a dead end. Our critique is compatible with thinking that exercises of constancy mechanisms, when part of larger systems, suffice for perceptual representation. In particular, our critique is compatible with thinking that exercises of constancy mechanisms, when part of a system of social interactions that are nonderivatively beholden to norms of accuracy, suffice for the resulting sensory states to be genuine perceptual representations. Thus we believe our critique points the way forward in thinking about the problem of perceptual representation. The key question now is this: what more is needed, beyond exercises of constancy mechanisms, for sensory states to be nonderivatively beholden to norms of accuracy?

Notes

1. We use "indicates" and its cognates roughly as Dretske does in his 1988, which builds on his 1981. As Dretske there observes, indication is essentially the same as "natural meaning" or "meaning$_N$" in Grice 1957. We follow Shannon and Weaver 1949 in referring to these sorts of theories as "information-theoretic" theories.

2. For discussion of the varieties, see Godfrey-Smith 1992 and Neander 2012. For our purposes here, we are overlooking the subtle but important distinctions between these varieties.

3. More precisely, a token of type X's being F indicates a token of type Y's being G when tokens of type X being F are caused by tokens of type Y being G. Although it can matter in other contexts, unless we say otherwise, we will generally abstract away from the distinction between types and tokens.

4. This objection is normally attributed to Fodor (1990). For Burge's formulation of it, see Burge 2010, 299.

5. Not everyone accepts this conclusion. Fodor (1990) offers a sophisticated version of IT theory, one that strives to make room for the possibility of error.

6. The most influential TS theories, those of Millikan and Dretske, draw on etiological theories of biological functions. See Wright 1973; Millikan 1989; Neander 1991; Griffiths 1993; Walsh and Ariew 1996; Buller 1998.

7. This is a rough gloss on functions. For further discussion, see Sober 2000 and Sterelny and Griffiths 1999.

8. Dretske 1988 is the best example of this specific sort of TS theory. For a contrasting sort of TS theory, see Millikan 1984.

9. See, e.g., Trimmer et al. 2008.

10. Burge 2010, 301.

11. As with Burge's criticism of IT theories, not everyone accepts this criticism of TS theories. See Neander 2012 for a sophisticated defense of TS theories.

12. Burge 2010, 317.

13. This is not to say that we perceive these images on our retinas. Visual experience is produced by the light that is absorbed by our retinas, not the light reflected by our retinas. That said, speaking of the "images on our retinas" is useful shorthand for speaking of the patterns of proximal sensory activity that take place when we visually perceive something. But we should not be misled by this shorthand: we simply do not perceive the curved, inverted images in our retinas.

14. For Burge's formulation, see Burge 2010, 388. For formulations by other authors, see Palmer 1999, 312–313, 723; Byrne and Hilbert 1997, 445; and Goldstein 2009, 218, 410.

15. For Burge's formulation, see Burge 2010, 344. See also Palmer 1999, 23, 716.

16. Palmer (1999) gives an especially clear introduction to this tradition.

17. Burge summarizes what he takes the dominant view to involve in Burge 2010, 101–103.

18. For Burge's formulation, see Burge 2010, 345.

19. As Burge puts it, "Formation principles describe processes that begin with selective filtering of the initial sensory registration" (2010, 92). See also Burge 2010, 274, 282, 285, 371. In some respects, filters are a useful metaphor for beginning to explain how formation principles solve the underdetermination problem. However, this metaphor should not be taken too literally or without qualification. Consider, for example, a standard filter, such as a HEPA air filter, which blocks the passage of airborne particles that are larger than 0.3 microns. HEPA filters do this by being presented with airborne particles of a variety of sizes and allowing only those that are smaller than 0.3 microns to pass through. Filters like this are very different from formation principles, because formation principles are not presented with all the possible distal stimuli. Possibilities, as such, do not exist for formation principles, so it is misleading to say that formation principles "filter" some of them out. Nevertheless, such metaphorical talk can be helpful shorthand for the way in which formation principles take an input that is indeterminate between many distal causes and generate an output that tracks just one distal cause.

20. Burge 2010, 100.

21. We ourselves think that the extent to which perceptual constancy mechanisms are modular is more contested in contemporary vision science than Burge himself seems to think, but for the purposes of this essay, we will follow him in characterizing this as the consensus view. Our own criticisms of Burge do not depend on contesting the extent to which perceptual constancy mechanisms are modular.

22. Burge 2010, 397.

23. Similarly, Strawson argues that genuine perceptual representation involves an ability to draw a "distinction between oneself and one's states on the one hand, and

anything on the other hand which is not either oneself or a state of oneself, but of which one has, or might have, experience" (1959, 69).

24. For Burge's formulation, see Burge 2010, 346.

25. Burge 2010, 98, 344.

26. Burge 2010, 303, 309.

27. See, e.g., Hemingway 1963; Jensen 2001.

28. This claim is neutral with regard to debates about whether appearances are exhausted by their representational contents.

29. For a useful summary of research on frog size-distance constancy mechanisms, see Ingle 1998.

30. For a useful historical survey of studies of underconstancy, see Wagner 2012.

31. Palmer says that underconstancy "is incomplete *perceptual constancy* in which the perception is a compromise between *proximal* and *distal* matches" (1999, 733). Palmer also characterizes underconstancy as the way in which constancy mechanisms exhibit "systematic deviations from accurate perceptions of objective properties in the direction of proximal image matches" (314). As will become clear, we think that Palmer's use of "accurate" in this characterization is ambiguous between two readings of what it is for a sensory state to be beholden to a norm of accuracy: namely, between being originally and derivatively beholden.

32. Another way to put this point is to note that many perceptual illusions are the direct result of the ways in which constancy mechanisms solve the underdetermination problem. Here's how Palmer puts this point:

Somewhat surprisingly, the same processes that usually result in constancy—that is, veridical perception—sometimes produce illusions. The reason is that ... veridical perception of the environment often requires heuristic processes based on assumptions that are usually, but not always, true. When they are true, all is well, and we see more or less what is actually there. When these assumptions are false, however, we perceive a situation that differs systematically from reality: that is, an illusion. In several cases we will find that in the perception of object properties, constancy and illusion are therefore opposite sides of the same perceptual coin. (Palmer 1999, 313)

Once again, as with "accurate," we think Palmer's use of words like "veridical" and "true" is ambiguous (see the previous note). It is understandable that Palmer would use such words, because they are far and away the easiest way to describe constancy and underconstancy. Nonetheless we think using words like "accuracy," "veridical," and "true"—without disambiguating what one might mean by them—can be misleading.

33. This is not to say that they cannot be part of such a system. Burge's account of perceptual representation treats constancy mechanisms as modular, self-contained

representational systems. Our goal in this essay is to show that such an account fails. We ourselves think that constancy mechanisms are often parts of representational systems. We describe the conditions under which this obtains in Adams and Maher 2012.

34. Burge refers to this sort of hyperintellectualized account of perceptual representation as *individual representationalism*, the view that "an individual cannot empirically and objectively represent an ordinary macro-physical subject matter unless the individual has resources that can represent some constitutive conditions for such representation" (2010, 13).

35. Seyfarth, Cheney, and Marler 1980; Bradbury and Vehrencamp 2011.

36. See Caro and Hauser 1992 for a summary of the possible evidence; but see Thornton and Raihani 2008 for some reservations.

References

Adams, Z., and C. Maher. 2012. Cognitive spread: Under what conditions does the mind extend beyond the body? *European Journal of Philosophy* 20 (4): 420–438.

Bradbury, J., and S. Vehrencamp. 2011. *Principles of Animal Communication*, 2nd ed. Sunderland, MA: Sinauer.

Buller, D. 1998. Etiological theories of function: A geographical survey. *Biology and Philosophy* 13 (4): 505–527.

Burge, T. 2010. *Origins of Objectivity*. New York: Oxford University Press.

Byrne, A., and D. Hilbert, eds. 1997. *Readings on Color*, vol. 2. Cambridge, MA: MIT Press.

Caro, T. M., and M. D. Hauser. 1992. Is there teaching in nonhuman animals? *Quarterly Review of Biology* 67:151–174.

Chirimuuta, M. 2015. *Outside Color: Perceptual Science and the Puzzle of Color in Philosophy*. Cambridge, MA: MIT Press.

Cohen, J. 2015. Perceptual constancy. In *Oxford Handbook of Philosophy of Perception*, ed. M. Matthen, 621–629. Oxford: Oxford University Press.

Dretske, F. 1981. *Knowledge and the Flow of Information*. Cambridge, MA: MIT Press.

Dretske, F. 1988. *Explaining Behavior*. Cambridge, MA: MIT Press.

Evans, G. 1996. Things without the mind. In *Collected Papers*, by G. Evans. New York: Clarendon.

Fodor, J. 1990. *A Theory of Content and Other Essays*. Cambridge, MA: MIT Press.

Godfrey-Smith, P. 1992. Indication and adaptation. *Synthese* 92 (2): 283–312.

Goldstein, E. B. 2009. *Sensation and Perception*, 8th ed. New York: Wadsworth.

Grice, H. P. 1957. Meaning. *Philosophical Review* 66:377–388.

Griffiths, P. 1993. Functional analysis and proper functions. *British Journal for the Philosophy of Science* 44 (3): 409–422.

Haugeland, J. 1998. *Having Thought*. Cambridge, MA: Harvard University Press. (Abbreviated as *HT*.)

Hemingway, A. 1963. Shivering. *Physiological Reviews* 43 (3): 397–422.

Ingle, D. 1998. Perceptual constancies in lower vertebrates. In *Perceptual Constancy: Why Things Look as They Do*, ed. V. Walsh and J. Kulinkowski. New York: Cambridge University Press.

Jensen, C. 2001. *Temperature Regulation in Humans and Other Mammals*. Berlin: Springer.

Millikan, R. 1984. *Language, Thought, and Other Biological Categories*. Cambridge, MA: MIT Press.

Millikan, R. 1989. Biosemantics. *Journal of Philosophy* 86 (6): 281–297.

Neander, K. 1991. The teleological notion of "function." *Australasian Journal of Philosophy* 69:454–468.

Neander, K. 2012.Teleological theories of mental content. In *Stanford Encyclopedia of Philosophy*. http://plato.stanford.edu/entries/content-teleological/ (accessed September 1, 2012).

Palmer, S. 1999. *Vision Science*. Cambridge, MA: MIT Press.

Schulte, P. 2012. How frogs see the world. *Philosophia* 40 (3): 483–496.

Seyfarth, R. M., D. L. Cheney, and P. Marler. 1980. Monkey responses to three different alarm calls: Evidence for predator classification and semantic communication. *Science* 210:801–803.

Shannon, C., and W. Weaver. 1949. *The Mathematical Theory of Communication*. Urbana-Champaign: University of Illinois Press.

Sober, E. 2000. *Philosophy of Biology*. 2nd ed. Boulder, CO: Westview Press.

Sterelny, K., and P. Griffiths. 1999. *Sex and Death*. Chicago: University of Chicago Press.

Strawson, P. F. 1959. *Individuals*. New York: Routledge.

Thornton, A., and N. J. Raihani. 2008. The evolution of teaching. *Animal Behaviour* 75:1823–1836.

Trimmer, P., A. Houston, J. Marshall, R. Bogacz, E. Paul, M. Mendl, and J. McNamara. 2008. Mammalian choices: Combining fast-but-inaccurate and slow-but-accurate decision-making systems. *Proceedings of the Royal Society B: Biological Sciences* 275 (1649): 2353–2361.

Wagner, M. 2012. Sensory and cognitive explanations for a century of size constancy research. In *Visual Experience: Sensation, Cognition, and Constancy*, ed. G. Hatfield and S. Allred, 63–86. New York: Oxford University Press.

Walsh, D. M., and A. Ariew. 1996. A taxonomy of functions. *Canadian Journal of Philosophy* 26 (4): 493–514.

Wright, L. 1973. Functions. *Philosophical Review* 82:139–168.

9 Recording and Representing, Analog and Digital

John Kulvicki

In the final analysis, however, a definition should be more than merely adequate to the intuitive data. It should show that the cases cited are instances of a theoretically interesting general kind, and it should emphasize the fundamental basis of that theoretical interest.

—John Haugeland, "Analog and Analog" (*HT*, 75)

How does aboutness reside in witless processes? Questions like these are often cast in terms of worries about naturalization. How could matter be formed—over evolutionary time, in a network of norms—so as to be about something else? Alongside these worries are concerns about how to understand distinctions between representational kinds. What makes something an image, graph, piece of language, analog, or digital? In some logical sense, these challenges are shallower than those posed by naturalization, but in another sense they are closer to home. Representations shape the world we have made, and nonintentional processes shape those representations. The following isolates a nonintentional process, recording, and shows that it is central to understanding the varieties of aboutness.

Haugeland (*HT*, chap. 8) built an account of representational kinds around the structures of their contents. Icons, which include pictures, images, graphs, and diagrams, differ from logical representations like linguistic expressions because iconic contents are essentially patterns of dependent and independent variables. Such structures are absent in language. Recording is a witless, replayable process that Haugeland deployed to calm worries about his content-wise account of representational genera. The following elevates recording from the supporting role Haugeland gave it to star of the show.

The main claim I defend here is that some kinds of representation—pictures, images, graphs, diagrams—are *modeled by recording processes*, while others, like languages, are not. Extensionally, this distinction is close to Haugeland's, but it is intensionally subtler, more plausible, and, as I hope to show, more useful. This approach abandons Haugeland's goal of distinguishing representational kinds exclusively in terms of their contents, but there are many advantages to doing so.

Section 1 explains the distinction between recording and representing. Section 2 then introduces Haugeland's distinction between skeletal and fleshed-out content. This allows section 3 to express what it means for representations to be *modeled by recording processes*. This takes a good bit of setup, because the main thesis uses many concepts that Haugeland developed in a manner distinct from how he used them. The remainder of the paper puts this notion to work.

Section 4 unpacks Haugeland's account of representational genera, and section 5 considers a new way of understanding them. Section 6 shows that recording sheds light on the analog-digital distinction. Representations modeled by recordings have features characteristic of analog representations, while those that are not must be digital. This is interesting because it grounds the difference between analog and digital in the interpretive demands of users. Section 7 looks more carefully at accounts of digital representation and shows how they are built around satisfying the interpretive demands outlined in the section 6. Section 8 wraps up the discussion by considering how we ought to think about digital representation in digital devices like computers and smartphones.

1   Recording versus Representing

Anything can represent anything else, or so they say. Being a representation is having the status of standing for something, and anything can be granted that status. The kitchen table can represent the Almighty, an Assyrian king, or a wombat. Representation might fail absent an agreeable audience, or the right kind of baptismal song and dance, so it's an interesting question just what has to happen to make the kitchen table represent something, but it's not the question that matters here. The point here is that the qualities of the kitchen table play little role in such enfranchisement; it's as good as anything else. Imperceptible things might be poor choices as representations, but nothing in principle prevents us from agreeing that Earth's magnetic field represents a Jovian moon. Representation is cheap, at least once creatures like us are on the scene. The right kinds of intention readily collaborate to confer such status.

Not anything counts as a representation of the Almighty in English or in German, however, and graven images are more than arbitrary scribbles. Norms govern depiction, languages, diagrams, and all other representational families, genera, and species. Conferring representational status in a one-off fashion is easy, but representing in a useful system is hard. Very hard. And the only representations we care much about are those that belong to useful systems. This is exactly what one would expect, and as it turns out, the notion of recording will help us understand what makes systems of representation useful.

Recording is difficult, like depicting something or describing it in grammatical English, because recording builds in standards for production and use. These standards depart from those in place for representing. While representations have an intentional character, recordings are *relational*. The relation between a recording and what it records is *witless*, and it allows *playback*. Let us consider each of these three features in turn, with a couple of examples. One is a recording of Whitman's voice, etched on Edison's wax wheel. The other is a daguerreotype of Lincoln.

Recordings are related to events. You don't have a recording absent something that it records. The source might cease to exist—Whitman and Lincoln have long since passed—but status as a record nevertheless requires a source. Lincoln stood before the daguerreotype that bears his likeness, and Whitman could have touched the wax cylinder as he spoke. This is just one condition on being a recording, and the following conditions will require being careful about what exactly this condition requires.

The process whereby something is recorded is *witless* in that it is "oblivious to content (if any) and ignorant of the world" and thus does not depend "crucially and thoroughly on general *background familiarity* with the represented contents—that is, on worldly experience and skill" (*HT*, 180). Cameras and Edison's cylinder phonograph make recordings. The process is causal, and as long as everything is working properly, no wits are required. The camera doesn't care whether it's Lincoln or a landscape out there, and it doesn't try to get things right. The photographer might care about her daguerreotypes, but once the cap is off the lens, she is taken out of the picture. Wait, and hope Lincoln doesn't sneeze. Wits might be prerequisite to making such machines, but recording processes don't require those wits. Smart machines, if that is what most of us are, can also make recordings. For present purposes, not much rides on the subtle boundary between witless and mindful processes, because recording processes are clearly on the witless side.

Among the many witless processes to which things can be subjected, recordings are those that allow *playback*. This feature is what makes recording distinctive, and as such it is important for what follows. Playback is a witless process whereby what is recorded can be reproduced. Turn the wax cylinder under a needle, and the result is a replica of the pattern's cause. It's like hearing Whitman speak. A digital camera saves a file, which can then be used to create an image. The image is a replica of its cause much as the audio recording is. In daguerreotypes, the pattern burned into a sheet of silver records a pattern of light and dark and also serves as a playback of that pattern, because it *is* the pattern of light and dark that was recorded. Just look, and you see, reproduced, the pattern that caused it.

Sometimes the recording is one step away from playback, as with wax cylinders and JPEGs. The record-*ing* in these cases is not the same pattern as the one that gets recorded, but the right witless apparatus allows one to reproduce the recorded pattern. Sometimes the recording itself constitutes playback of the pattern, as with the daguerreotype. *Copying* happens when the recording and playback processes coincide in this manner. Consider another example. Copying an inscription is producing a recording of it, in that one can easily produce another such inscription based on the copy. But the copy itself is an instance of the inscription, so it is in effect a playback of a recording too, which can itself be recorded and played back over and over: "a copy of a copy is a copy" (*HT*, 180). Along these lines at least, copying an inscription is akin to making an image with the daguerreotype. The recording process is also a playback process. I will return to the distinction between copying and merely recording in the following sections. For now, we need to get clearer on the possible scope of recordings.

Some things cannot be recorded, because there is no way to capture their features via a witless, replayable process. Being a U.S. dollar bill requires the consent of the U.S. Treasury. One can easily record the pattern of light and dark that constitutes a bill's surface. It might even be possible to record the paper's characteristics. But it is impossible to record dollarhood. No witless

process can take a recording and produce a dollar bill. Similar considerations apply to humans and wombats, if being such requires the right
phylogeny and ontogeny.

This limit on recording matters quite a bit because it means that particular individuals, like Whitman, cannot be recorded. We record the voice in
the sense that we record the sound pattern he produces. That can be played
back, but Whitman cannot. Neither can Lincoln. So while recordings are
relational in that they require a source, that source, in its particularity, is
not what is recorded. The recording captures features of something—an
instance of a kind—that can be reproduced: an audible pattern in Whitman's case, and a visible pattern in Lincoln's. Recording, in this sense, is
primarily concerned with witlessly transmissible patterns, not connections
to particular individuals.

Witlessness does not entail practicability. It might be impressively difficult to produce a recording of something, and even more difficult to play it
back. The key is that the mess is a witless mess. The recording and playback
processes we employ, of course, tend to be practicable.

Recording, in this technical sense, fits with some of the everyday things
we are willing to say about recording, but not with all of them. Whitman
can, in the everyday sense, be recorded, even though the man cannot be
played back. The photo is a record of Lincoln's presence in a photo studio,
even though that particular event cannot be reproduced by the image. In
the technical sense, we can record patterns of features that can enter into
causal relations and thus participate in witless processes. The recording
allows reproduction of an instance of a kind. Those kinds that can witlessly
be produced are the kinds that can, in that sense, be recorded.

Recording also fails to coincide with two closely related technical
notions: information and Gricean natural meaning. It is possible to carry
information about particular individuals, like Whitman and Lincoln
(Dretske 1981). In fact, it's likely that the recordings carry information
about what Whitman said, what Lincoln wore. A family photo carries information about one's ancestors on their wedding day. The right kinds of
counterfactual support allow information to flow, but they are helpless
when faced with demands for playback. Similar remarks apply to Gricean
natural meaning. Grice suggests that a photo naturally means something
about a tryst (1957, 382–383), but there is no witless way to reproduce such
a thing, at least not from a photo.

Recording is not offered as a replacement for information or natural meaning. It's just another nonintentional process that relates in interesting ways to representation. Information and natural meaning are used to illuminate (Grice) or serve as a reduction base (Dretske) for intentionality of all sorts. Recording will help us distinguish kinds of representation from one another in a manner neither of the other notions is suited to doing. It does this, surprisingly, because of the two-way street required by playback.

Finally, recording has been introduced here using two examples that are also *representations*. At a minimum, it is common to think about audio recordings and daguerreotypes in that way. But nothing has been said about the relationship between what these things record and what they represent. As we will see, not everything they record figures in their representational contents, and not everything they represent is recorded. But recording nevertheless plays an interesting role in representations, which will become clear in section 3. Before getting there, we need one more idea from Haugeland's tool kit.

2   Skeletal Content

Before I can show how recording can help us understand representation, we need to understand the distinction between skeletal and fleshed-out content. A representation's skeletal content is "not augmented or mediated by any other" (*HT*, 185), and its fleshed-out content is the augmented mess.

A simple remark can convey worlds of content. "The letter is on the table" can be a simple answer to a simple question or can suggest that the utterer is too busy to fetch the letter herself, that someone forgot to post it, that the spy is dead, and so on. Its boring skeletal content is that the letter is on the table. The complete message fleshes out these bare bones.

Only if "implication" is stretched to the full gamut of what can be expected from conversational skills, topical associations, critical judgments, affective responses, and so on, can it hope to encompass the sort of content routinely conveyed in the newspaper, in the boardroom, and over the back fence. (*HT*, 187)

The distinction between skeletal and fleshed out content is supposed to apply to all kinds of representation and help to distinguish kinds of representation from one another.

The skeletal contents of photographs, for example, are nothing but "variations of incident light with respect to direction" (*HT*, 189). Rarely do we interpret a photo as representing nothing but variations of incident light—it's Lincoln!—but such patterns are the core contents of any picture. The formula for showing this is simple. First, peel off the layers of implication, in its broadest sense, by imagining alternative, acceptable fleshed-out contents. Then, with these alternatives in mind, ask what they all have in common. The portrait depicts Lincoln, but it's possible that an exactly similar daguerreotype could have been the result of focusing on a wax model of Lincoln, an alien impostor, or even another daguerreotype of Lincoln. These disparate objects have in common only the way that they project onto a surface to form the picture. That common ground is the content's skeleton, that on which all the fleshings-out must build.

It's easy to suggest that photos, audio recordings, and daguerreotypes have skeletal contents of the sort just mentioned. The mechanical nature of their production lends itself to such abstractions. Even handmade pictures, in this view, have skeletal contents, and their skeletal contents are akin to those of photos (see Kulvicki 2006, chap. 3). The photo's skeletal content does not rely on its being a recording, though as the next section shows, these skeletal contents reveal a distinctive relationship between photos and recording processes.

3   Representations Modeled by Recordings

Section 1 ends asking after the connection between what some things record and what they represent. Section 2 articulates a distinction between skeletal and fleshed-out content. This section suggests, first, that many representations are interesting because their skeletal contents are exhausted by what they record. Not everything recorded need show up in skeletal content, but everything in skeletal content is recorded. Second, another class of representations is such that every aspect of their skeletal contents is record-*able*, even if these representations are not themselves recordings. All together, these representations form an interesting class corresponding to what Haugeland called icons and what I have called the images, generally speaking (Kulvicki 2014). These representations are all, as I will explain, *modeled* by recording processes. The relationship between recording and this class of representations has not heretofore been noticed.

The previous section suggests that a representation's skeletal content is quite limited. In the case of photographs, for example, it amounts to little more than a pattern of color, or in some cases a pattern light and dark. Many scenes are projectively compatible with these patterns, and so such photos can be fleshed out in a number of ways. Similar remarks apply to audio recordings. They capture certain sound patterns skeletally, though they are understood to represent much richer scenes than that. So how does recording relate to what such things represent?

First, notice that everything included in the skeletal contents of photos and audio recordings can be recorded. Skeletal contents, in those cases, amount to patterns of features that can be multiply instantiated. Second, these features are all plausibly features that such representations in fact record in Haugeland's sense of the word. The photo is witlessly connected to a pattern of light and dark, and when we view the photo, we encounter just that pattern. The wax cylinder witlessly registers a sound pattern, and playing it back delivers just that pattern again. It might be that audio recordings and photos record a bit more than we take to be included in their skeletal contents, but the point for now is that everything that's a good candidate for being included in skeletal content is also recorded.

Even some handmade images fit this pattern. Chuck Close's paintings after photos of his friends were so carefully done that they might be just as much recordings as the photos on which they are modeled. That's not to say Close is witless, so much as that his process, mindfully adopted, left his wits to one side. Drafting techniques like lithic illustration are governed by norms so strict that if properly executed they record prehistoric tools' features (Lopes 2009).

The connection between recording and representing in these cases is striking. Not only is some part of the photo's content also recorded, but an important part of it is: its skeletal content. The same holds for audio recordings. Skeletal content forms the core of any representation's content, and in these cases, it is wholly the result of a recording process.

Other kinds of pictures, like most paintings and handmade drawings, fail to be recordings. Tenniel's illustrations of Carroll's *Through the Looking Glass* (1871) do not record what they represent. They are expressions of his imaginative efforts and drafting skill, but they are not related, for example, to the Jabberwock via a witless process. They represent, but do not record, looking-glass beasts. Nevertheless the norms governing these pictures' interpretations are consistent with them having been recordings, in the following sense. Even imaginative drawings have skeletal contents, and these skeletal contents—patterns of light and dark—could have been recorded in such a way that the recording would be indistinguishable from Tenniel's picture. Put differently, the skeletal contents of these representations are recordable, even if his images are not recordings.

Some representations, like photographs and audio recordings, are built around recording processes. Others, though not built around recordings, are such that their skeletal contents are recordable. This turns out to be an interesting family of representations, all of which are *modeled by recording processes*. The sense of modeling I have in mind is not terribly complex. The idea is just that such representations can be understood as built around a skeletal core that is recordable. It might not be that the skeletal contents of all such representations are recordings, but they are all scaffolded by contents that can be recorded.

There are interesting and uninteresting cases of representations modeled by recording processes. A description like "the red round thing" has a fairly straightforward skeletal content, which could be recorded. That is, one can set up a witless process that produces "red round object" in response to red, round things. That description can then be fed into a machine that produces red, round things in response. But this is not generally true for linguistic representations. Some expressions, like those involving names, pick out particular individuals, and we have already seen that they cannot be recorded. Even if we stick to descriptions that pick out recordable qualities, there is no particularly systematic process at work. There could not be, it seems, if only because any recordable property is readily represented linguistically, and any number of words can stand for the same quality.

The interesting cases of representations modeled by recording processes meet the following two conditions. First, the representation is part of a system, all of whose representations have skeletal contents that can be recorded. Second, the recording process in question is not terribly gerrymandered. These two conditions interact. Simple skeletal contents can be the result of simple recording processes. The next two sections unpack these claims in more detail. It turns out that the distinction between representations modeled by recording processes and the rest is quite helpful.

4   Recording and Haugeland's Genera

The previous section distinguishes representations based on whether they are modeled by recording processes. Haugeland distinguished kinds of representation, which he called *representational genera*, exclusively in terms of their contents. This section presents Haugeland's view and connects his distinction between representational kinds to recording. The next section offers an alternative that does justice to the motivations behind his view while evading worries about his approach (Kulvicki 2006, chap. 6).

For Haugeland, *iconic* representations, which include pictures, images, and many graphs and diagrams, have skeletal contents structured in terms of dependent and independent variables. The graph represents a series of times with temperatures attached. No time is represented as having two temperatures, though the same temperature might show up at more than one time. "Iconic contents might be conceived as variations of values along certain dimensions with respect to locations in other dimensions" (*HT*, 192). These contents can be complex—think of all the variations of color possible across any two-dimensional expanse—but they are in another sense simple for being built out of a limited number of dimensions.

The contents of *logical* representations are different: "The primitive elements of logical contents ... are always identifiable separately and individually. That is, they can enter into atomic contents one by one, without depending on their concrete relations to one another" (*HT*, 191). The complexity countenanced in logical representations is of a different sort than we find in icons. Any collection of elements can come together to form the content of a linguistic expression, for example, and in that sense their skeletal contents can be indefinitely variegated: Marco the elephant happily ice-skating on Mars while wearing a pink tutu, and thinking of Jerry Falwell. The fine detail of icons might exceed the typical reach of logical representations, but the dimensions of iconic skeletal content are always limited. The indefinite variability of logical content will be important for the discussion of digital representation in sections 6 and 7.

Extensionally speaking, Haugeland's proposal is rather good. Common practice distinguishes items in his class of icons from their logical cousins like descriptions, and that ordinary distinction has something to do with dimensionality. Icons represent patterns of variation in some dependent dimensions along other independent dimensions.

What Haugeland's proposal fails to capture is the way in which such things not only represent but *manifest* such patterns in a way deeply connected to their role as representations (Kulvicki 2006, chap. 6). A graph of temperature over time need not manifest an interesting pattern of temperature over time, of course, but it is a pattern of heights along an axis. The photo is a display of light and dark, and it represents such a pattern too. Once we see this, we can also see that most icons are linked to their (skeletal) contents via a fairly straightforward recording process. And if they are not recordings, they are at least modeled by relatively simple recording processes.[1] In fact, the aspect of iconic content that Haugeland placed center stage—variations of features along a few dimensions—means that all such representations are modeled by recording processes. This suggests an interesting way to distinguish kinds of icons from one another, which is the task of the next section.

5   New Representational Genera

By drawing his distinction between representational genera exclusively in terms of content, Haugeland ignored the fact that useful examples of such representations have syntactic qualities that are also patterns of features along multiple dimensions (Kulvicki 2006, chap. 6). Noticing this helps sort the icons into three finer classes, or new representational genera. They are genera because each of them also accommodates many species. What Haugeland called the icons are more like a family that itself includes many genera, and I will talk about family matters in the next section. The generic distinctions here depend on whether icons copy none, some, or all of the features that figure in their skeletal contents.

Some representations modeled by recording processes instantiate none of the qualities they represent. A graph of temperature over time can be any temperature you like: it's the form of line that counts. Two spatial dimensions have little to do with temperature and time, but they are excellent ways of recording temperature over time. One dimension maps onto time, the other onto temperature, and they do so in a structure-preserving manner. A location between two others along the time dimension stands for a time between the ones represented by the others. A location above two others along the other dimension represents a warmer temperature than the others. Similar remarks apply to diagrams and graphs of many sorts. They don't necessarily share qualities with what they represent—the graph's temperature has nothing to do with its content—but they do share a structure with it. This class of representation corresponds to what we usually call diagrams and graphs (e.g., Swoyer 1991; Barwise and Etchemendy 1995; Stenning 2002).

Another genus of representations modeled by recording processes instantiate some of the qualities they skeletally represent. The fMRI image has a shape of uniform color that stands for a region of activity shaped just that way in a subject's brain. The brain scan preserves structure, but it is also partly *mimetic* (Kulvicki 2006, chap. 4). Shapes of uniform color represent similarly shaped regions of uniform activity. In that sense, shape is not just recorded but copied, in that the recording itself manifests, and thus serves as a playback of, the shape it records. The color of the scan, by contrast, has nothing to do with colors in the brain, even though relations between the colors map onto relations between activities in a structure-preserving manner. Representations like this are commonly called images, though most people seem reluctant to call them pictures. They are, I suggest, the nonpictorial images.

At the extreme end of this spectrum, we have representations that instantiate all the qualities that constitute their skeletal contents. *Pictorial* representations exemplify this condition. Haugeland suggested that "all the photos 'strictly' represent is certain variations of incident light with respect to direction" (*HT*, 189). In fact, that they represent patterns of incident *light* is unclear. Closer to the truth is that they represent colored patterns of brightness, saturation, and hue. But this is exactly what the photos *are*. Both the scene depicted and the picture itself instantiate the qualities that constitute the skeletal content. In that sense, the entirety of the skeletal content is copied: the recording serves as a playback. Such representations are modeled by *copying* processes.

Haugeland distinguished representational genera exclusively in terms of their contents. The finer, threefold distinction just offered corresponds roughly to what we call graphs and diagrams, nonpictorial images, and pictures.[2] Such distinctions become possible when we notice, first, that all these representations are modeled by recording processes, and second, that there are different ways of modeling a system of representation with a recording process. In "Representational Genera," Haugeland wasn't after these fine distinctions, but his notion of recording offers a way to formulate them.

6   Family Matters

Representations modeled by recording processes can be distinguished based on whether they instantiate none, some, or all of the features that constitute their skeletal contents. Logical representations, exemplified by languages and their arbitrary pairings of words with things, also bear an interesting relationship to recording, and especially to copying. To see this, we need to consider a few points of contrast between logical and iconic representation.

First, representations modeled by recording processes have limited arbitrary links with their contents, corresponding to their syntactically significant dimensions of variation. For example, colors might represent temperatures, regions of enhanced brain activity, or porosity. Relative heights can represent relations between any values you like. In fact, any readily grasped range of properties can serve well as a stand-in for some other range of qualities, but representations modeled by recordings have a limited number of dimensions that do this. As a result, iconic representations are *generative* (Schier 1986): once one understands the few dimensions that matter, it's easy to interpret indefinitely many novel representations in the system.[3]

Second, even with such a limited palette of dimensions, representations like this can have impressively rich contents. Indefinitely fine differences in hue indicate fine differences in temperature, for example, beyond even what a careful observer could notice. Displayed in a surface map, subtle variations in hue matter across indefinitely small distances. The richness of such contents can hide the small number of arbitrary links that tether such representations to the world.

Third, representations modeled by recording processes are forgiving, interpretively speaking. The details of hue across space might be lost to the casual user of the temperature map, but she is nevertheless able to interpret it accurately and effectively. Often the fine details of the map don't matter, because the interpreter's interest is at some remove from them. Which quadrant is warmest? Is the pattern of hot and cold regular or haphazard? These questions can be answered even if the subtleties of hue across space are ignored. Such representations are readily interpretable even if their syntactic details are elusive. This is the mark of *analog* representation (Kulvicki 2010, 2015).

The situation is quite different with linguistic and other logical representations. First, their syntactic qualities are not easily analyzed in terms of a small set of dependent and independent variables, so they do not map readily onto dimensions of variation in content. Recording processes do not model such representations, and indefinitely many links tie logical representations to the world. The lexemes of a language must be learned by rote, so such representations fail to be generative.

Second, logical representations can have impressively detailed contents, but they purchase their subtlety with many parts, each arbitrarily paired with what it represents. Detail is captured not by subtle variation along very few dimensions but rather with many unrelated syntactic types, sometimes, as in language, shaped by grammar.

Third, under these circumstances, confusion over syntactic type has vast and unpredictable semantic consequences. Logical representations are thus profoundly unforgiving, interpretively speaking. "You wanted me to clean the *house*? I hosed down the horse!" "Moses didn't have horns?" Likeness, luckless, lackluster. Syntactic indeterminateness is as bad as straight-up confusion. Being unsure whether a token is this or that syntactic type leaves one semantically at sea. Approximations to the syntactic identities of these representations are in no way approximations to their semantic identities, so syntactic mistakes can be semantically disastrous.

Systems of logical representation must place a premium on reliable syntactic identification, lest they be useless. If you want an interpretable system tethered in indefinitely many ways to contents, then instances of its syntactic types must be readily identified. A related constraint is that it should be easy to produce instances of each type. This suggests that, in logical systems, instances of each syntactic type are readily recorded and indeed copied. In brief, systems of logical representation must be digital. "Digital devices are precisely those in which complex form is reliably and positively abstractable from matter" (*HT*, 82). This isn't an account of digital representation, but it shows the need that the digital satisfies, and it gestures at why copying is so important in such systems. The next section looks more closely at accounts of digital representation and makes the importance of copying clear.

7   Copying and Digital Representation

Goodman suggested that digital representational systems are, among other things, syntactically disjoint and syntactically finitely differentiated. A *disjoint* system is just one in which no mark, say, an inscription, actually belongs to more than one syntactic type (Goodman 1968, 132–133; see also *HT*, 78). So if the Roman alphabet is disjoint, then there is no mark that is simultaneously an *A* and a *B*. Some scribbles fail to be any letter, some manage to be one, and none manage to be two or more. A disjoint system is *finitely differentiated* when:

> For every two characters K and K* and every mark *m* that does not actually belong to both, determination that *m* does not belong to K or that *m* does not belong to K* is theoretically possible. (Goodman 1968, 135–136; italics removed)

Characters, in Goodman-speak, are syntactic types, like the first two letters of the written alphabet, the written word "dog," and so on. Differentiation concerns whether users can in principle determine the character membership, if any, of marks. If the Roman letters are finitely differentiated, for example, then for any mark it will be possible to determine either that it is not an *A* or that it is not a *B*. This condition holds "for every two characters," so for any mark it's possible to determine that it is not one of at least twenty-five letters. Most scribbles fail to be any letter at all, and we can determine that this is so, but (1) all marks fail to be twenty-five (disjointness), and (2) we can determine which twenty-five any given mark fails to be, at least in principle (finite differentiation).[4]

It is not always possible to determine that a mark does not belong to any of the twenty-six letters, even if it belongs to none of them. This is because it will always be possible to produce marks that are indefinitely close to being *A*s, for example.

> There is no way of preventing this infiltration at the borders, no way of ensuring that due caution will protect against all mistakes in identifying a mark as belonging or not belonging to a given character. (Goodman 1968, 134)

The important point is that there is clear blue water between the types. Anything that one might confuse with an *A* had better not be something one might also confuse with a *B*, or any other letter, for that matter. Similar remarks apply to written words.

In "Analog and Analog" (*HT*, chap. 4), Haugeland agrees that "in digital devices, the main thing is eliminating confusion over the type of any token" (*HT*, 82). He departs from Goodman in focusing on the production of marks, *writing*, as well as their consumption, *reading*. Digital systems are such that it is easy to produce marks that are members of each syntactic type. Instances of syntactic types must be producible and legible via positive and reliable procedures: "one which can succeed absolutely and without qualification" and "which, under suitable conditions, can be counted on to succeed virtually every time" (*HT*, 77). The procedure for writing *A*s is *positive* in that the result of writing can be an *A*, not just something that approximates an *A*. The procedure is also *reliable* in that ordinary individuals, in appropriate circumstances, can carry it out. The same is true for reading, in that it's possible to identify something as an *A*, and most of us can do so under appropriate circumstances.

Haugeland's conditions are stricter than Goodman's in one sense and more forgiving in another (*HT*, 79–80). First, what is in principle possible might escape us in practice. For Goodman, positive reading procedures suffice, even if they are unreliable in that the intended users of such representations might have trouble reading them. Haugeland is thus stricter than Goodman as far as reading goes, and he also insists on conditions that govern writing. But bringing writing into the mix actually loosens Goodman's in-principle constraints on reading, in the following way.

In digital systems, it is easy to identify members of a type, in part *because* it's easy to produce them. Easy production reduces the preponderance of infiltrating border cases and also diminishes the importance of finite differentiation. Reasonable effort is all that is required to write Roman letters. We succeed in making *A*s when we want them, virtually every time. Since most *A*-attempts succeed, most things that seem to be *A*s are *A*s, not borderline nonsense. Also, there are marks that come uncomfortably close to being *A*, *B*, and *R* all at once. The Roman alphabet might not be finitely differentiated. But writers can readily avoid making those marks. Contested borders between syntactic types are of little practical concern as long as no one produces tokens anywhere near the border. By insisting on a positive and reliable "write-read cycle" (*HT*, 79), Haugeland avoids the need for finite differentiation.

A simple virtue of Haugeland's approach is that standards for reading affect standards for writing and vice versa (*HT*, 79–80). For example, the

letter identification game requires more care with the pen than word identification does. Written words are complexes of letters in most writing systems. It's a problem for letter identification if an *A* looks a lot like an *R*, but
much less of a problem if a borderline letter appears in a word, because
word identification is typically easy even absent a clear sense of each letter.
In isolation, the scribble might look somewhere between a script *m* and *n*,
but in the word "rhyming" that mess is unimportant: it's easy to read the
word. Syntactic misidentification of words is typically disastrous, but this
does not mean that each letter need be carefully crafted. In this way, the
syntactic complexity of written words makes them relatively easy to write.
But this ease of writing is only possible because of how easy it is to read
words written from many letters (cf. Goodman 1968, 137–139).

We can now easily see the role that copying plays in digital representation. First, these representations are the kinds of things that can be recorded.
That is, the features that make something the word or letter it is can be
parts of the witless replayable processes required by recording. Syntactic
misidentification is disastrous in systems that are not modeled by recording
processes. But syntactic identification is precisely the kind of thing that can
be done witlessly. Combine with this the need to produce such tokens, and
we get the requirement that digital representations are readily copied. An
inscription is both a representation in language and a record from which
one can reproduce another inscription of the same kind.

A survey of representational kinds typically locates pictures and written
languages about as far from one another as possible. For Goodman, pictures
were syntactically and semantically dense and relatively replete, while
notational systems, the antipodes of pictures, were syntactically and semantically finitely differentiated and not relatively replete. Ordinary languages
failed the semantic test of finite differentiation but were nevertheless quite
far from depiction. Diagrams, graphs, ordered lists, and the like are hybrid
forms that occupy the space between.

Pictures, as we saw earlier, are modeled by copying processes. In an
important sense, they reproduce what they represent. And that's a long way
from written words, arbitrarily paired with content. But copying plays an
important role in written languages too. While copying places pictures
close to what they represent (they reproduce parts of their contents), it
places interpreters close to words (we readily produce tokens of any given

syntactic type). Difficulty identifying and producing words makes a system, already tenuously tethered to its contents, completely unusable.

8   A Digital Revolution?

The last two sections motivate digital representation by appeal to what happens when indefinitely many representations pair arbitrarily with contents. Under such circumstances, the high cost of syntactic misidentification forces representations to be readily recognized and produced, or *copied*. Writing was thus the first great digital technology.

Today, increasingly dense collections of transistors have reshaped the world in what some regard as a digital revolution. This digital revolution is not about the ordinary tasks of writing and reading, however. No one reads the electrical signals that make computers work, and production is always a highly mediated affair. The bit stream bears at best a tenuous relationship to the swipes and clicks constitutive of electronic input, not to mention the text, audio, and video outputs that constitute the most appealing aspects of fancy electronic technology.

In these cases, both reading and writing are taken out of our hands. We read text and interpret the photos, videos, and audio tracks these machines present, of course, but the machines work with a different code. "In digital devices, the main thing is eliminating confusion over the type of any token; and a primary motivation (payoff) for this is the ability to keep great complexity manageable and reliable" (*HT*, 82). When we write with pen and paper, the complexity is the lexemic mess characteristic of any language. With computers, it is the vast number of files and instructions that must frequently be reproduced. We might be deeply confused about whether one photo, video, or audio recording is the same as another, of course, even though these things are digitized. The precise identification of the photos that a computer stores is not important for ordinary interpreters, but it is essential for the computer to do what we need it to do.

For example, it's no easier to recognize all the details of a five-megapixel picture than it is to notice those of a daguerreotype. Both are beyond our powers of discernment, and practically speaking, no one is in a position to notice all the details of either one. Digital photos are important not because we interpret them differently from film photos or daguerreotypes but because making them digital allows their "great complexity" to be managed. We can store, retrieve, and share digital photos much more straightforwardly than their old-school kin. The consumers of these digital things, qua digital representations, are the computers, not their users.

From a user standpoint, one of the great advantages of the electronic revolution is that it makes analog representations like photos, videos, and audio recordings readily stored, copied, and reproduced. That is, it confers on analog representations many of the virtues of their digital kin without thereby removing the virtues of analog representation. This isn't so much a digital revolution—that happened with the invention of writing—as a *digitizing* revolution. As I've argued elsewhere (Kulvicki 2015), it's even reasonable to understand digital photos and audio tracks as analog representations, although they are also digital.

9 Conclusion

Recording processes play a heretofore unnoticed role in representation. Some representations, Haugeland's icons, are modeled by recording processes. They can have impressively detailed contents, but they are tethered to the world with relatively few arbitrary links. Some of them—pictures— are even modeled by copying processes: they instantiate their skeletal contents. They are interpretively forgiving, in part because of their limited arbitrary links to content. Representations not modeled by recording processes are linked in many arbitrary ways to contents. Such representations are uninterpretable absent precise syntactic identification because simple syntactic mistakes have vast semantic consequences. Such representations must themselves be readily recorded. Recording thus shapes our representational practices.

It has been common to link the demand for perfect copies with digital representation. This paper shows the need such copying fills, and it reveals the role that copying and, more generally, recording play in those representations that are often not readily copied. It is interesting that the representations most distant from the digital—pictures—are the ones modeled by copying processes. This is just what we would expect, however, once we see the dual roles copying plays in pictures, on the one hand, and written language, on the other. Among representations, pictures bring us closest to their contents by, in part, reproducing them. We cannot perfectly copy most pictures, but that is of little consequence because minor failures to notice pictorial detail have at best minor interpretive consequences. Linguistic representations leave us at the farthest remove from content because of their many arbitrary links. Here the potential for copying representations is essential because even minor syntactic mistakes leave us at a complete loss interpretively.

Acknowledgments

I thank audiences at Aarhus University and Lingnan University for helpful comments on earlier drafts of the paper, and thanks to Zed Adams and Jacob Browning for detailed comments on a late draft.

Notes

1. Some simple pairings of dimensions are impossible to interpret. Represent ranges of brightness with saturations, or conversely, and you will be lost. It's difficult to distinguish those two in the first place, and almost impossible to focus on variations in one independently of variations in the other. Other pairings fail to be user-friendly while also being awfully complex. Link temperatures haphazardly with heights along the y-dimension of a plane, and do the same for times along the x-dimension. The result is uninterpretable, at least for ordinary mortals, and the process is also a mess. There are, unsurprisingly, many ways to make useless representations. The ones that succeed pair easily noticed syntactic dimensions in a simple manner with the semantic dimensions of interest.

2. Elsewhere (2006, chap. 4) I unpack this set of distinctions in different but complementary terms. The point here is that we can relate these distinctions to recording, which is something that has not been mentioned before.

3. This is, as far as I know, a new way to think about why certain kinds of representation are generative. Both Schier's (1986) explanation and Lopes's (1996) happen at a much higher level. They suggest that pictorial representations exploit our abilities to recognize things in the wild, so basic familiarity with depiction naturally generates the ability to interpret pictures of anything one can already recognize.

4. The Roman alphabet probably fails this test, at least in general. It's close enough to passing it, however, to remain useful, as we will see hereafter.

References

Barwise, J., and J. Etchemendy. 1995. Heterogeneous logic. In *Diagrammatic Reasoning*, ed. J. Glasgow, H. Hari Narayan, and B. Chandrasekaran, 179–193. Menlo Park, CA: AAAI/MIT Press.

Carroll, L. 1871. *Through the Looking Glass, and What Alice Found There*. London: Macmillan.

Dretske, F. 1981. *Knowledge and the Flow of Information*. Cambridge, MA: MIT Press.

Goodman, N. 1968. *Languages of Art*. Indianapolis: Hackett.

Grice, H. P. 1957. Meaning. *Philosophical Review* 66 (3): 377–388.

Haugeland, J. 1998. *Having Thought*. Cambridge, MA: Harvard University Press. (Abbreviated as *HT*.)

Kulvicki, J. 2006. *On Images: Their Structure and Content*. Oxford: Clarendon Press.

Kulvicki, J. 2010. Knowing with images: Medium and message. *Philosophy of Science* 77 (2): 295–313.

Kulvicki, J. 2014. *Images*. London: Routledge.

Kulvicki, J. 2015. Analog representation and the parts principle. *Review of Philosophy and Psychology* 6 (1): 165–180.

Lopes, D. 1996. *Understanding Pictures*. Oxford: Clarendon Press.

Lopes, D. 2009. Drawing in a social science: Lithic illustration. *Perspectives on Science* 17 (1): 5–25.

Schier, F. 1986. *Deeper into Pictures*. Cambridge: Cambridge University Press.

Stenning, K. 2002. *Seeing Reason*. Oxford: Oxford University Press.

Swoyer, C. 1991. Structural representation and surrogative reasoning. *Synthese* 87:449–508.

IV Two Dogmas of Rationalism

10   Two Dogmas of Rationalism

John Haugeland

Rationalism and empiricism were both inspired by the scientific revolution—but with different emphases. Empiricists were most impressed by the new experimental method: the idea that all scientific results are ultimately supported by observation (via "the senses"). Rationalists, by contrast, were more impressed by the mathematical form of those results—hence their amenability to formal constructions, derivations, and proofs.

What follows is an attempt to expose two covert "dogmas"—tendentious yet invisible assumptions—that underlie rationalist thought, both modern and contemporary. Though neither term is perfect, I will call these assumptions *positivism* and *cognitivism*. For historical continuity, my focus will remain on science.

1   Positivism

As I intend the term, positivism is a metaphysical assumption—a view about the essential structure of the world as such. Picturesquely, it is the view that reality is "exhausted" by the facts—that is, by the true propositions. (Note that "positive" and "propose" stem from the same root.) Thus an elegantly austere expression of positivism is the opening line of Wittgenstein's *Tractatus*: "The world is everything that is the case" (1974, 5).

Now, this needn't mean that there are no things, properties, or relations, but only that these are intelligible only as constituents of true propositions. Things-bearing-properties and things-standing-in-relations just *are* facts—the facts that F(a), R(a,b), and so on. In the same spirit, but the formal mode: if to be is to be the value of a bound variable, then to be is to be an argument for which the value of some propositional function is "true."

Clearly, if this equation is intended as an essential conception of the world, then it requires an independent understanding of propositions. That understanding is invariably discursive: a proposition is something that could (in principle) be said—that is, articulately "pro-posed" or "put forward" by somebody as the content of an explicit claim. And that articulability, in turn, is cashed out in terms of finitely specifiable interpreted formal systems—of which natural languages are approximations, and the symbolic notations of mathematics and logic are exemplars.

The germ of this idea goes back to the dawn of modern science and rationalism. Thus Galileo said already in 1623:

> Philosophy [that is, physics] is written in that great book, the universe, which is always open, right before our eyes. But one cannot understand this book without first learning to understand the language and to know the characters in which it is written. It is written in the language of mathematics, and the characters are circles, triangles, and other figures. (1623/1968, 232)[1]

It isn't the "book" image per se that's significant here, but how Galileo intends it. For astrologers and alchemists had long undertaken to decipher the "book of nature"—by which they meant finding and decoding covert signs and hidden messages about other things that they were independently interested in.

Galileo, by contrast, is looking not for meaning in the world but rather for the world's own intrinsic structure—not just in space but also in time. Thus he achieved an unprecedented abstraction in allowing the length of a line to represent elapsed time, the slope of another line to represent rate of acceleration, and the area of the enclosed triangle to represent distance traversed, and then being able to prove, geometrically, a general relation among them. That's the point of saying that the universe *itself* is an open book, written in the language of mathematics (geometry).

But though Galileo spoke of a "language" and achieved an important innovation in abstract representation, he did not have our conception of a formal system, with syntactical rules and arbitrary symbol types. Thus, if you systematically permute his "characters"—lines, circles, and triangles, say—the constructions won't work.

So it was more nearly Descartes who could first understand and defend positivism. For he showed that a symbolic formalism—algebra—could express "analytically" (formally) the axioms, theorems, and proofs of traditional geometry. Therefore, his doctrine that the essence of the real world is extension amounts to the claim that reality as such is formally describable in principle. And that, close enough, is the idea that later became dogmatic positivism—an idea that Newton, Leibniz, and many others then greatly extended the scope and plausibility of.

It remained to Kant, however, to appreciate that positivism is a substantive metaphysical thesis, and so to argue for it—at least as regards the empirical world. For the transcendental deduction of the categories is nothing other than an attempt to prove that there can be nothing to the empirical world except such as can be the content of a judgment.

Now, to those thinking backward from the twentieth century, deeming Kant a hero of positivism may come as a surprise. For the Vienna school—the so-called *logical* positivists—often defined themselves in opposition to Kant. But their objection was not to Kant's positivism—they took that so much for granted that they no longer saw it as a substantive thesis, never mind in need of proof. Rather, they protested the transcendentalism and, especially, the idea of synthetic knowledge a priori. Encouraged by recent developments in physics and mathematical logic, they held that all synthetic knowledge is a posteriori, and all a priori knowledge is analytic.

Thus, they also thought of themselves as rehabilitating a scientific empiricism. But that could never have been an empiricism in the eighteenth-century sense, because impressions and ideas can't support the logical form required for the predicate calculus, nor could association based on similarity and constant conjunction ever amount to valid inference in it. Yet if sensory impressions are epistemologically forfeit, what remains of empiricism, after all?

Remarkably, it was several decades before Sellars brought the issue into focus and pointed the way out. If experience is to provide rational grounds for empirical knowledge, then it must itself already be in "the logical space of reasons"; that is, it must already be conceptually articulated with the logical form of the factual. That left it to McDowell to extend the argument beyond experience to the entire world, insofar as it is knowable at all. "The realm of the conceptual," McDowell says, "is unbounded on the outside"—a self-conscious reformulation of Wittgenstein's positivist slogan (1996, 157).

2   Scientific Know-How

The "everything" in "everything that is the case" means both "all" and "only." It's not just that all facts (and their constituents) are included, but also that nothing else is. The world is *exclusively* things bearing properties and standing in relations. In the present section, I will question this latter "nothing but" clause. That is, I will try to show—specifically in connection with empirical science—that there must be more to the world than what is the case.

As empirical, science depends essentially on learning or finding out about the world by way of observation, measurement, and, above all, experiment. As Sellars showed, if the results of these endeavors are to be intelligible as evidence—for or against theories, for instance—then they must themselves be in the space of reasons. In other words, only facts (true propositions) can be rationally probative.

Gathering evidence, however, is not the only way of learning or finding out about the world that is essential to empirical science. Scientists must also, for instance, learn or find out *how* to make or perform the observations, measurements, and experiments that yield that factual evidence. Yet if that is so, then that learning-how cannot itself be just more evidence gathering, on pain of regress. On the face of it, therefore, what is learned or found out about the world, in learning or finding out how to do something, must be different in kind from learning or finding out what is the case— that is, the facts. But it will be worthwhile to spell this out.

The result of learning or finding out how to do something—how to gather evidence, say—is (or at least includes) what Ryle called knowing-how, as opposed to knowing-that. These are alike, of course, in that they are species of knowing—that is, ways of being onto the world, and vulnerable to error. So, to cement a case against positivism, two points need to be secured: first, that empirical know-how embodies genuine knowledge of the world; and second, that what is known in such know-how is not exclusively factual.

Empirical scientific know-how is a cultural-historical achievement: it is the product of great effort by many individuals over many years. Such a history implies not only that know-how is difficult to develop but also that not just anything will work. Reliable laboratory procedures are highly constrained, and just how they are constrained is what must be learned or found out, by the relevant community, in learning or finding out how to design and perform them. The same point can be cast comparatively, in terms of learning or finding out what will work better, as opposed to not so well. But what will work or not in actual practice, or work better than something else, is a function of the world. Therefore, in learning what will and won't work, or what will work better—that is, in acquiring the relevant know-how—scientists are learning something about the world.

It remains to show that what is known in such practically acquired know-how is not exclusively factual. Insofar as the skillful know-how of empirical technique is embodied in a cultural-scientific heritage, it must be handed down to each succeeding generation—from experts to novices. But this handing down cannot be entirely verbal: the heritage of expert practice cannot be reduced to a text. Rather, as every science teacher knows, the students must themselves "practice," in the sense of performing hands-on exercises. First they watch while the tutor demonstrates, and then they try it themselves, repeatedly, under critical supervision. As their skills improve, the exercises get harder, and their standards go up. No amount of lecturing or assigned reading can substitute for such firsthand laboratory experience—which is why science education always includes both. Thus, phenomenologically, what is known in know-how cannot be reduced to verbally articulable facts.

It might be held, however, that the phenomenology is not decisive. Perhaps the limited articulability of practical skills merely reflects a limitation of available vocabulary and/or other linguistic resources. It could be, in other words, that a more refined and developed scientific language would enable scientists to make their practical know-how fully explicit, even though they can't actually do that now. Then educational science labs wouldn't really be necessary—lectures and textbooks would suffice, after all.

But this suggestion merely moves the bump in the rug, for language itself is an essentially skillful capacity—one that has to be learned and that, once learned, embodies knowledge of the world. To take the simplest example, learning to tell a duck when you see one is (part of) learning how to use the word "duck"—an instance of verbal know-how that embodies, as we might put it, knowing what a duck looks like. Without a great deal of such learning, students wouldn't even be able to read the textbooks, never mind use what they read in their own future practice.

Thus, supervised and ever-more-sophisticated laboratory exercises are indispensable not only for training the hands and eyes but also in teaching the very language and vocabulary in which empirical work can be described and reported.

Yet if such prerequisite skillful know-how, linguistic or otherwise, embodies a kind of worldly knowledge that cannot itself be expressed in words or formulas, then what is known of the world in that knowledge-qua-know-how is not facts. I tentatively conclude, therefore, that empirical scientific know-how constitutes a counterexample to metaphysical positivism.

Why only tentatively? Because the argument so far has presupposed a fundamental distinction, the credentials of which have not been established. In introducing know-how as a species of knowing, I pointed out that it is vulnerable to error, and elaborated by noting that, in practice, not just anything will "work." These are hardly controversial claims. Yet, in a discussion of the essence of science as science—that is, as knowledge—they cannot simply be taken for granted. Before this issue can properly be addressed, however, several other points will need to be spelled out. For the moment, therefore, we are left with an unanswered question: How is experimental failure to be understood?

3   Scientific Laws

By positivist lights, scientific laws are facts. That is, statements or mathematical formulations of them state or formulate facts. And from one point of view—the point of view from which these statements or formulas merely tell us what is, was, or will be the case—no one could disagree. Yet even positivists agree that laws are not ordinary facts. In the present section, I consider the ways in which laws might be extraordinary.

The usual view is that statements of laws do not specify merely what happens to be the case, whether locally or globally, but rather what must be the case, always and everywhere. Thus, what's extraordinary about laws is that they are necessary facts. This doesn't mean that they are anything more than (or other than) facts, but only that they are a special kind of fact.

The idea that necessity is the distinctive feature of a kind of fact is also implicit in standard systems of modal logic. If a proposition is to be represented as necessarily true, its formula will begin with a "necessity operator," typically printed as a small box prefixed to the relevant clause. What is important is that the operator is not separate from the formula— situated to the left of it, like Frege's "assertion sign"—but rather part of the formula itself, for that is what represents the necessity as belonging to fact being formulated, or, as we might also say, to the propositional content of the proposition expressed.

What alternative could there be? Note first that the very question is an index to a sort of blindness. Part of what distinguishes a dogma from a general belief, or even conviction, is the extent to which it leaves no imaginable space for alternatives. For, once the spell is broken, coherent alternatives may not be that hard to find.

Consider, by way of preparation, the difference between negation and denial. Both have to do with opposition and contrariety, but they are not the same. Negation is a logical operator that, in the simplest case, transforms a given proposition into another with the opposite truth value. Denial, on the other hand, is what Austin called an "illocutionary act" by which (for instance) a speaker takes a stand against some specified proposition. Thus one can explicitly "oppose" some given assertion in either of two ways: one can negate the asserted proposition and assert the contrary proposition instead; or one can leave the proposition itself unchanged but deny it instead of asserting it.

Suppose, now, we look at "modalizing" in the light of those alternative ways of "opposing." It is immediately clear that conventional approaches to modality all follow the model of opposing an assertion by negating the asserted proposition. That is, they all understand the "modalization" of an assertion as an assertion of a modalization of the propositional content originally asserted. This is not only the only alternative compatible with positivism; it's the only alternative that positivists can see. But as the example of denial suggests, it's not the only intelligible alternative.

For, just as denial is a distinctive illocutionary stand that effectively opposes an assertion, we can easily imagine—or invent—distinctive illocutionary stands that effectively modalize assertions. So, for instance, the illocutionary counterpart to "I assert that necessarily p" might be something like "I insist (or require) that p." Likewise, the explicit stand of "allowing" or "declining to reject" might be the illocutionary counterpart of asserting possibility, and so on.

Now, some may find this an alien or even bizarre suggestion. So, before proceeding, I would like to say a few soothing things about its strengths and weaknesses. First, it does not prevent the articulation of a modal logic, though it does impose an important limitation on that logic: nested or embedded modalities are ruled out. The reason is that illocutionary acts must, qua acts, have widest scope. They cannot be within the scope of any others and still themselves be acts. For instance, one cannot say "I assert that I assert that p" without equivocating on the word "assert." The first occurrence is an explicit performative—an acting or doing—whereas the second, as embedded, can only be a characterization of something done.

On the other hand, subject to that limitation, all of classical modal logic can be recast in performative terms. Thus, there are clear performative notions of contradiction and entailment, as *logical* relations among illocutionary acts or stands. For instance, a speaker cannot, without self-contradiction, both affirm and deny the same proposition; and that contradiction is clearly between the acts, not the contents (which, by stipulation, are the same). Similarly, the performance "I insist that p" surely entails (in a performative sense) affirming that p and allowing that p. (Note that these obvious points couldn't even be made if logic were restricted to propositions.)

Finally, it's worth mentioning that the logic of performatives extends beyond acts or stands directed upon propositions. For we also affirm and deny, insist and repudiate, our faith, our love, our loyalty, our responsibility. Simultaneously affirming and denying any of those would likewise be a performative contradiction. And affirming, say, one's loyalty to something entails standing up for it, much as making a promise entails an undertaking to keep it. Again, one can insist or require—as a condition of staying in it, perhaps—that a job or personal relationship live up to certain understandings or essential values; and that entails not only allowing it to have the character in question, but also disallowing it not to.

Scientific laws do, of course, have propositional contents. Our concern, however, has been with their distinctive modal character and, in particular, whether it must be understood as being built into to those contents. The main obstacle to thinking otherwise, I claim, has been the apparent absence of any intelligible alternative. But, drawing on speech-act theory, I have tried to bring a cogent alternative into view. This alternative is logically limited in that it does not allow for embedded modalities. But that limitation is irrelevant when the topic is scientific laws, since their modal character always has widest scope (laws do not contain embedded modalities).

What I have not done, however, is offer any considerations in favor of this new alternative as an understanding of lawlikeness, nor am I yet in a position to do so. Accordingly, this section, like the last, ends with a question that cannot be taken up until later: what advantage could there be to treating the distinctive character of laws on the model of performative acts or stands?

4 Cognitivism

Cognitivism—the second dogma of rationalism—is an assumption not about the world but rather about the mind. It is the view enshrined in Descartes's definition of us as thinking things—which has evolved into the idea that reason is to be understood in terms of cognitive operations on cognitive states.

A cognitive state is something with propositional content (such as a mental representation) together with a cognitive attitude toward that content (such as deeming or wanting it to be true). So, paradigm cognitive states are knowings and willings—or, in more anemic contemporary terms, beliefs and desires. A cognitive operation is a modification of a given body of cognitive states in accord with a procedure that is reliably truth- and/or success-maximizing—in other words, a rational inference. Thus the most straightforward cognitivist view, common to Hobbes and Fodor, is that cognitive states are just sentences *in foro interno*, and that reasoning is but reckoning (i.e., computation). As with positivism, Kant was the first to spell the idea out with clarity and precision.

More recent versions have liberated the core theses of cognitivism from extraneous modern commitments to mind–body dualism, mentalist internalism, and even basic individualism. Quine, Davidson, and Dennett, for example, have argued that cognitive states and processes are determinate only relative to the interpretability of manifest behavior (both verbal and nonverbal). This is still cognitivist, however, inasmuch as what the interpretations attribute are cognitive states and processes, subject to an overall constraint of rationality.

More recently still—though with roots in Hegel and Dewey—Brandom has argued that cognitive states and processes cannot be ascribed to individuals in isolation at all. Rather, the former are to be understood ontologically as norm-governed social statuses, defined in terms of interlocking social commitments and entitlements. And reasoning itself is correlatively explained (with a bow to Sellars) as making moves in a socially normative "game of giving and asking for reasons." Finally, the governing norms themselves are grounded in systematic social sanctions on deviant behavior. It's not that individual thought and ratiocination *in foro interno* are denied but that they are reconceived as derivative internalizations of the more basic public practice.

All these variants, however, betray a pinched and shallow conception of human life and personhood. Though this is clear in many ways, I will try to illustrate it via a consideration of individual integrity and responsibility. For concreteness, I begin with a comparison of Brandom and Nietzsche on promising.

Brandom, of course, understands promising as undertaking a socially instituted commitment or obligation, and thereby conferring similarly constituted entitlements on others (1994, 163–165). Those entitlements include not only innocent reliance on the promiser to make good but also invocation of communal sanctions if she does not. Needless to say, such an institution, if consistently sustained, is of great value, both to individuals and to organized society as a whole.

Nietzsche sketches a similar though less-developed account in *The Genealogy of Morals* (early in the second essay). He is rather more colorful than Brandom in his characterization of social sanctions ("mnemotechnics," Nietzsche calls them), but the effect is the same (1887, §2:3). Promising and many other valuable institutions, he agrees, are made possible by socially imposed norms—or, in his more vivid phrase, the "social straightjacket" of "slave morality" (§2:2). By such means, the naturally erratic animal, man, is made regular and calculable, even to himself.

What sets Nietzsche apart, however, is that this account is only preparatory for another and quite different understanding of what promising can be and mean: the self-responsible exercise of an autonomous protracted will—the act of a sovereign individual with the *right* to make promises. Social sanctions, while prerequisite for the genesis of this capacity, are left entirely behind in its maturity. A will that endures the vagaries of the world by its own willful law is no mere inculcated habit or persisting desire, no social institution or status, but something new that has emerged out of them and transformed the possibilities for mankind. Or so, at any rate, Nietzsche claims.

Brandom and cognitivists more generally have no room for any such phenomenon. Indeed, social pragmatism is just a more sophisticated account not only of slave morality but of Plato's marketplace, Heidegger's *das Man*, and Kuhn's normal science. The point here is not to endorse any of their particular alternatives but to suggest that something important is at stake—something incompatible with cognitivism, something that can be indicated by terms like "integrity" and "responsibility," something therefore related to courage, loyalty, and the ability to love, and something that, as I argue later, is essential to science.

5   Scientific Understanding

Despite a hoary tendency to assimilate them—as old as the word *epistêmê*, and still conspicuous in Kant—understanding is not the same as, or any species of, propositional knowledge. Indeed, it is not a propositional attitude at all but something quite different that is, in fact, prerequisite for them. Though I think this is obvious, there's also an easy argument for it. One cannot have a cognitive attitude toward a proposition that one does not understand, but one can certainly understand a proposition without having any attitude toward it at all.

If understanding is not a cognitive state in any canonical sense, then what is it? In science, the gold standard of understanding is the ability to explain. In all scientific explanations, what is explained is shown to be compatible with (often entailed by) its particular boundary conditions, given what is possible and necessary in general. Accepting the connection between modality and lawlikeness, this clearly subsumes familiar deductive-nomological explanations as a special case.

But not all scientific explanations are deductive-nomological. Explanations are also cast in terms of functional organization, statistical selection processes, causal mechanisms, rational decision making, and so on. What these have in common is showing how the actual can be seen or grasped in the light of the modal. While I maintain that this connection with modality is quite general, I must here confine myself to a single illustration—an explanation in terms of functional organization. The most famous such explanation in science may be the double-helix account of DNA replication. But I will stick to something more mundane: an explanation of how a pendulum clock works—that is, how it keeps and shows the time.

One doesn't explain the working of a clock in terms of the physics of rigid bodies—even though one does (and must) take for granted that most of its internal parts are rigid. And, of course, one needs to assume (what only physics explains) that the period of a pendulum is (nearly) constant. But what matters for understanding the clock *as a clock* is not so much the physics as the particular configurations, arrangements, and interactions of all those physical parts.

A brief account might go as follows (preferably with an actual clock or picture to point at). Given that the pawl is connected as it is to the pendulum, its two ends *cannot but* rock up and down once per swing; given the shapes and relative positions of the pawl and ratchet wheel, the latter *can* advance one—but *only* one—notch per rocking of the pawl; given that, plus the torque that *cannot but* be imparted to it by the weight on the chain passing over a sprocket on the same shaft, the ratchet *cannot but* so advance; and, given all those, plus the way the ratchet shaft is geared to the hands, and the constant period of the pendulum (plus a few other things), the hands *cannot but* rotate at certain constant rates—thereby keeping and showing the time.

Now, even though various laws of physics are presupposed in that account, none are actually cited—nor need they be, since they are not what provide the relevant insight. Nevertheless, as my emphases indicated, the notions of what can, can only, and cannot but occur, given the mechanical configuration, play a crucial role. The reliable ability to keep time depends on the possibility and inevitability of the interactions that enable it. (There is no explanation in mere recital of what happens to happen.) Thus, even in a mechanical or engineering explanation, without any mention of laws, an ability to see the actual in the light of the modal is fundamental.

So far, we have examined understanding through the lens of explanation—on the grounds that, in science, the ability to explain is the "gold standard." But understanding as such does not presuppose the ability to explain. If explaining is showing how the actual can be grasped in the light of the modal, then understanding by itself is just grasping or being able to grasp the actual in that light. What we need, to spell this out, is to discharge those metaphors. In the present context, "grasping the actual" means noticing, or at least tracking, some nontrivial pattern or structure of relationships in the phenomena. And grasping that pattern or structure "in the light of the modal" means appreciating that it is possible and/or necessary, in a way that others, superficially similar to it, would not be.

That formulation, it bears mentioning, is intended to accommodate the fact that not all understanding is articulate. Obviously, for example, our understanding of our own words and sentences cannot itself presuppose an explicit account expressed in the same vocabulary and grammar. Yet it does presuppose grasping them in the light of norms and proprieties of usage—which are the relevant sort of modality. And, indeed, a great deal of our everyday and scientific understanding is largely inarticulate—and no less legitimate for it.

All of the foregoing, however, leaves a surprising question about understanding quite unanswered: Why does it *matter*? If understanding is qualitatively distinct from knowledge, then why is the latter alone not sufficient? What does understanding add?

6   Science without the Dogmas

When God died—by most accounts, sometime in the nineteenth century—so did the only conceivable guarantee that there is any complete true description of the world, or even any language in which such a description could be formulated. In other words, the cognitivism implicit in the very idea of an omniscient intellect was modern rationalism's only compelling basis for the metaphysics of positivism. Yet forfeiting these familiar anchors might seem to float objective truth and understanding free of any solid mooring in "how things really are"—and, perhaps worse, out to sea on some postmodern relativism. In this epistemic debacle, science too would be at stake, for it is nothing other than the systematic pursuit of general objective knowledge and understanding.

The appropriate response is not to denigrate scientific objectivity, but rather to reconceive it in thoroughly posttheological—hence postrationalist—terms. But that will mean rethinking epistemology from the ground up. So I will close by trying to sketch how that might go (tying up my earlier loose ends along the way).

The essential move must be to understand truth itself as a distinctively human achievement. Before there were people, there was no such thing as truth (at least not on Earth). By this, I don't mean just that there were no truth bearers—sentences, cognitive states, or whatever—but that there was no truth for anything to "bear." So, for instance, though it may be true *now* that there were frogs a million years ago, it was not true a million years ago that there were frogs *then*. Truth and objectivity as such had to be opened up and enabled by people, and that happened a lot more recently than a million years ago. Until we can understand them as in this way our own, philosophy will have yet to recover from the death of God.

Material truth is nothing without falsehood. The crucial opening up, therefore, was really of the true–false dichotomy. We ask, accordingly: what else has to be in play to make that opposition intelligible? One obvious prerequisite is that distinct, formally compatible claims can nevertheless conflict—that is, be such that not all can be true. That's what forces and gives sense to the acknowledgment that at least some are materially false. So our question becomes: How can logically compatible scientific claims be nevertheless scientifically incompatible? What kind of *non*logical constraint is there on the mutual compatibility of, say, empirical results?

The empirical findings of a science are clearly constrained, in their relations to one another, by the laws of that science—as a simple example illustrates. If you and I each measure the speed of the Mississippi River, but at different times and places, we could get quite different answers and still both be right. But if we measure the speed of light in a vacuum, no matter when or where, any discrepancy at all will show that at least one of us got it wrong. The difference, of course, is that a law of physics says the speed of light is constant, whereas there are no such laws for the speeds of rivers. This is how, at the most basic level, experimental failure is to be understood.

But the positive side of that same point is equally fundamental. If substantially different measurements and experiments consistently give results that are related just as the laws require, then they are mutually supporting. And, importantly, it's not just the results that support one another but also the respective techniques and procedures by which they are obtained. Perhaps it is a cliché by now to point out that theories are "holistic," but it is less often noted that this essential holism must extend also to established scientific practice—including, in particular, the inarticulate know-how, and even the specialized equipment, on which researchers invariably rely.

For, ultimately, it is only such integrated theoretical-practical "packages" that can confront nature in a way that might genuinely fail. If there are no experimental results, obtained by recognized means and expressed in the terms of the theory, then that theory is detached from any external constraint. Yet if the theory, in turn, does not impose constraints of its own, expressed in the same vocabulary, then what happens in the laboratory is irrelevant to it. It's only via the potential collision between these two sets of constraints that anything can actually go wrong. And only if something actually could go wrong can the fact that it doesn't be of any significance.

The constraints imposed by a scientific theory are its laws. If, however, these are construed merely as further propositions, added to those accepted on the basis of experiment, then all you get is a larger set of propositions— possibly inconsistent. But a set of propositions, consistent or otherwise, is utterly inert. And adding "necessity" to some of their contents can't change that. Propositions themselves, whatever their contents, don't do anything at all.

It is only people—in our case, scientists—who act or take stands. It is they, and only they, who do anything with propositions; and what they do depends in part on the "status" they accord them. In particular, if scientists accord some propositions the status of laws, then they will not put up with, will not credit, incompatible empirical claims (as the speed-of-light example illustrates). In other words, to take a proposition as a law just is to insist or require that empirical results be compatible with it. And that is why I suggest that the "modal character" of scientific laws might be understood better in terms of performative stands than in terms of propositional operators.

By the same token, if the essence of understanding is the ability to grasp or see the actual "in the light of the modal," and explanation is making that ability explicit, then it becomes clear why they matter to science. For that ability is nothing other than the success condition on the foregoing insistent stands. In the most basic case, explaining just is *showing* how actual empirical results are compatible with modal laws—which is to say, showing how they are intelligible.

Yet science is intrinsically hard—and precisely because it must make itself hard, to be what it can be. If it is to yield nontrivial knowledge and understanding, it must assiduously maximize what I called the potential for "collision" between theoretical and empirical constraints. Only by leaving itself wide open, as wide as it can, to such threatening difficulties can it ever claim to have discovered anything by discerning a narrow path that avoids them. Only on these terms can meeting the above "success condition" amount to anything epistemically substantial.

Taking—nay, not just taking but actively pursuing—this fundamental risk, a risk even to career and profession, demands something like personal responsibility and integrity—if not of all scientists always, at least of some scientists sometimes. For the ultimate threat is not merely to this research project or that but to the whole integrated "package" in terms of which alone it and many others are scientifically viable. History shows that this threat is not idle.

The elaborate institutional structure that makes such a threat so much as possible is what gives sense to determinate scientific claims—that is, to assertions that can be determinately true or false. And giving sense to that distinction, exclusively in terms of our own resources, is what I mean by the *human* opening up and enabling of truth. As I have tried to show, however, those necessary resources are incompatible with the modern assumptions of positivism and cognitivism. I conclude, therefore, that, inasmuch as science itself flourishes, neither dogma of rationalism can survive the death of God.

Note

1. Translation mine. Page citations are to the 1968 edition edited by G. Barbera, vol. 6.

References

Brandom, R. 1994. *Making It Explicit: Reasoning, Representing, and Discursive Commitment.* Cambridge, MA: Harvard University Press.

Galilei, G. 1623/1968. Il Saggiatore. In *Le opre de Galileo Galilei: Nuova ristampa della Edizione Nazionale*, ed. G. Barbera. Rome: Ministero della Pubblica Istruzione.

McDowell, J. 1996. *Mind and World: With a New Introduction*. Cambridge, MA: Harvard University Press.

Nietzsche, F. 1887. *Zur Genealogie der Moral: Eine Streitschrift*. Leipzig: C. G. Neumann.

Wittgenstein, L. 1974. *Tractatus Logico-Philosophicus*. Trans. D. F. Pears and B. F. McGuinness. Introduction by B. Russell. London: Routledge & Kegan Paul.

11 Rationalism without Dogmas

John McDowell

The question to be answered is whether the prospect [raised in §13], that objects might be presented to *our* intuition (in space and time, as described by the Aesthetic) that are not subject to the conditions of the understanding (that is, are unintelligible), is a genuine one. And the answer to that question is: No!

—John Haugeland, "The Transcendental Deduction of the Categories" (this vol., 342)

1. John Haugeland's "Two Dogmas of Rationalism" has his characteristic forthrightness and eloquence, but I think it partly misses its target. Haugeland implies that the dogmas he attacks are characteristic of rationalism as such. Only so can his argument seem to establish what he suggests at the end of the paper: that discarding the dogmas requires us to rethink epistemology in postrationalist terms, which implies that we should discard the rationalist impulse altogether. I think the result is that rationalism at its best goes missing in Haugeland's treatment.

It is an abiding regret for me that I did not get to discuss Haugeland's paper with him. I am grateful to Zed Adams for suggesting I might respond to it in this volume.

I make no apology for celebrating Haugeland by taking issue with him. I think anyone who knew him would acknowledge that this is a fitting way to honor his memory.

2. The first of Haugeland's two dogmas is what he calls "positivism": the idea that reality, or the world, is exhausted by what can be truly said to be the case.

Why call that "positivism"? The idea is that reality can be completely captured by true propositions. And the label is meant to exploit the etymological link between "positive" and "proposition."

Now, how should we interpret the equation of the world with what is the case? Haugeland writes: "Clearly, if this equation is intended as an essential conception of the world, then it requires an independent understanding of propositions" (294).

But this ensures, I think, that the best interpretation of the thought Haugeland has in his sights goes unconsidered. We can find insight in linking the idea of reality with the idea of a discursive capacity, a capacity for articulable thought, without needing to look for an *independent* understanding of the idea of a discursive capacity. Haugeland does not consider the possibility that the point of the equation is this: the idea of reality and the idea of a discursive capacity can be properly understood only *together*, not one of them in terms of an independent understanding of the other.

He cites the first sentence of Wittgenstein's *Tractatus*—"The world is everything that is the case"—as "an elegantly austere expression" of positivism in his sense (293). The citation might suit his purpose if, in the conception of language and the world that Wittgenstein at least goes through the motions of elaborating in the *Tractatus*, there were a clear indication of the directionality of understanding offered by positivism as Haugeland understands it, with propositions explained independently and the idea of the world explained in terms of the independent understanding thus provided. But that is certainly open to dispute.

And anyway, the *Tractatus* is not the best context in which to consider positivism as a doctrine. We should be alive to the possibility that Wittgenstein is *only* going through the motions of setting out a conception of world and language. Famously, near the end of the book he says that anyone who understands him (not "understands what he has been saying" or something to that effect) recognizes his "propositions" as nonsense. And he instructs us to throw away the ladder we have climbed in making as if to understand them.

3.   Haugeland thinks positivism, in the sense of the dogma, came to maturity with Kant. He writes (295):

It remained to Kant ... to appreciate that positivism is a substantive metaphysical thesis, and so to argue for it—at least as regards the empirical world. For the transcendental deduction of the categories is nothing other than an attempt to prove that there can be nothing to the empirical world except such as can be the content of a judgment.

I would not dream of disputing that Kant thought the empirical world is exhausted by what can be contents of judgments. But I think Haugeland is wrong in taking Kant to have conceived that as a substantive thesis, in need of argument. That is not what the Transcendental Deduction is for. To explain this, I need to say a fair bit about the Deduction.[1]

4. It is fundamental to Kant's thinking that a finite understanding— "an understanding whose whole power consists in thought" (B145)—can have its power actualized in knowledge only thanks to its partnership with sensibility, which is needed to supply the finite or (equivalently for Kant) discursive understanding with objects for its knowledge.

The question addressed by the Transcendental Deduction is this: if the whole power of the finite understanding consists in thought, how can we be entitled to suppose that acts of that power include knowledge? In the basic case of empirical knowledge, as Kant sees things, objects are sensibly presented to knowing subjects in a way that conforms to requirements whose source is the understanding. But if the requirements originate in the understanding independently of sensibility, and sensibility is needed to supply the understanding with objects, what licenses us to think conformity to those requirements has anything to do with objectivity? It is a live question how it can be knowable a priori that putative pure concepts of the understanding—putative concepts of objects as such whose content reflects requirements that originate from the pure power of thought—are indeed what they purport to be, concepts of objects as such; or, as Kant puts it, that pure concepts of the understanding have objective validity.

5. In the first part of the B Deduction,[2] Kant works up to a conception of empirical intuitions, cases of objects being sensibly present for knowledge, that is supposed to entitle him to the following: "All the manifold, ... so far as it is given in a single empirical intuition, is *determined* in respect of one of the logical functions of judgment, and is thereby brought into one consciousness. Now the categories are just these functions of judgment, in so far as they are employed in determination of the manifold of a given intuition. ... Consequently the manifold in a given intuition is necessarily subject to the categories" (B143; this is what, at B144, he calls "a beginning ... of a deduction of the pure concepts of understanding," and the

big question about the structure of the B Deduction is why it is only a beginning).

Let me sketch how Kant entitles himself to that preliminary result.

It helps to focus on the implication, in this statement of it, that the role of determination by the functions of judgment, in the conception of intuitions he is putting into place, is to provide for the manifold given in an intuition to be *brought into one consciousness*. An empirical intuition as Kant is conceiving it here is, we might say by definition, something in which multiple sensibly given "representations"—elements in a sensory manifold—figure not as a mere multiplicity, an aggregate or heap, but as a unity self-consciously experienced as such. It may not be too much of an exaggeration to say we can understand the "beginning of a deduction" as no more than an elaboration of that.

For me to experience a multiplicity of representations self-consciously as a unity, they must be mine together in a sense that is spelled out by saying that the "I think" must be able to accompany them (B131–132). And of course it is analytic that for representations to be mine together, they must conform to whatever is required for representations to be mine together: that is, whatever is required for them to be able to be accompanied together by the "I think."

If it is to be possible for the "I think" to accompany a multiplicity of representations that count as jointly mine by virtue of that possibility—if it is to be possible for them to be brought into one consciousness—then they must be combinable with one another in accordance with the unity that is characteristic of thought (or we might say "at least one of the unities that are characteristic of thought"; cf. "one of the functions of judgment" in the passage I quoted from B143). That specifies a bit further the requirement for a multiplicity of representations to constitute a unitary intuition that is mine. It is still analytic (it is the "I *think*" that makes it explicit that the representations belong to one consciousness), even though the unities that are characteristic of thought are synthetic. Thus, after saying that "the first pure knowledge of understanding ... upon which all the rest of its employment is based ... is the principle of the original *synthetic* unity of apperception" (B137), Kant remarks (B138):

Although this proposition makes synthetic unity a condition of all thought, it is, as already stated [he is harking back to B135], itself analytic. For it says no more than that all *my* representations in any given intuition must be subject to that condition under which alone I can ascribe them to the identical self as *my* representations, and so can comprehend them as synthetically combined in one apperception through the general expression, "*I think.*"

We can follow Kant in considering this idea in abstraction from the question what these synthetic unities, the unities characteristic of thought, are. Of course Kant has an answer, the table of the categories, but the Deduction does not exploit that answer. The Deduction is concerned with how such unities—whatever they are, we might say—figure in the presence of objects to finite knowers.

So far, then, what we have is this: it is analytic that what is required for representations to be mine together is the synthetic unity of apperception; there must be synthetic unity if representations are to be brought into one consciousness. And, in a specific application of that: an empirical intuition, a case of being sensibly presented with an object for knowledge, is constituted by a multiplicity of sensible representations (a sensibly given manifold) with a togetherness that exemplifies the unity, or one of the unities, characteristic of thought. As I have suggested, we can conceive that as true by definition; it spells out the conception of empirical intuitions that Kant is working with, and it does no more than elaborate a basic feature of the way he conceives finite knowledge.

On these lines, it emerges as analytic that objects are sensibly given to me (sensibly given in representations that are mine, in the sense that the "I think" can accompany them) only in intuitions so understood, that is, only in episodes or states in sensibility that instantiate the synthetic unity of apperception. It is analytic that if an object is sensibly present to me for knowledge, my consciousness in having it present to me exhibits a synthetic unity.

The passage I quoted from B138 says it is analytic that synthetic unity is a condition of all thought. Kant might have said (more immediately to the point) what I have just said on his behalf: it is analytic that synthetic unity is a condition of all sensible givenness of objects to subjects. It is no more than an application of the claim that synthetic unity is a condition of thought to say, as he said earlier in B138: "It [the synthetic unity of consciousness] is not merely a condition that I myself require in knowing an object, but is a condition under which every intuition must stand in order *to become an object for me.*"[3]

6. In the first part of the B Deduction, then, Kant puts in place, as no more than an analytic elaboration of the fundamental structure of his conception of finite knowledge, a claim on the following lines: objects can be sensibly present to subjects, sensibly present for knowledge, only in intuitions, understood as sensory manifolds unified by modes of synthetic unity whose source as requirements lies in the understanding.

Now why is that only a beginning of what he needs to do, as he says at B144? Why does the B Deduction need its second part?

The Deduction's task is to establish that the categories (the pure concepts of the understanding: concepts whose contents consist in modes of the synthetic unity that is characteristic of thought) have objective validity. Kant's aim is to establish that by showing, in effect, that it is the categorial unity of intuitions that provides for the directedness of sensory consciousness at a reality independent of it. And he has not done that yet, even though the first part brings out, as analytic, that synthetic unity is a condition under which sensory consciousness must stand in order to present an object to a subject.

If there could be episodes or states in which there was presence to my senses without conformity to the requirement of synthetic unity, they would not present objects to me. They would not be cases of my being sensibly conscious of objects. That is an implication of what we have as analytic from the first part of the Deduction.

If there could be such episodes or states, the idea of presence to my senses could be detached from the complex of analytic truth spelled out in the first part. Presence to my senses that was not presence of objects to me would not require a synthetic unity of consciousness. And it is tempting to think Kant's conception of finite knowledge allows for such presence. It can seem that his account of our human sensibility, in the Transcendental Aesthetic, is meant to provide for presence to the senses on its own, without any need to invoke requirements that originate from the understanding.

But any presence to the senses would be at least a minimal case of sensory consciousness being directed at something other than itself. So if we think the Aesthetic explains presence to the senses independently of any need to invoke the understanding, we are unable to reject this idea: what accounts for the fact that sensory consciousness is directed outside itself is just that it is *sensory*, and so exemplifies the form of our sensibility. On this supposition, the directedness of sensory consciousness does not require the synthetic unity of thought.

After the Deduction's first part, we have to grant that episodes or states of sensibility "in me" that did not exemplify the unity of apperception would not make objects present to me. But if such episodes or states could, even so, be cases of presence to my senses, it would be left open that categorial unity secures at most that some cases of sensory consciousness—which, on this supposition, has its directedness anyway by virtue of exemplifying the form of human sensibility—are describable as presence of objects to subjects.

And now we cannot exclude the possibility that if we describe them like that, we are illicitly crediting merely subjective requirements with objective significance. We have not ruled out the possibility that, contrary to how Kant describes the Deduction's aim (B138, quoted earlier), the synthetic unity of consciousness is merely a condition that I myself require in (what, without justification on this supposition, I describe as) knowing an object.

In the second part of the Deduction, then, Kant needs to exclude this apparent possibility of presence to my senses that, because it does not conform to the requirement of synthetic unity, does not count as presence of objects to me. He needs to show that synthetic unity is already a condition for presence to my senses, not a merely additional condition for what (illicitly on this view) I count as presence of objects to me. As he says early in §26, when he is on the brink of finishing the execution of his project: "We have now to explain the possibility of knowing *a priori*, by means of *categories*, whatever objects may *present themselves to our senses*" (B159); without that, "there could be no explaining [as there needs to be] why everything that can be presented to our senses must be subject to laws which have their origin *a priori* in the understanding alone" (B160). The first part seemed to leave open a possibility of presence to our senses without categorial unity. (It would not have been presence to *us*, and it would not have been presence of *objects* in a demanding sense that is connected with that; but, as I have tried to explain, that does not help.) The second part (which he is near finishing at that point) needs to exclude that seeming possibility.

This makes Kant's statement of what he needs to do in the second part of the Deduction exactly right (B144–145):

In what follows (cf. §26) it will be shown, from the mode in which the empirical intuition is given in sensibility ["in our sensibility" would have been better], that its unity [the unity of the mode: the unity that is implicit in the idea that our sensibility is spatially and temporally formed] is no other than that which the category (according to §20) prescribes to the manifold of a given intuition in general. Only thus, by demonstration of the a priori validity of the categories in respect of all objects of our senses, will the purpose of the deduction be fully attained.

He needs to argue that the unity implicit in the idea that our sensibility is spatially and temporally formed is not (as we might have thought when we first read the Aesthetic) a self-standing unity, independent of the unity characteristic of the understanding—that it is not something that could by itself provide for presence to our senses, if not for presence of objects to subjects. He does that, in §26 as he promises in that passage from B144–145, by exploiting the fact that "space and time are represented *a priori* not merely as forms of sensible intuition, but as themselves intuitions which contain a manifold [of their own], and therefore are represented with the determination of the unity of this manifold" (B160). Synthetic unity is presupposed already in the requirement of spatial and temporal order that first came into view, in the Aesthetic, as required by our sensibility. There is only one unity, the synthetic unity that is intelligible only in terms of the unifying power of the spontaneous intellect. And the apparent possibility of presence to the senses that is not presence to the understanding is unmasked as the mere appearance of a possibility.

7.   The task of the Deduction's second part, then, is to rule out the apparent possibility of a presence to the senses that is not an availability of objects for discursive knowledge, because it does not exemplify categorial unity.

We might formulate what needs to be shown like this: there cannot be "rogue appearances," cases of presence to the senses without the synthetic unity required by acts of the understanding.[4] That certainly has the look of a substantive thesis. And if we formulate what Kant needs in those terms, we may seem to be committed to agreeing with Haugeland that, in the Deduction, Kant treats his version of positivism as a substantive thesis, since he thinks he needs to argue for it. No doubt something on those lines is what Haugeland has in mind.

But this involves misconstruing the threat posed by the apparent possibility that Kant excludes in the Deduction's second part. Or, what comes to the same thing, it involves underestimating the extent of the analyticity Kant brings out in the first part.

The threat posed by the apparent possibility of "rogue appearances"—the apparent possibility of items that, though not within the reach of our power of empirical knowledge, which is essentially discursive, are nevertheless present to our senses—is not that there may be more to a reality that is empirical, in that it impinges on our senses, than what we can judge. If we could not rule out a supposed presence to the senses that was not presence to the understanding, we would not even be entitled to the idea of presence to the senses. The implication would not be that the reality that impinges on our senses transcends what we can think, as if the understanding operated as a filter on what we are anyway entitled to conceive as given to our senses, admitting into our apperceptive consciousness only what conforms to its requirements. If conformity to the understanding's requirements is not what licenses the very idea that sensory consciousness is directed at something other than itself, we are not entitled to suppose that those requirements have anything to do with objectivity. We are not entitled to exclude the idea that when we count sensory manifolds that exemplify the unity required for thought as intuitings, as states or episodes in which objects are present to us, we are imposing subjective requirements on what is really only a manifold of sensation, in which nothing is given from outside our sensibility. The upshot would not be that we have to suppose empirical reality might outrun our discursive capacity. The upshot would be that we lose any entitlement to the idea of an empirically accessible reality, objective in being distinct from our manifolds of sensation.

The conceptual connections Kant elaborates in the Deduction's first part are analytic, and the analyticity extends to this: it is analytic that anything we could mean by "empirical reality" is thinkable. The apparent possibility Kant excludes in the second part would, if actual, make that analyticity amount only to this: it is analytic that empirical reality would be thinkable if we could make sense of the idea of empirical reality. Eliminating that apparent possibility is not arguing, as if for a substantive thesis, that there can be nothing to the empirical world except such as can be the content of a judgment, as Haugeland has it. For Kant that is analytic. What the Deduction needs to exclude is the specter of our being unable to mean anything by "empirical reality."

8. Kant's version of the positivist equation of reality with what can be thought is distinctive because as he sees things, the understanding, the power of discursive thought, has its formal character independently of sensibility, which is needed to provide it with a subject matter. That is why the Deduction's question is a question for him (and why his version of the equation needs to be restricted to empirical reality, as Haugeland notes). To repeat (see §4 earlier), the question is this: if the understanding needs sensibility to have objects at all, how can concepts whose content is given by unities required by the understanding, conceived in abstraction from sensibility, be guaranteed to be concepts of objects?

Things change if we stop supposing that the formal character of the power of thought can be understood in abstraction from something that plays the role of sensibility in Kant: something other than the power of thought, taken to be needed for the acts of the power of thought to have a subject matter. This opens up the possibility of a version of the positivist equation for which no analogue to the Deduction's question arises. I think that fits Hegel's conception of logic. For Hegel, logic's unfolding of the forms of thought is an unfolding of the forms of reality. And as with Kant's version of this idea, the equation of the forms of thought with the forms of reality is not a substantive thesis. It is not that, as in the dogma Haugeland considers, logic is supposed to yield an independent understanding of the forms of thought, from which we are to derive an understanding of the forms of reality. On the contrary, Hegel does not think it needs arguing that the idea of the forms of reality and the idea of the forms of thought are just two guises of a single idea. Here we have an especially unqualified version of the positivist strand in rationalism, interpreted in what I suggested is the best way: that is, without the directionality, the order of understanding, that figures in positivism as Haugeland considers it.

9. Of course Haugeland might be wrong about the significance of the positivist equation in rationalist thought at its best, but still right to reject the equation, even interpreted in the way I have urged, as an untenable dogma.

His main case against positivism turns on scientific know-how.[5] He argues that when one learns how to conduct scientific investigations, one learns something about the world, but what one acquires in such learning cannot be knowledge about the world as it is conceived by positivism. Haugeland writes (297):

> What will work or not in actual practice, or work better than something else, is a function of the world. Therefore, in learning what will and won't work, or what will work better—that is, in acquiring the relevant know-how—scientists are learning something about the world.

He is surely right that what will work or not is a function of the world. For instance, what one can do with a material is a function of its physical properties, which are part of the world. One can know how to work with a material without having explicit concepts of properties like malleability or ductility. But we can count knowing how to work with a material as a distinctively practical way of knowing relevant things about the world. In knowing how to do things with the material, one knows that it can be worked with in such and such ways.[6]

Analogously, a skilled scientific investigator knows that one can acquire knowledge and understanding of worldly phenomena by following such and such investigative procedures. So, as Haugeland argues, when one acquires a bit of know-how, one learns something about the world.

In these exemplifications of the point, what one comes to know about the world when one learns how to do something is specified by a "that ..." clause. Even so, Haugeland claims that such knowledge "is not exclusively factual" (297). He means that by the lights of positivism, it cannot count as knowledge about the world.

His ground for this is that such knowledge "cannot be entirely verbal," "cannot be reduced to a text," "cannot itself be expressed in words or formulas" (297). When, in a passage I quoted earlier (§2), he claims that positivism requires an understanding of propositions independent of the idea of the world, he goes on (294):

> That understanding is invariably discursive: a proposition is something that could (in principle) be said—that is, articulately "pro-posed" or "put forward" by somebody as the content of an explicit claim. And that articulability, in turn, is cashed out in terms of finitely specifiable interpreted formal systems—of which natural languages are approximations, and the symbolic notations of mathematics and logic are exemplars.

And in his discussion of the worldly knowledge one acquires in learning how to do things, it becomes clear that in positivism as Haugeland understands it, interpreted formal systems are not just a model or an ideal case but figure in the very definition of discursive capacities, in the sense in which positivism claims that the world can be completely captured by acts of discursive capacities. As Haugeland understands positivism, it conceives what can be truly said as restricted to what can be made explicit by words (or other symbols) without help from anything besides themselves.

This cries out to be questioned. A possessor of know-how might say, "Here is a way one can work with this material," while displaying a technique that could not be fully specified in a text; the display enters into determining what it is that is being said. And a scientific educator might say, "Scientific knowledge can be acquired by investigating the world in the way you will be acquiring familiarity with as you work with me"; the training that is under way enters into determining what it is that is being said. Haugeland's point is that these expressions of meaning could not be replaced by pure strings of symbols that by themselves made their meaning explicit. But why should that debar rationalists from counting them as cases of speaking the truth? We can agree with Haugeland that there is more to the world than what can be expressed in exploitations of finitely specifiable formal systems. But on a sufficiently liberal conception of capacities to say things, that is no threat to the idea that the world is exhausted by what can be truly said.

10.   Haugeland's second dogma is cognitivism: "the idea that reason is to be understood in terms of cognitive operations on cognitive states" (301). He spells this out as follows (301):

A cognitive state is something with propositional content (such as a mental representation) together with a cognitive attitude toward that content (such as deeming or wanting it to be true). So, paradigm cognitive states are knowings and willings—or, in more anemic contemporary terms, beliefs and desires. A cognitive operation is a modification of a given body of cognitive states in accord with a procedure that is reliably truth- and/or success-maximizing—in other words, a rational inference.

He mentions a number of variations on this theme. But he claims that all the variants "betray a pinched and shallow conception of human life and personhood."

He illustrates this claim by comparing Brandom and Nietzsche on promising. In both there is a picture of promise keeping as an obligation instituted by social practices, centrally involving sanctions for noncompliance. But in Nietzsche, and not in Brandom, such a picture stands in unfavorable contrast with "another and quite different understanding of what promising can be and mean: the self-responsible exercise of an autonomous protracted will—the act of a sovereign individual with the *right* to make promises" (302). Haugeland goes on (302–303):

Social sanctions, while prerequisite for the genesis of this capacity, are left entirely behind in its maturity. A will that endures the vagaries of the world by its own willful law is no mere inculcated habit of persisting desire, no social institution or status, but something new that has emerged out of them and transformed the possibilities for mankind. Or so, at any rate, Nietzsche claims.

As with positivism, when he explains what he means by "cognitivism," Haugeland cites Kant as its first clear expositor. But Kant's conception of respect for the moral law bears striking similarities to Nietzsche's conception of the capacity to bind oneself by promising, on the understanding that Haugeland suggests is out of reach for rationalism. Haugeland casts Kant as a proponent of the second dogma of rationalism as well as the first, but surely Kantian respect for the law is an instance of just the sort of thing Haugeland says cognitivism, and therefore rationalism, are doomed to leave out of the picture: "something that can be indicated by terms like 'integrity' and 'responsibility,' something therefore related to courage, loyalty, and the ability to love" (303). Haugeland notes that different conceptions of human life or aspects of human life come under attack by different thinkers, in effect as pinched and shallow, in the way he exemplifies with his criticism of Brandom's social pragmatism. He declines to endorse any of the proposed alternatives to pinched and shallow conceptions that he mentions: Nietzsche's alternative to slave morality, Plato's alternative to the marketplace, Heidegger's alternative to *das Man*, Kuhn's alternative to normal science. Similarly, we do not need to embrace Kantian morality to recognize that the moral will, as Kant conceives it, can be described in the terms Haugeland uses to express a Nietzschean idea that is supposedly unavailable to rationalism: "a will that endures the vagaries of the world by its own willful law" (302–303).

And Kant's conception of the self-legislating will is surely rationalist in the sense that is relevant to Haugeland's conception of a pervasive strand in philosophy. It is not rationalism as such that is to blame for conceptions of human life from which such things as self-legislating wills are absent. If a particular version of rationalism saddles itself with a pinched and shallow conception of reason, as the mark of a distinctively human way of living, it is not surprising that the result is a pinched and shallow conception of human life and personhood. The remedy is not to abandon rationalism but to liberalize our conception of reason.

It is not just in respect of the absence of will as ethical character that cognitivism, as Haugeland describes it, embodies a conception of reason that is needlessly restricted, and so need not be conceived as exemplary of rationalism as such. There is something questionable about the way cognitivism restricts acts of reason to inferences conceived as transitions from one cognitive state to another. Consider intentional agency. Cognitivism would limit reason, in its practical guise, to inferences starting from cognitive states that register grounds for acting, and concluding in something on the lines of practical commitments. This leaves no room for the idea that intervening in the world can itself be practical reason in operation. And there is no ground to saddle rationalism as such with being inhospitable to that idea.

It seems right to say that some version of positivism in Haugeland's sense—the idea that the world is fully accessible to reason—would have to be part of any outlook deserving to be described as rationalism. The problem with Haugeland's first dogma is not that a defensible rationalism contains nothing of the sort, but that some specific features of the version Haugeland attacks—the idea that the acts of reason that are said to be capable of embracing the world must be understood independently of the notion of the world, and in particular must be glossed in terms of exploitations of interpreted formal systems—are at best optional. So there is room for an improved positivism, which we can see as characteristic of an improved rationalism. But the second dogma is different. Cognitivism reflects a needlessly restricted—indeed, a pinched and shallow—understanding of reason. A proper rationalism should not seek to modify cognitivism, as I have suggested in the case of positivism, but should simply disown it.

11.   Our grasp on the world includes not just knowing worldly phenomena, but also understanding them. Haugeland insists, convincingly, that a bit of understanding is not just another bit of knowing.

He glosses understanding in terms of "see[ing] the actual in the light of the modal" (304). The modality in question can be the necessity that belongs to a law of nature. But one sees the actual in the light of the modal not only in full-blown scientific explanation, but also in a familiar kind of understanding that does not traffic (directly, at any rate) in laws of nature. Haugeland illustrates this beautifully with a sketch of how we can understand the workings of a pendulum clock (304):

Given that the pawl is connected as it is to the pendulum, its two ends *cannot but* rock up and down once per swing; given the shapes and relative positions of the pawl and ratchet wheel, the latter *can* advance one—but *only* one—notch per rocking of the pawl; given that, plus the torque that *cannot but* be imparted to it by the weight on the chain passing over a sprocket on the same shaft, the ratchet *cannot but* so advance; and, given all those, plus the way the ratchet shaft is geared to the hands, and the constant period of the pendulum (plus a few other things), the hands *cannot but* rotate at certain constant rates—thereby keeping and showing the time.

He explains persuasively how the capacity of science to "open up" truths about the world depends on its containing "theoretical-practical 'packages' that can confront nature in a way that might genuinely fail" (307). It is an essential part of what makes such confrontation possible that "if scientists accord some propositions the status of laws, then they will not put up with, will not credit, incompatible empirical claims. ... In other words, to take a proposition as a law just is to insist or require that empirical results be compatible with it" (307).

He thinks this talk of insisting or requiring validates a suggestion he made earlier in the paper, that we should understand modality in a distinctive way. In contemporary logic it is usual to recognize only one way of rejecting a claim: asserting its negation. But in older traditions, which we should perhaps not be too quick to dismiss, denying a claim is distinguished from asserting its negation, and denying is understood as an illocutionary act in its own right, distinct from and on a level with asserting. Analogously, modal claims are usually understood as assertions of modally modified propositions, but Haugeland proposes that we understand them instead as performances of special illocutionary acts, distinct from and on a level with asserting: insisting (or requiring) and allowing (or declining to reject).

It seems open to question whether this explanation of modal talk in terms of what those who engage in it are doing—an explanation specifically designed to contrast with one in terms of how they are saying things are—can accommodate the idea of seeing the actual in the light of the modal. Surely when one understands an actual phenomenon by seeing it in the light of (say) a necessity, one's understanding is genuine to the extent that the insistence one expresses, by all means, when one affirms the necessity in the light of which one sees the phenomenon, is warranted. If there is really no such necessity, the supposed understanding is illusory. It seems doubtful that Haugeland's account of understanding can do without a conception on which claims of necessity or possibility purport to be true, and do their distinctive explanatory work only if they are indeed true—that is, only if they have the perfection that is proper to claims about how things are.

Diverting attention from questions about how one is saying things are, when one makes a modal claim, to questions about what one is doing—from something about the world (if one's claim is true) to something about the subject who makes the claim—is in the spirit of Haugeland's picture of "science without the dogmas," the topic of his culminating section. He writes (306):

The essential move must be to understand truth itself as a distinctively human achievement. Before there were people, there was no such thing as truth (at least not on Earth). By this, I don't mean just that there were no truth bearers—sentences, cognitive states, or whatever—but that there was no truth for anything to "bear." So, for instance, while it may be true *now* that there were frogs a million years ago, it was not true a million years ago that there were frogs *then*. Truth and objectivity as such had to be opened up and enabled by people, and that happened a lot more recently than a million years ago. Until we can understand them as in this way our own, philosophy will have yet to recover from the death of God.

But, properly understood, the positivism that Haugeland thinks he needs to avoid by saying this is only a truism. Truths are just facts. Conceiving them as truths locates them in relation to the idea of our discursive capacities; conceiving them as facts locates them in relation to the idea of the world. Those are two angles on a single kind of thing, each intelligible only in the context of the other. Certainly truths—that is, facts—had to be opened up by people, but that is not to say that acts of human intelligence were required to bring them into being. Haugeland's suggestion to the contrary is an ontologically inflected counterpart to the conceptual directionality—discursive capacities independently understood, and the idea of the world explained in terms of them—that figures in the way he interprets positivism.

And if we discard that suggestion, we can read Haugeland's picture of how science opens up the world as sketching the core not of a postrationalist epistemology but of an improved rationalism, free of the dogma of cognitivism but still characterized by a version of positivism. Positivism remains, in that we recognize our understanding of the very idea of the world as interdependent with our understanding of the idea of discursive capacities. What we learn from Haugeland's account of understanding is that for discursive capacities to play that role, they must include capacities for the modal thinking that enters into inquiry into the laws of nature.

What about Haugeland's proposed understanding of modal thinking? As he notes, it diverges from more usual understandings in not making room for iterated or embedded modalities. But it matches ordinary understandings in that those special illocutionary stands are conceived as directed toward what are already full-fledged propositions in their own right.

For an alternative that might be congenial to the project of rereading Haugeland's picture of science as a sophisticated positivism, let me invoke Kant once more. For Kant, modality is in the first instance a dimension along which judgments differ in form. Judgments that differ modally differ in how their contentful elements hang together. This conception of modality shares the first of those two features of Haugeland's proposal; there is no room for iterated or embedded modalities. But it does not share the second; a necessary truth is not an ordinary truth with a special feature, but a truth with its own distinctive form.[7]

This conception makes room, as Haugeland's does not, for modal forms of judgment to be forms of distinctive ways for things to be. That makes Haugeland's conception of understanding unproblematic. Now we can recognize that seeing the actual in the light of the necessary or the possible is seeing a way things are with one such form in the light of a way things are with another.

Notes

1. What I am going to say about the Deduction traces back to working through it with Haugeland himself and James Conant. It is a bit surprising that a reading that I thought the three of us had worked out together, at least in outline, seems to me to contradict what Haugeland says about the Deduction in the remark I have just quoted. This feature of Haugeland's paper is one that I especially wish I (and Conant)

could talk to him about. One way of conceiving the next few sections of this paper is as a program for that sadly impossible conversation.

2. I take Kant to think, rightly, that the B version improves on the A version. I quote from the translation of Norman Kemp Smith (Kant 1781/1929).

3. "Intuition" (*Anschauung*) here clearly means, as often in Kant, not "act of intuit-*ing*" but "item intuit*ed*."

4. The label "rogue appearances" is suggested by something Kant says when he is explaining why a transcendental deduction of the categories is needed: "That objects of sensible intuition must conform to the formal conditions of sensibility which lie *a priori* in the mind is evident, because otherwise they would not be objects for us. But that they must likewise conform to the conditions which the understanding requires for the synthetic unity of thought, is a conclusion the grounds of which are by no means so obvious. Appearances might very well be so constituted that the understanding should not find them to be in accordance with the conditions of its unity" (A90/B122–123).

5. *Scientific* know-how because, for continuity with his introduction of rationalism as one response to the scientific revolution, Haugeland concentrates on the meta-physics and epistemology of science in particular. But I think he means his argu-ments to apply more generally.

6. This need not be framed as *equating* know-how with suitable knowledge that. I think we can appreciate Haugeland's point without addressing the much-discussed questions in that area. Haugeland does not commit himself on those questions.

7. Of course that is consistent with recognizing, for instance, entailments between judgments with different modal forms: for instance, between a judgment whose content is a necessary truth and a judgment whose content is a contingent truth with the same contentful elements.

References

Kant, I. 1781/1929. *Critique of Pure Reason*. Trans. N. K. Smith. London: Macmillan.

Wittgenstein, L. 1974. *Tractatus Logico-Philosophicus*. Trans. D. F. Pears and B. F. McGuinness. Introduction by B. Russell. London: Routledge & Kegan Paul.

12 "Two Dogmas of Rationalism": A Second Encounter

Mark Lance

The appropriate response is not to denigrate scientific objectivity but rather to reconceive it in thoroughly posttheological—hence postrationalist—terms.

—John Haugeland, "Two Dogmas of Rationalism" (this vol., 305)

The project is to understand the objecthood of objects—their standing as criteria for objective skills—in terms of their constitutedness.

—John Haugeland, "Truth and Rule-Following" (*HT*, 326)

I first encountered John's reflections on rationalism when he presented "Two Dogmas of Rationalism" at Georgetown more than a decade ago. I was excited by the ideas but had a few objections, and we explored briefly some ways around these objections, as this generally pragmatist approach to epistemology and modality is something I am also committed to. The latest, still sadly unfinished, version of the paper that is included in this volume incorporates some of the objections but does not develop our thoughts about ways around them. In this brief response, I indicate the general direction of these ideas in the hope of a fuller development in the future.

1 On Understanding versus Knowledge and the World of Facts

John begins by defining positivism as a metaphysical assumption: "the view that reality is 'exhausted' by the facts—that is, by true propositions" (293). I'm not sure I understand this definition.[1] I am heartened to be told that "this needn't mean that there are no things, properties, or relations"—being a simple Appalachian boy at heart, I would otherwise be inclined to point to my dog, who is no proposition, as a counterexample—but my confusion returns when I read "but only that these are intelligible only as constituents of true propositions" (293). Of course, John is going to reject positivism so defined, but presumably the view is meant to be plausible enough to be worthy of extended refutation, and in my own simple way, I just don't see how it is. Again, why isn't my dog a counterexample?[2] Even if I were to grant—in a fit of Russellian sympathy—that she is a constituent of propositions, I have no idea what it could mean to say that she is only understandable as such. I know, at least roughly, what it means to say that I can understand her only by *grasping* propositions—that is, my understanding consists in my having beliefs with a propositional content, and so on. But that is not to understand *her as* a constituent of a proposition. The propositions that I deploy in understanding my dog are, for example, propositions like "dogs are animals," "dogs are far nicer on average than humans," and so on. But to hold those beliefs is to understand her as a constituent of the class of animals, or the class of nice beings, not as a constituent of a proposition. (The *concept* "dog" is a constituent of the propositions in question, but Lacy is not a concept.)

Happily, for purposes of my own engagement with this essay, the bulk of John's criticism of positivism focuses on just the epistemological claim that I more or less understand. That is, what he argues against is precisely the claim that there is some set of propositions such that a complete understanding of the world would be constituted by knowledge of each element of the set. John seems to target this as a consequence of the metaphysical doctrine of positivism, but for my purposes this doesn't matter. The epistemological doctrine is interesting in its own right, and worth considering whether one takes it to be a consequence of an independently interesting metaphysical thesis or not.

So what exactly is the argument against this epistemological version, or consequence, of positivism? First, note that we should understand "epistemological positivism" as more than just the claim that there is a set of propositions such that if one knows all of them, one understands all there is to understand. Rather, the claim is that the knowledge of the various propositions *together constitutes* complete understanding. Thus one might think that there are forms of understanding that take a very different form than knowing that P, but such that one can only count as knowing various propositions if one also has lots of this other sort of understanding. In such a view—an example hereafter—one could still think that it was a necessary truth that anyone who knows that P for every truth P would understand all that there is to understand, but deny that all understanding is amenable to formulation as S knows that P.

So what is the argument against this epistemic exhaustion thesis? Essentially it is an argument by counterexample, to wit: "scientific know-how." John writes:

> As empirical, science depends essentially on learning or finding out about the world by way of observation, measurement, and, above all, experiment. … Scientists must also, for instance, learn or find out *how* to make or perform the observations, measurements, and experiments that yield that factual evidence. Yet if that is so, then that learning-how cannot itself be just more evidence gathering, on pain of regress. (296)

Now in one sense there certainly is a regress in the offing. If one's goal is to explain know-how, then there is know-how involved just in understanding sentences; no need to move to fancy things like scientific evidence gathering. So one certainly can't explain know-how in general, for example, how one knows the sentences that describe empirical results, in terms of one's understanding further sentences. Indeed, John makes this very point:

> For language itself is an essentially skillful capacity—one that has to be learned and that, once learned, embodies knowledge of the world. To take the simplest example, learning to tell a duck when you see one is (part of) learning how to use the word "duck"—an instance of verbal know-how that embodies, as we might put it, knowing what a duck looks like. Without a great deal of such learning, students wouldn't even be able to read the textbooks, never mind use what they read in their own future practice. (297)

But this point undercuts John's main argument. The epistemological claim in question, recall, was that a total understanding of the world could be constituted by an understanding of some range of sentences—the complete description of the world. Even in this epistemological guise, if this claim is to have anything to do with the metaphysical claim that the world is constituted by propositions, it cannot be one that is falsified by a claim about the way that understanding language is constituted by skillful know-how. That I need perceptual skill to understand "duck" does not, in itself, imply an incompleteness in my description of the situation in front of me when I say that there is a duck swimming in the pond. My skills are not a part of the total description of the pond, of the totality of what is going on there.

They are, to be sure, a part of the total description of things in the broadest possible sense, but there is no vicious circularity in supposing that such skills can themselves be fully characterized as the content of a proposition. The claim is not the obviously false one that some list of sentences all on its own constitutes understanding of anything, but the claim that if one does understand all these sentences—whatever that understanding consists in, including the sentence that describes just what this understanding consists in—one understands all that there is to understand about the world.

If one is to argue that a complete understanding of the world is not constituted by knowing some total set of facts, one needs to show that there are particular aspects of this understanding that are not linguistically representable as propositional contents. Put another way, one must show that no complete understanding could be constituted by grasping a set of declarative assertions. As it happens, I think that there are such arguments. In Kukla and Lance 2009, we argue in detail that declaratives are not autonomous. That is, a social practice that makes only declarational assertions about an empirical world is not intelligible. Rather, the very existence of empirically meaningful declaratives presupposes both Lo! speech acts that serve to bring into the space of reason acts that express first-personal uptake through perception and at the same time call for intersubjective uptake on the part of others, and Yo! speech acts that express recognition of other speakers and call for symmetrical recognition from the one recognized. Since these other speech acts are, or so we argue, not reducible to declaratives, I take John's counterexample to hold in the end. Full understanding is not constituted by knowledge of propositions alone. Though I reject his specific argument for the conclusion, I think that John has identified a key false presupposition of at least one strand of rationalism.[3]

2   Modality

The second of John's two counterexamples concerns the understanding of modal claims. Here he begins with a distinction between two ways of understanding the function of a bit of discourse. Suppose that we have some grammatical operator * that applies to declarative sentences. The typical "semantic strategy" is to understand *P by taking this to be a candidate content to be asserted. Thus, presupposing that we have a background account of assertion and of the content of P, we explain the difference in content between P and *P—the meaning of *—and apply the same notion of assertion. Alternatively, one can suggest that what looks to be an assertion—"*P"—is in fact a distinct pragmatic act. That is, we see the * as functioning not primarily on the content P but on the act of asserting that P; we understand an act of producing "*P" as a *-assertion of P.

To illustrate, John takes up the example of negation.

Consider, by way of preparation, the difference between negation and denial. Both have to do with opposition or contrariety, but they are not the same. Negation is a logical operator that, in the simplest case, transforms a given proposition into another with the opposite truth value. Denial, on the other hand, is what Austin called an "illocutionary act" by which (for instance) a speaker takes a stand against some specified proposition. ... Suppose, now, we look at "modalizing" in the light of those alternative ways of "opposing." It is immediately clear that conventional approaches to modality all follow the model of opposing an assertion by negating the asserted proposition. That is, they all understand the "modalization" of an assertion as an assertion of a modalization of the propositional content originally asserted. ... But, as the example of denial suggests, it's not the only intelligible alternative. For, just as denial is a distinctive illocutionary stand that effectively opposes an assertion, we can easily imagine—or invent—distinctive illocutionary stands that effectively modalize assertions. So, for instance, the illocutionary counterpart to "I assert that necessarily *p*" might be something like "I insist (or require) that *p*." (299)

I think this approach is exactly right, and in what follows I want to suggest both that taking this line does not require the theoretical concession that John accepts—and why it is important not to—and also to add a bit of detail to the position. Both moves depend on radicalizing John's methodology: that is, taking this pragmatic approach to the function of language to apply not merely to certain specific locutions but to linguistic practice in general.[4]

John's concession is the following: "First, it does not prevent the articulation of a modal logic, though it does impose an important limitation on that logic: nested or embedded modalities are ruled out. The reason is that illocutionary acts must, qua acts, have widest scope. They cannot be within the scope of any others and still themselves be acts" (300). Later he suggests that this is not a particularly troubling limitation: "But that limitation is irrelevant when the topic is scientific laws, since their modal character always has widest scope (laws do not contain embedded modalities)" (301).

I disagree on both counts: the limitation is entirely relevant when the topic is scientific laws, but not, in the end, necessary. It is relevant because while it may be true that laws do not contain embedded modalities—though this is hardly uncontroversial—it is certainly not true that nested law claims are irrelevant to science. "If it is a law that P, then it is a law that Q" is certainly something that is intelligible and often important in the context of scientific discourse. Further, one sees structural principles of the form "Any law of domain D is such that F" and other compounds. So it does seem important that any account of the function of law-talk incorporate nested constructions. Thankfully, John's sketch of an obstacle for this is not as devastating to the project as he thinks.

It is true enough that a subsentence Q of a given sentence P that is produced in a speech act S is not typically itself S-ed. If, for example, I assert "If P then Q," I have not asserted P, or Q. But it does not follow from this that we cannot understand the pragmatic function of a sentence in various speech acts in terms of the pragmatic function of that sentence's subsentences in various speech acts.[5] Indeed, John's example of negation is a clear one.

Here we might begin with the two pragmatic performances of asserting and denying that John identifies. Now how would we understand the content of the complex sentence ~P? Well, the natural suggestion, following Brandom, is to begin with a set of material inferential proprieties across the atomic sentences of the language—themselves understood pragmatically in terms of commitment or entitlement. Then we say that to assert ~P is to deny P, and to deny ~P is to assert P. For other complex constructions C, what we need is to show how the set of assertions and denials one undertakes in uttering C depends on the assertions and denials of various simpler sentences. Thus to assert P&Q is to assert P and to assert Q (and hence to commit oneself to anything one is committed to in either assertion). To deny P&Q is to commit oneself to every sentence that follows from the denial of P and also from the denial of Q. There are details to work out, but this has been done.[6] So John's own example shows that a pragmatist approach to the function of language need not preclude understanding nested constructions.

Turning to modality, the key idea is to notice that contexts of reasoning can be initiated hypothetically. The idea of a hypothetical context is simple and crucial for understanding the force of reasoning.[7] Thus we can introduce a new context within the game of giving and asking for reasons by saying "suppose P." The norms governing that new hypothetical context are a simple emendation to those of the normal one, namely, that we reason as if everyone is committed to P, and then if we conclude Q within the context of a hypothetical context introduced by "suppose P," we are allowed to conclude in the nonhypothetical context "If P, then Q."

But "suppose P" is said in many ways. In a paper with Heath White (Lance and White 2007), we consider a distinction between two important sorts of hypothetical contexts, contexts governed by subjunctive and indicative norms of material inferential propriety. These correspond to contexts in which the resulting → introduction rule corresponds to a subjunctive and an indicative conditional. Thus one might reason as follows:

1. Suppose (we discover) that Booth did not kill Lincoln. [Introducing an indicative hypothetical context in which one is given free entitlement (and commitment) to "Booth did not kill Lincoln."]
2. But Lincoln died of a gunshot wound that night at the theater.
Therefore,
3. Someone else must have killed him. So [now returning from the hypothetical context]
4. If Booth did not kill Lincoln, then someone else did.

Alternatively, one might reason:

1. Suppose (we stipulate) that Booth did not kill Lincoln. [Introducing a subjunctive hypothetical context in which one is given free entitlement (and commitment) to "Booth did not kill Lincoln."]
2. The only reason Lincoln died that night is that Booth killed him.
Therefore,
3. Lincoln lived to finish his term.
So [now returning from the hypothetical context]
4. If Booth had not killed Lincoln, then he would have lived out his term.

Details are not important for current purposes (see Lance and White 2007). What matters is that by applying a pragmatic methodology generally, we make sense of hypothetical contexts and the function of speech acts within them. This allows us to make sense of both various forms of conditional reasoning and also various conditionals that, in nonhypothetical contexts, make explicit proprieties governing such hypothetical inference.

What's more, introducing subjunctive hypothetical contexts allows us to develop the pragmatic function of modal "insisting" that John offers as an alternative to semantic accounts of modality. To commit oneself to "necessarily P" is to commit oneself to the propriety of asserting P within all discursive contexts: both hypothetical and nonhypothetical. Specific forms of modality—logical, physical, and so on—will correspond to contextual variations in the class of relevant contexts. The best way to think of this is in terms of the notion of a set of situations that are "onstage" in the sense of Lange (2007). The idea is that each scientific project determines a range of contexts that are relevant to lawlike claims made within the development of that project. (Lange speaks in terms of worlds, but we can easily translate that into the pragmatic terms of hypothetical contexts.) To make a lawlike assertion is to make a claim that applies to reasoning in each onstage context.[8]

Once one thereby accounts for the pragmatic significance of freestanding claims of the form "necessarily P," one is in a position to understand the significance of embedding them within conditionals as well by adapting the general account of conditionals as commitments to the propriety of inferences, or even to nest them if one allows that introducing a subjunctive hypothetical context can change the rules of acceptable hypothetical reasoning. It is a substantive empirical question whether such allowances will invoke scientific inferential practices that succeed in confronting the world, but the formal apparatus that begins with practically induced statuses in no way precludes this.

Clearly many details remain to be worked out, and my goal here is not to offer a formal semantics or even a partial formal pragmatics of modal claims. Rather, I aim merely to show that we have no principled reason not to push John's pragmatist methodology as far theoretically as any "semantic" approach. To do so, however, requires that we push the pragmatism consistently, understanding the function of "if-then" and other important logical devices by way of the same pragmatist methodology that we apply to modal notions.

3 Conclusions

Beyond these specific criticisms of positivism, John offers in "Two Dogmas of Rationalism" another variant on two themes that became prominent in his later work. The first is the idea that by subjecting our engagements with the world to the right kind of normative constraint, we "let things be" (*HT*, 325)—that is, we open up the possibility of being normatively constrained by the way the world is. Our disclosive activity thus comes to involve a commitment to getting things right, to truth and objectivity. The second is the idea that such normative commitment must ultimately call for a sort of engagement that is not capturable as a declarative content, even if that content concerns social reality. That is, in the crucial notions of integrity, love, and commitment, we see the fundamental ontological problem in Heidegger's sense, the issue that is ontologically grounding of the very possibility of objective disclosure and at the same time essentially a matter of taking a stance on our own being. Ontology, on this Heideggerian reading, is a matter of calling us to a sort of engagement with the world and our social practices of disclosing us. Both of these are deep and powerful ideas that have inspired a generation of students, and I am certainly pleased to count myself among them. It is one dimension of the tragedy of John's early death that these ideas did not receive a full development in his own voice and, for many of us, a personal loss that we did not get to engage with him fully around them.

At times in this essay, as in others, John flirts with a sort of individualist existentialism ("A will that endures the vagaries of the world by its own willful law is no mere inculcated habit or persisting desire, no social institution or status, but something new that has emerged out of them"), a tendency that Rebecca Kukla and I have criticized recently (302–303).[9] As for the idea that forthright engagement in social norms and forthright subjection to the resulting worldly exposure are not constituted by any social fact, I have argued for just such a view in recent work with Andrew Blitzer.[10]

Neither of these papers nor much of the rest of my own work in developing pragmatic understandings of language would have been possible without John's teaching and collegiality. John leaves us not only a profound and fertile picture of being-in-the-world but a huge range of detailed philosophical arguments with which to grapple. It has been my own great good fortune to be a small player in the project of developing this vision, from my student days, learning of these ideas from John at the University of Pittsburgh, to long conversations at Asilomar conferences and other venues. It is my hope that this "second encounter" will give rise to further valuable conversations, although John's own voice is no longer with us.

Notes

1. I often fear that I do not understand any strictly metaphysical claims. On my more confident days, I attribute the defect to metaphysics, and on my less confident days to myself.

2. She is a wonderfully friendly pit bull mix, in case you are curious.

3. Andrew Blitzer and I develop a different line of argument that has as a corollary this same conclusion. In "The Question of Being: Existential Pragmatics and Critical Ontology" (work in progress), we argue that what Heidegger calls ontological claims—claims that express the transcendental structure of being—must be seen as nondeclaratival.

4. It seems to me that such a global pragmatic approach to language is exactly in line with John's Heideggerianism. (See my other paper in this volume for one development of that line of thought.) For more on the systematic pragmatic approach to language, see Kukla and Lance 2009, 2013.

5. I grow more skeptical by the year of the idea that there is a complete compositional theory of language, whether based on pragmatics or semantics, but (a) I'm much more skeptical about systematic semantic theory, and (b) this argument is completely independent of the current considerations. I have no doubt that certain pragmatic functions of language are compositional.

6. The general idea and much of the underlying pragmatist motivation are developed in Brandom 1994. I work out details of the underlying interactions of commitment, entitlement, and inference in Lance 1995 and develop formal commitment-based semantics for a language with &, v, $\sim$, and a conditional expressing entailment in two papers with Philip Kremer (Lance and Kremer 1995, 1996). The general idea of beginning with two statuses undertaken in the acts of commitment and denial generates a natural four-valued logic for complex claims. This was first studied by J. M. Dunn (1971) and is systematically related to a pragmatically induced notion of incompatibility in Lance 2001.

7. I first discussed this in the context of explaining how a pragmatism of roughly the sort developed in this paper should understand quantification in Lance 1996. The move also makes an appearance in my other paper in this volume.

8. Lange offers one other important development of the ideas that John sketches in this paper. In addition to asserting P as a law having the implication that P is true across subjunctive contexts, it alters the structure of acceptable indicative inference within those contexts. To treat a claim as a law is, in Lange's terminology, to put forward an inductive strategy. Thus, to say that it is a (putative) law that sugar dissolves in water is to endorse an inductive strategy according to which testing this bit of sugar to see if it dissolves in this bit of water is an act that, if successful, raises the probability for any bit of sugar and any bit of water that the former will dissolve in the latter. By contrast, testing whether there is a quarter in Marc's pocket in no way changes the probability of there being a quarter in John's pocket.

9. See Kukla and Lance 2014. We did get to present an early version of that paper to the Asilomar conference with John present, and at the time he happily, loudly, and profanely disowned this individualism.

10. "The Question of Being" (work in progress).

References

Blitzer, A., and M. Lance. Work in progress. The question of being: Existential pragmatics and critical ontology.

Brandom, R. 1994. *Making It Explicit: Reasoning, Representing, and Discursive Commitment.* Cambridge, MA: Harvard University Press.

Dunn, J. M. 1971. An intuitive semantics for first degree relevant implication. *Journal of Symbolic Logic* 36:362–363.

Kukla, R., and M. Lance. 2009. *'Yo!' and 'Lo!' The Pragmatic Topography of the Space of Reasons.* Cambridge, MA: Harvard University Press.

Kukla, R., and M. Lance. 2013. Leave the gun; take the cannoli! The pragmatic topography of second-person speech. *Ethics* 123 (3): 456–478.

Kukla, R., and M. Lance. 2014. Intersubjectivity and receptive experience. *Southern Journal of Philosophy* 52 (1): 22–42.

Lance, M. 1995. Two concepts of entailment. *Journal of Philosophical Research.* 20:113–137.

Lance, M. 1996. Quantification, substitution and conceptual content. *Noûs* 30 (4): 481–507.

Lance, M. 2001. The logical structure of linguistic commitment III: Brandomian scorekeeping and incompatibility. *Journal of Philosophical Logic* 30 (5): 439–464.

Lance, M., and P. Kremer. 1995. The logical structure of linguistic commitment I: Four systems of non-relevant commitment entailment. *Journal of Philosophical Logic* 23 (4): 369–400.

Lance, M., and P. Kremer. 1996. The logical structure of linguistic commitment II: Systems of relevant commitment entailment. *Journal of Philosophical Logic* 25 (4): 425–449.

Lance, M., and H. White. 2007. Stereoscopic vision: Reasons, causes, and two spaces of material inference. *Philosophers' Imprint* 7 (4): 1–21.

Lange, Marc. 200. *Natural Laws in Scientific Practice*. New York: Oxford University Press.

13 Appendix: The Transcendental Deduction of the Categories

From the *Critique of Pure Reason*, B edition (an outline and interpretation)

John Haugeland

Editors' introduction: While at the University of Pittsburgh, John Haugeland participated in a long-running reading group on Kant's *Critique of Pure Reason* with James Conant and John McDowell. The following is an outline of the second edition Transcendental Deduction that Haugeland put together as the result of this reading group. Although the outline is incomplete, the overall thrust of the reading is clear and easy to follow.

Background and overview
The Aesthetic argues that, if anything is to be an object for us, it must be a spatiotemporal object, intuited under the forms of space and time.
The first half of TD-B [Transcendental Deduction, B edition—Ed.] argues that if anything is to be an object of intuition (at all) it must be a possible object of judgment—the condition of the OSUA [objective synthetic unity of apperception—Ed.]—thus subject to the categories.
The second half of TD-B argues that being presented in space and time is a sufficient condition for being an object of judgment.
The Aesthetic (§§1–8) argues that objects can be presented to *our* senses only in accord with the formal conditions of space and time. The legitimacy of such notions as "intuition" and "sensibility" is simply taken for granted in this discussion.
Chapter 1 of the Analytic of Concepts (§§9–12) identifies the categories—the pure concepts of the understanding—by identifying them with the logical functions of judgment; but it does not justify this identification or show the legitimacy of the categories as *a priori* concepts.
The first segment of the deduction chapter (§§13–14) explains the problem of legitimating the pure concepts by broaching the possibility that objects might be presented to our senses (*a la* the Aesthetic) whether they accord with the pure concepts (categories) or not. To put the worry another way: Why couldn't there be sensible (intuitable) objects that aren't subject to the conditions (pure concepts) of the understanding (that is, aren't intelligible)?

The first half of the deduction in B (§§15–20) addresses the question: What are the conditions on anything's being a *representation* at all? It argues that the ability to represent presupposes the ability to judge. The argument is this: to be a representation is to have an object; but an object is that in which more than one representation can be united; and judgment is just the ability to unite representations (in an object). To say that more than one representation can be united in an object is to say that the object itself has (must have) a *synthetic* unity—which amounts to saying that there can be (synthetic) judgments about it. The *categories* are nothing other than specifications of the possible forms of such synthetic unity.

Hence, the ability to represent at all is essentially two-sided: it is *both* an ability to be given a manifold (receptivity) *and* an ability to ascribe unity (spontaneity). *Sensibility* is the ability to represent insofar as it is receptive; and *understanding* is the ability to represent insofar as it is spontaneous— that is, the ability to judge.

Think of this as a rejoinder to, for instance, Descartes and Hume. Descartes said that ideas differ from bodies in that they have objective reality (that is, they are representations); but he had no account of *how* they could have objective reality, or even what that is—he just took it for granted. Hume said that ideas are just faint copies of impressions (items that are somehow impressed upon us), thus leaving no options but to take representationality for granted (as did Descartes) or to deny that there is any such thing— between which he vacillated.

The first half of the B deduction considered representation as such—in the "abstract." The second half (§§22–26) lifts that abstraction, and undertakes to show that what *we* have are representations in this sense—in particular, that the "intuitions" discussed in the Aesthetic really deserve that title. (An *intuition* is, by definition, an immediate *representation* of a single object.) So, in effect, the question to be answered is whether the prospect raised in §13, that objects might be presented to *our* intuition (in space and time, as described by the Aesthetic) that are not subject to the conditions of the understanding (that is, are unintelligible), is a genuine one. And the answer to that question is: No!

The reason is that the forms *of our* intuition, space and time, are themselves intuitable objects (formal intuitions), and so are subject-to the conditions of the understanding. In other words, they must have the synthetic unity prerequisite to being the objects of synthetic judgments (namely, the synthetic a priori judgments of geometry and arithmetic). But that means that any object presented *in* space or time must, as conforming to the forms thereof, *ipso facto* also participate in that synthetic unity—which is none other than the unity that the categories specify. Therefore, any object that is so much as presented to our intuition is subject to the categories— the conditions of the understanding. (Hence the use of the terms "intuition" and "sensibility" in the Aesthetic was legitimate all along.)

> Here's another way to see essentially the same dialectic. Someone (a defender of the prospect raised in §13) might acknowledge that to be a representation is to have an object, and that to be an object is to have a synthetic unity (that is, to be representable by more than one representation), but maintain that the resources of the Aesthetic alone provide for this. In particular, to conform to the forms of space and time is already to have all the unity that an object needs in order to be an object. But obviously, the unity of space and time themselves cannot be accounted for in this way. So the second half of the B deduction can argue that any unity that an object could inherit from the forms of intuition would just be the unity that the categories specify, but at one remove.
> *There is only one unity.*

Analytic of Concepts

Chapter II: The deduction of the pure concepts of the understanding

> The aim of the deduction is to show that every object apprehensible by us (every *empirical* object) is determined in respect of the categories (and is, therefore, subject to laws).
> Thus, it will be shown that the categories apply to such objects, hence have objective validity.

Section 1 [common to the first and second editions]
13. *The principles of any transcendental deduction*

> To *deduce* a concept is to show its *objective validity*—that is, that it applies to *objects*.
> A deduction is *transcendental* if it is shown *a priori* that the concept applies to objects.
> It's no particular problem to deduce the a priori concepts of space and time, since objects can't even appear to us except under the pure forms of intuition (= space and time).
> But deducing the pure concepts of the understanding is tougher, since it seems [prima facie] that intuited objects need not accord with the conditions of understanding (= be intelligible).

- Jurists distinguish the question of right *(quid juris)* from the question of fact *(quid facti)*.

  —Proof of the former (the right) is called a deduction.

- To deduce a concept is to show the legitimacy of using it in regard to objects.

  —That is: to show that it has *objective reality* or *meaning*.

> The terms *objective reality* and *objective validity* are notoriously hard to interpret; but there's a reading that works fairly well. First, both are terms for the "objective purport" (intentionality, meaningfulness) of representations—without which, of course, they wouldn't be *representations* at all.
> *Objective reality* is used for all representations *other than judgments;* so, to a first approximation, it means what it did for Descartes.
> *Objective validity*[1] is used for *all* representations, *including judgments.* For the latter, it means: having a truth value (that is, purporting to represent an objective fact). This is important, because Kant's account of judgment lies at the heart of the philosophical breakthrough in the first *Critique* (and hence lies behind his radically new accounts even of concepts and intuitions).

—For most empirical concepts, no deduction is needed (though some are problems).

—For pure a priori concepts, however, whose legitimacy can never be proven empirically, an explicit deduction is always needed—to show how they can relate to objects at all.

—Showing how such concepts can relate a priori to objects is a *transcendental* deduction.

—We already have two kinds of concepts that relate to objects completely a priori:

  —The concepts of space and time, as forms of sensibility; and

  —The categories, as concepts of understanding.

—So, if they need deductions [which they do], these will have to be transcendental.

• We are indebted to Locke for opening up the field of psychological investigation of how we come to have the concepts we have, including general concepts, as a result of experiences.

  —But such derivation is not deduction [since it establishes only origins, not legitimacy].

• But is a deduction (transcendental, to be sure) of pure a priori concepts really *needed?*

  —For instance, we have given [in the Aesthetic] a transcendental deduction of the concepts of space and time—that is, explained and determined their a priori objective validity.

  —But geometry doesn't *need* philosophy to be (and know that it is) legitimate a priori.

—That, however, is only because a priori intuition provides immediate evidence of the pure form (= space) of the outer sensible world.

—And geometry, applying, as it does, only to that world, and indeed only as regards its pure form (spatiality), is grounded in that immediate intuition.

—It is, however, quite different with the *pure concepts of understanding*.

—Since they speak of objects through predicates of pure a priori thought (not of intuition or sensibility) they apply to objects [*Gegenständen*] universally.

—That is, they are not limited to objects given under the conditions of *our* sensibility.

—Hence, there can be no a priori [formal] intuition that grounds their synthesis—that is, establishes their objective validity.

—In fact, this very generality of the pure concepts of understanding also leads to a tendency to employ the concept of space beyond the conditions of our sensibility.

—Which is why a transcendental deduction of space and time was needed after all.

• It wasn't that hard to explain why the concepts of space and time, though a priori, must nevertheless relate to [empirical] objects, and even enable synthetic knowledge of them.

—For, if any object [*Gegenstand*] is going to appear [to us] at all, it's going to have to be an Object [*Objekt*] of empirical intuition, and thus to accord with the a priori conditions of [our] intuition—namely, space and time.

Kant seems to distinguish (in the B edition, anyway, not in A) between *Objekt* and *Gegenstand* (both of which are rendered *as* "object" by Kemp Smith). Again, it's not very clear—but here's a reading that works fairly well.

Objekt[2] is a formal term for that which a representation purports to represent. It means: "whatever is represented" (if anything). To be a representation at all is to have such an *Objekt*—so the two terms are coordinate.

A *Gegenstand*[3] is an object that is (or could be) *given to us*—that is, one that is (or could be) intuitively apprehended *by us,* and thus be known. In *particular,* to be an *empirical* object (in space and time) is to be a *Gegenstand.*[4]

I will mark the distinction by capitalizing "Object" when it is *Objekt.*

> To show that a representation has objective validity, it suffices to show that
> it has an *Objekt*—which is just to show that it is indeed a representation.
> But to show that a *synthetic* representation has objective validity *a* priori—
> hence universally—it must be shown that all possible *Gegenstände* would be
> among its *Objekte*.

> It is worth bearing in mind that, in the preface to the B Edition, Kant
> explicitly says (twice, at Bxvii) that things as objects [*Gegenständen*] of
> experience (appearances) are the very same things as things in themselves.

—So the synthesis that rakes place in such objects [*Gegenstände*], qua
appearances, has objective validity.

• But it's going to be a lot harder to deduce the categories of the understanding, for they do not [ostensibly] represent conditions under which
objects can so much as appear to us.

> The bracketed qualifiers in this and the next few items flag an
> interpretation of the point of this paragraph (A89f/B122f)—namely, that it
> sets up the task of the deduction by putting forward a seeming possibility
> which the deduction will then show not to be genuine. This
> possibility could be called the possibility of *rogue appearances*—that is, the possibility
> of appearances (objects of intuition) that are *unintelligible*, in the sense that
> they do not accord with the conditions of the understanding (that is, the
> forms of the unity of thought or judgment—namely, the categories).

—So it's hard to see how *subjective conditions of thought* could be conditions of the possibility of all knowledge of objects [*Gegenstände*], and
[so] have a priori objective validity.

—For [one would suppose] appearances can certainly be given in intuition independently of the functions of the understanding.

—The concept of cause, for instance, signifies a special kind of synthesis
whereby two quite distinct appearances must relate to each other in
accord with a rule.

  —But it's certainly not obvious a priori why they should have to do
  that.

  —(Nor could experience help, since what we're after is a priori objective validity.)

  —Appearances might very well [it seems] be in such a confusion that
  nothing presented itself that would answer to the concept of cause
  and effect.

  —But, since intuition as such [on this assumption] doesn't need the
  functions of thought, those appearances would still present objects
  [*Gegenstände*] to our intuition.

- In the meantime, it would be useless to try to evade the issue, by suggesting that the concept of cause might instead be abstracted from experience, since that concept itself demands strict necessity and universality—which can never be established through experience alone.

14. *Transition to the transcendental deduction of the categories*

> The basic strategy of the deduction will be to show that the categories are conditions of the possibility of experience (empirical knowledge of objects [*Gegenstände*]) after all.

- There are only two cases in which synthetic representations and their objects [*Gegenstände*] can make contact, relate to one another in a *necessary* way, and, so to speak, meet each other:

  —Either the object alone makes the representation possible;

  —Or the representation alone makes the object possible.

  —In the first case (which includes appearances, in regard to that in them that belongs to sensation), the relation is only empirical, and the representation cannot be a priori.

    —In the second case, the representation itself would not *produce* its object, in the sense of causing it to exist (since causality by the will is not in question).

    —Rather, the representation determines the object a priori, in the sense that, only through the representation, is it possible that anything is *an object to be known.*

- Now, there are two conditions under which alone knowledge of an object is possible.

  —First: *intuition*, through which an object is given (though only as appearance); and

  —Second: *concept*, through which an object corresponding to this intuition is thought.

> Don't think of these as two independent conditions. Rather, they are two essential aspects of any finite knowing—two sides of one coin.

  —As has been shown, the first condition—under which alone objects can be intuited—does indeed lie a priori in the mind as the ground of Objects [*Objekten*] as far as their form is concerned [namely, in the a priori forms of intuition].

—So, the question is: Might a priori *concepts* likewise be conditions under which alone anything can be, if not intuited, at least *thought* as an object [*Gegenstand*] at all?

—If so, all empirical knowledge of objects would have to conform to them, since they would be presupposed by the possibility of anything being an *object of experience.*

—To be sure, all experience does contain, in addition to the intuition in which something is given, also a *concept* of an object as being thereby given—that is, as appearing.

—Thus, concepts of [what it is to be] an object at all [*überhaupt*] do lie at the basis of all empirical knowledge as its a priori conditions.

—So, the a priori objective validity of the categories rests [as will be shown] on the fact that experience becomes possible only through them.

• The transcendental deduction of all a priori concepts has therefore a guiding principle: these concepts must be recognized as a priori conditions of the possibility of experience—whether of the intuition or of the thought. [That is, they must be recognized as conditions of *both*.]

• Locke and Hume both held that all concepts of objects had to be derived from experience.

—Locke included pure concepts of the understanding among these, and, not seeing that this would gravely limit those concepts, gave way to enthusiasm in their application.

—Hume, appreciating that pure concepts could not be so derived, despaired of finding any pure concepts applicable to objects at all, and so gave way to skepticism.

• The categories are concepts of [what it is to be] an object [*Gegenstand*] at all, and through which the intuition of an object is regarded as determined in respect of one of the logical functions of judgment.

—So, for instance, from a merely logical ["formal"] point of view, either concept [either term] of a categorical judgment could be a subject, and the other a predicate.

—But, if a concept is brought under the category of substance, it is determined such that its empirical intuition can only be considered as subject, and never as mere predicate.

Section 2 [as restated in the second edition]

Transcendental deduction of the pure concepts of the understanding

15. The possibility of combination in general

Here begins the "first half" of TD-B (the transcendental deduction in B). The aim of this half is to show that, insofar as anything is an Object [*Objekt*] of a representation at all, it is determined in respect of the categories (that is, the categories apply to it).
§ 15: Combinedness of representations is never given, but must be spontaneously supplied.
The *act* of combining is called *synthesis*; the *ability* so to combine is called *understanding*.
Such synthesis presupposes [hence implies] a synthetic unity, which is *very* fundamental.

• Though a manifold can be given in intuition, its combinedness cannot.

—Thus, the combinedness cannot be already contained in the pure form of sensible intuition either.

—The form of intuition, though it lies a priori in our ability to represent [our faculty of representation], is no more than the mode in which we are affected.

—Combining, on the other hand, is a spontaneous act of our ability to represent.

—This *ability* to represent, insofar as it is thus spontaneous, we call *understanding*.

The same ability to represent, insofar as it is receptive, is called *sensibility*. So, like, concept and intuition, understanding and sensibility are two sides of one coin—namely, the spontaneous and receptive side, respectively, of the ability to represent (or, as we could also say, of the *finite* ability to know).

—The *act* of combining we call *synthesis*.

—Combining cannot be given through objects, but is only ever executed by the subject.

To *combine* (synthesize, conjoin, unify) representations is not to gather, merge, or fasten them together, but rather to *take* (conceive, think of) them as *belonging* together, as a function of what they represent.

• The concept of combining includes [presupposes] the concepts of:

—the manifold [what gets combined];

—its synthesis [the act of combining]; and *also*

—the [synthetic] *unity* of the manifold [as so combined or combinable].

• Combining is a representing of the *synthetic* unity of the manifold.

  —Thus, the representation [concept] of this unity can't arise from the combining.

  —Similarly, this unity cannot be [what is meant by] the *category* of unity, since the combinability in question is presupposed by all concepts, hence all the categories.

• Therefore, we must seek this unity higher up, in whatever contains the ground of the unity of all sorts of concepts in judgment (and, hence, of the possibility of understanding at all).

16.   The originary synthetic unity of apperception [OSUA]⁵

> In order for anything to be a representation *of mine,* it must be possible for me to be conscious of it (for my "I think" to accompany it). This is the *analytic* unity of apperception.

> But I can be conscious of representations as mine only insofar as I can consciously conjoin them—unite them—with one another in one self-consciousness (one apperception).
> The implied unity (in the synthetic unifi*ability*) of all *my* representations *as mine* is the *originary synthetic unity of apperception—the OSUA.*⁶

• It must be possible for the "I think" to accompany any of my representations [insofar as they are anything to me at all].

  —Intuitions are representations—namely, ones that can be given prior to any thought.

> Intuitions are representations by definition. It remains to be shown, however, whether—or, rather, *how*—we can actually have such representations.

  —Thus, all intuitions given to any subject bear a necessary relation to its "I think."

• But the "I think" is an act of spontaneity.

  —This [the act, the "I think"] is called *originary* or *pure apperception.*

  —It is called that because it is *self-consciousness*—namely, the self-consciousness that:

    —*generates a priori* the "I think" (which can accompany any other representation), and

    —cannot itself be accompanied by any further representation.

- The *unity* of this apperception is transcendental, in the sense that a priori knowledge can arise from it.

  —The many representations given in intuition, if they are to be all *mine*, must belong to one self-consciousness.

    —Thus, insofar as they are mine, they must conform to whatever conditions there may be on standing together in one consciousness.

  —From this [prerequisite, hence necessary] originary combinedness [or: combinability], many consequences follow.

- The identity [one-ness] of the apperception of a given manifold [many-ness] contains [presupposes] a synthesis, and is possible only through the consciousness of that synthesis.

  > This *numerical identity* or *one-ness* of apperception—of the *one* "I think" that can accompany *many* representations—is later called the *"analytic* unity" of apperception (as opposed to its *synthetic* unity: the *OSUA*).
  > The presupposed synthesis will turn out to be the transcendental = productive synthesis of imagination (see §24, pp. B151f; and cf. A118f and A124f).

  —For, in what way are the many representations to be related to the identity of the subject?

    —Not through any accompanying empirical consciousness—that's too inconstant.

  —Rather, the many representations are related to one and the same subject only insofar as that subject *can* conjoin them with one another, and be conscious of so doing.

    —Only insofar as I *can* unite [actively synthesize] many given representations in one consciousness, *can* I so much as represent the one-ness of that consciousness.

  —Hence, the *analytic* unity of apperception presupposes a *synthetic* unity [or: unifi*ability*].

- Thus, the thought that the many representations given in intuition all belong to me—are all *mine*—is equivalent to the thought that *I can unite* them in one consciousness.

  —So, (a priori) synthetic unity of the manifold of intuitions is the ground of the [analytic] identity of apperception—which, in turn, precedes a priori all *my* determinate thought.

• The principle of [the necessary originary synthetic unity of] apperception [*POSUA*] is the highest principle of human knowledge.

—Though it is itself an analytic proposition, it reveals the necessity of [i.e., it presupposes] a synthesis without which the identity of self-consciousness could not be thought.

—This necessary synthesis is called the *originary synthetic unity of apperception* [*OSUA*].

—All representations given to me must stand under this unity of apperception—but they have to be brought there by a synthesis.

> The originary synthetic unity of apperception is the necessary objective unifi*ability* of the manifold given in intuition; it replaces what was called the transcendental *affinity* of the manifold in TD-A (compare pp,~ B134f to pp. A113f and, especially, A122f). In each case, the affinity/unifiability is inferred from the fact that, otherwise, I would be able to have representations of which I *could not* be conscious—grasp as "mine"—which is incoherent.
>
> That is, the synthetic unity is inferred from the analytic unity (because the latter presupposes it). Thus, the synthetic unity is really more basic—the synthetic unity *grounds* the analytic unity.

17.   The POSUA is the supreme principle of all employment of the understanding

> The principle of the synthetic unity of apperception [POSUA] is this: in order for anything—in particular, [anything presented in] any intuition—to be a knowable Object [*Objekt*] for me, it must conform to the condition of the originary synthetic unity of apperception.

• There are two parallel supreme principles, which can be presented and compared this way:

The supreme principle of the possibility of all intuition, in its relation to	
sensibility	understanding
—that is, insofar as the manifold of intuition	
are *given* to us—	must allow of being *combined* in one consciousness—
is that all the manifold of intuition should be subject to	
The formal conditions of space and time	conditions of the OSUA [whatever those might be]

> Note that *both* principles concern the possibility of *intuition*—in its relations to sensibility and to understanding, respectively.

- *Understanding* is the *ability to know* [the faculty of knowledge].

  —Knowing consists in the determinate relation of representations to an Object [*Objekt*].

  > Determinate in the sense of determining. Thus, in a synthetic judgment, the predicate is not contained in the subject, and so further determines it.

  —An *Object* [*Objekt*] is that in the concept of which the manifold of a given intuition is *united*[7]

  —Since any unification of representations demands [*synthetic*] unity of consciousness [apperception] in their synthesis, that unity alone constitutes [*ausmacht*]:

  (i)   the relation of representations to an object [*Gegenstand*]

  (ii) the objective validity of representations, and

  (iii) the fact that they are cognitions [modes of knowledge] at all

  —Hence, upon it [OSUA] rests the very possibility of understanding [the ability to know].

- Accordingly, the mere form of outer sensible intuition, space, is not yet knowledge.

  —It is only a manifold of a priori intuition for a possible knowledge.

  —Even to know something purely spatial—a line, say—I must *draw* it, and thus synthetically bring to a stand [make "sit still"] a determinate combination of the given manifold.

  —The unity of this act = the unity of consciousness [= OSUA] (as in the concept of a line).

    —And, through this unity, an Object [*Objekt*] (a determinate space) is first known.

  —So, OSUA is an objective condition of *all* knowledge [even a priori knowledge of space].

- OSUA is the condition under which [anything presented in] any intuition must stand in order to *become an Object* [*Objekt*] *for me*. [This is the POSUA].

  [Repeats that, though the POSUA presupposes a synthesis, it is itself analytic:]

  [Points out that POSUA wouldn't apply to a creative understanding, such as God's.]

18.   The objective unity of self-consciousness

> The OSUA is *objective:* that is, the implied unity is in the *Objects* of
> knowledge themselves.

- The OSUA is not only transcendental but *objective,* in the sense that,
  through it, the manifold given in a [unitary] intuition is united into a
  concept of an Object [*Objekt*].

  —This must be distinguished from the empirical unity of consciousness,
  as in [Humean] association, which has merely *subjective* validity.

19.   The logical form of all judgments consists in the objective unity of the
apperception of the concepts which they contain

> To be a *judgment* is to be a cognition in which other cognitions are brought
> to the *objective* unity of apperception—synthesized in virtue of its necessary
> unity.
> That is, they are represented as combined *in the Object* (not just as
> associated *subjectively*).
> This, for instance, is what the copula "is" means.
> [The real question is how we can so much as *mean* that "... *is* ..." (about
> *empirical* objects).]

- To define a judgment as a representation of a relation between two concepts, is (whatever else is wrong with it) utterly hopeless in that it does
  not say what that relation consists in.
- The relation among the given cognitions [intuitions, concepts, or clauses]
  in a judgment is nothing other than the way of bringing them to the
  objective unity of apperception.

  —This relation is distinguished, as belonging to the understanding, from
  any relations according to laws of [Humean] association (which are
  merely subjective).

- This is what is intended by the copula "is":

  —It distinguishes objective unity of given representations from subjective unity, and

  —It indicates their relation to originary apperception and its *necessary
  [synthetic] unity.*

    —The latter is so, even if the judgment itself is empirical and therefore
    contingent.

    —The point is not that the representations in the judgment go together
    necessarily *with each other,* but rather that they go together at all [in
    a judgment] only *in virtue of the necessary unity of apperception* [= the
    OSUA].

—Only in this way does there arise from the relation among the representations a *judgment*—that is, a relation with *objective validity*.

> Thus, judgments, as representations of an *objective* unity of given cognitions, are objectively valid by definition. So the issue is not to show that judgments are or can be objectively valid, but rather to show that any object that can be given to our intuition—any *Gegenstand* (including any possible objects of experience)—must be a possible *Objekt* of a judgment.

• For example: consider the cognitions [concepts] "body" and "weight."

—If these cognitions were to become subjectively associated for me (due to constant conjunction in my perception), then all I would be in a position to say would be: "Were I to [represent] hold[ing] a body, I would feel [come to represent] weight."

—But I would *not* be in a position to say: "It (the body itself) *is* weighty" [=a *judgment*].

—To say [that is, to *judge*] the latter is not to say that the representations have always been conjoined *in me,* but rather that they are combined *in the Object* [*Objekt*] (regardless of what's in me).

20. *All sensible intuitions are subject to the categories, as conditions under which alone their manifold can come together in one consciousness*

> Summary and conclusion of the first half of TD-B.
>
> Therefore, since the manifold given in any intuition must be subject to the OSUA, it must also be determined in respect of a category—that is, subject to the categories.

• The manifold given in any sensible intuition is necessarily subject to the OSUA, since that's the only way it can be unified (combined in one consciousness). (§17: POSUA)

• But the act of understanding by which any manifold of given representations (intuitions or concepts) is brought under one apperception is just the logical function of judgment (§19)

—Therefore, insofar as any manifold is given in a unitary empirical intuition, it is *determined* in respect of one of the logical functions of judgment.

—And the categories are [by definition: A79/B105, B128] just these functions of judgment, insofar as they are employed in the determination of the manifold of a given intuition. (§13 [but does he mean §14?])

• So the manifold in any given intuition is subject to the categories.

21. Remark

Explanation of the relation between the first and second halves of TD-B.
It has been shown that the manifold in any given intuition [qua representation] must be subject to the categories [but it hasn't been shown that any such intuitions are in fact given *to us*].   The demonstration has relied, so far, only on the spontaneity of the understanding, its functions of unity in judgment (the categories), and the generic receptivity of sensibility.   No use has yet been made of the *modes* of givenness *to us* (the forms of *our* intuition), through which alone empirical objects [*Gegenstände*] can be presented to our senses.   Now it must be shown that insofar as any *Gegenstand* is presented to *our* senses, it too must be determined in respect of the categories—that is, it is something about which we can form judgments. This demonstration will depend on the *modes* of our intuition.   Thereby, it will be shown that, when *Gegenstände* are presented to our senses, we indeed *intuit* [represent] them, and that the categories actually function as *concepts* [representations] in us—that is, they have objective validity with regard to all objects of our experience.

- A manifold contained in an intuition that is mine is [that is, would have to be] represented as belonging to the *necessary* unity of self-consciousness [OSUA].

  —Such representing, by means of the synthesis of the understanding, is effected in terms of the category.

  —Hence, the empirical consciousness of a given manifold in a unitary intuition is subject

    (i) not only to a pure sensible intuition—which is a priori,

    (ii) but also to a pure self-consciousness—which is likewise a priori.

  —Hereby a *beginning* has been made of a *deduction* of the pure concepts of understanding.

- Since the categories have their source in the understanding, *independently of sensibility,* this deduction has had to abstract from any *mode* of givenness of a manifold of an intuition.

That mode of givenness—*for us*, anyway—is *our* forms of intuition (in space and time); but conceivably it could be different for other finite knowers.

  —In this abstraction, we have concentrated solely on that unity which enters into empirical intuitions by means of the understanding (and thus in terms of the category).

—Now, however, we must consider that mode of givenness itself, and show (§26) that its unity, too, is none other than that which the category prescribes (§20) to the manifold of any given intuition in general [i.e., without regard to any mode of givenness].

> The *forms* of space and time are themselves given as (formal) intuitions; but they are given *a priori,* not *sensibly* (not *in* space and time), so the argument of the first half of the deduction already applies to them. So they, at least, must have the unity prescribed by the categories (this is argued in §24).

—Only thus, by demonstrating the a priori validity of the categories in respect of *all objects* of our senses, will the purpose of the deduction be fully attained.

> In other words, we must show that there is no *other* source of unity for unitary empirical intuitions, apart from the categories; in particular, the pure forms of intuition do not provide an independent source of unity. The point being that objects cannot even so much as be presented to our senses (whether intelligible to us or not) except as determined in respect of the category—something that an independent source of unity might have seemed to make possible. This, then, will finally repudiate the possibilities bruited in the long paragraph at A89f/B122f: the categories are not merely conditions on what we can understand, but on what can be an object in our experience at all.

• Although the argument in the first half of the deduction abstracted from the *mode* of givenness of the manifold, it did not, and could not, abstract from the *fact that it is given* (that is, prior to and independently of the synthesis of understanding).

—The only counterpossibility would be God, to whose intuition nothing is given because it is productive of its own objects.

22. The category has no other application in knowledge than to objects of experience

> The aim of the second half of the deduction is to show that the, categories apply to all and only objects of experience. The "only" is §§22 & 23; the "all" is §§24 & 26.
> The structure of the argument is to show that the categories can apply to objects because and only because those objects are subject to the conditions of space and time. To put it another way, we can form judgments about objects because and only because they are spatiotemporal.

> §22: The categories enable knowledge *only* of objects of possible experience [empirical objects]. In particular, mathematics is knowledge *only* of empirical objects, so not a counterexample.

- To *think* an object [*Gegenstand*] and to *know* an object are not the same.

  —Knowing an object requires two factors:

  (i) the concept through which any object in general is thought (the category); and

  (ii) the intuition through which it is given.

  —If no intuition could be given, the concept would still be a thought (in form, at least), but objectless, and so not usable for knowledge of anything.

- Now, the only intuition possible to us is sensible (see the Aesthetic).

  —So thought can become knowledge for us only insofar as concepts of the understanding are related to objects of the senses.

  —Sensible intuition is either pure (space and time) or empirical.

- Through the determination of pure intuition, we can acquire a priori knowledge of objects (as in mathematics), but only in regard to their form (as appearances).

  —Whether there can be things which must be intuited in this form is left undecided.

  —Mathematical concepts are therefore not, by themselves, yet knowledge—not until it is shown that there are things that can be given in these forms: space and time.

- But *things in space and time* are given only empirically.

  —Even, therefore, with the aid of pure intuition (as in mathematics), the categories do not yet afford us any knowledge of things.

  —They afford knowledge of things only through their applicability to *empirical intuition*.

  —That is, they serve only for the possibility of *empirical knowledge* (= "experience").

- So: the categories, as yielding knowledge of *things*, are applicable only to things that could be objects of possible experience.

23.   [No title]

> Though, in principle, the categories are not limited to objects given under our forms of intuition, they are in fact—since, without given intuitions, they are empty and useless.

- Space and time, as conditions under which objects can be given to us, are valid only for objects of the [= *our*] senses—hence, only for experience.

  —In theory, the pure concepts of understanding are free from this limit, and extend to objects of intuition in general (whether like ours or not, so long as it's sensible [= finite]).

  —But this seeming freedom is useless.

    —For, beyond *our* sensible intuition, these concepts, as concepts of Objects [*Objekten*], are empty.

    —We couldn't even judge whether such Objects would be possible or not.

- Here's why: the OSUA is the whole content of the pure concepts of understanding.

  —So, insofar as there could be no intuition to which that unity could be applied (thus determining an object [*Gegenstand*], those concepts would be *mere* forms of thought.

    —They would thus lack objective reality [hence they wouldn't really even be concepts].

  —Only *our* sensible and empirical intuition can give them meaning and significance.

- Were [*per impossibile*] an Object [*Objekt*] to be given to us outside the forms of our sensible intuition, we could not apply a single one of the categories to it.

  —But more of this to follow ...

24.   Of the application of the categories to objects of the senses in general

Any object of any intuition whatever must *as such* be subject to the pure *intellectual* synthesis (the form of judgability), hence to the categories. [See first half of TD-B]. But how can this condition actually be satisfied *for us?* How can *we* in fact *have* any judgable intuitions?
The answer lies in the fact that, because all of our sensibility is subject a priori to the form of time, the manifold of inner sense [time as a whole] can be synthesized in a more basic way, called the *figurative synthesis* or the *transcendental* [*or productive*] *synthesis of imagination.*
This synthesis is the *most basic* application of the understanding to the *objects* of our intuition.

- Through the understanding alone, the pure concepts relate to objects [*Gegenstände*] of intuition in general (whether like our own or not, so long as it's sensible).

  —For that very reason [the generality], however, these concepts are [so far] mere *forms of thought*, through which (alone) no determinate object is known.

  —The synthesis in them, therefore, though transcendental, is also purely intellectual.

- However, there lies in *us* a priori, at the basis of everything, a certain form [time] of sensible intuition—one which depends on the receptivity of our ability to represent (i.e., sensibility).

  —Because of this, the understanding, *as* spontaneity, is able to determine inner sense [itself], through the manifold of given representations, in accord with the OSUA.

  —Thereby the understanding can think, a priori, a synthetic unity of the apperception of the manifold of *sensible intuition* [as a whole—a comprehensive horizon of all intuition].

  —And that [the a priori thinkable horizon of time] is the condition under which all objects of our human intuition must necessarily stand.

- By this means [as we shall see, §26], the categories—in themselves mere forms of thought—obtain objective reality.

  —That is, they obtain application to objects [*Gegenstände*] that can be given to us in intuition.

- This synthesis of the [whole] manifold of sensible intuition is possible and necessary a priori.

  —It can be called *figurative* synthesis—in order to distinguish it from [purely] *intellectual* synthesis [the mere forms of thought: what the first half of the deduction was about].

  —Both are *transcendental* syntheses, in that they make a priori knowledge possible.

  —However, the figurative synthesis, [even] if it concerns merely the OSUA—that is, the unity of thought in the categories—is [still] not the same as mere intellectual synthesis.

> The point is that, even though the figurative synthesis is concerned with
> no *unity* other than the unity of judgability—the OSUA—it is concerned
> with this unity within the specific horizon of *temporal* givenness; and that
> makes it different from the mere intellectual synthesis.
> I think the difference is something like this: the horizon of time provides a
> unique "locus" for each unification in a judgment. Thus, if I judge that the
> water is cold, these concepts must be unified *in the object*. But the object
> isn't given as an object independent of this unifiability of concepts in
> it—its being an *object* just is the unifiability of concepts in it in judgments.
> So, "which" water is said to belong with "which" coldness by this
> judgment? Well they must at least coincide in time; and, since this is an
> object of outer sense, they must also coincide in space. The *objective* unity
> that is the content of a judgment is possible only "at" a time (and,
> typically, place).

—To emphasize this, it should [also] be called the *transcendental synthesis of imagination*.

- *Imagination* is the ability to represent in intuition an object that *is not itself present*.

> Time and space, for instance, are not—at least, not entirely—present.

—Now, on the one hand, since all our intuition is sensible, imagination likewise can give an intuition to concepts *only* under the conditions that sensibility places on all intuition.

—So, to this extent, imagination belongs to *sensibility*.

—But, on the other hand, inasmuch as a synthesis in imagination, like any synthesis, is spontaneous, and spontaneity is determinative (unlike sense, which is determinable only), [transcendental] imagination is an ability to determine sensibility a priori, in accord with the categories.

—So, to this extent, imagination belongs to *understanding*.

- Transcendental synthesis of imagination is an action of the understanding on the sensibility.

—It is the first [in other words, the most basic] application of the understanding to the objects [*Gegenstände*] of our possible intuition.

> Presumably, this "most basic" application is the ability to judge temporal
> (and spatial) characteristics of objects: location, relation, extent, and so on.

—Thereby it is the ground of all other applications of understanding to such objects.

—It is figurative in that it involves sensibility [that is, *our* forms of it, *our* modes of givenness in intuition], whereas the intellectual synthesis involves understanding only.

* * *

25. [No title] [Section left blank.—Ed.]

26. Transcendental deduction of the universally possible employment in experience of the pure concepts of the understanding

> If any appearance is to be so much as perceptually apprehensible by us, it must conform to the forms of space and time. But space and time are themselves intuitable as (formal) *objects*—and as such must exhibit the unity prescribed by the categories. Consequently, any appearance, as having spatiotemporal form, must also exhibit this unity. That is, as determined in time and space, an appearance is also something about which we can form judgments. This means they can be *objects*—namely, of *intuitions*.
> Since this includes all appearances, the categories are valid a priori for *all* objects of experience.

· In the transcendental deduction of the categories [first half], we have shown their possibility as a priori cognitions of objects [*Gegenstände*] of any intuition in general (§§20 & 21) [that is, of *given* intuitions, but without regard to their *mode* of givenness].

  —(In the metaphysical deduction [§§9 & 10], their a priori origin had already been proved through their agreement with the logical functions of thought.)

  —We have now to explain the possibility of knowing, by their means, whatever objects [*Gegenstände*] may *present themselves to our senses* [that is, in *our* modes of intuition].

  —What's to be explained isn't knowledge of objects in respect of the form of their intuition [that was done in the Aesthetic], but rather in respect of the laws of their combination.

    —In effect, this prescribes the law to nature, and even makes nature possible.

> This doesn't mean, of course, prescribing the particular "laws of nature" that science discovers (those are empirical, as Kant says at B165). Rather, it means that, as intelligible (in terms of judgable unities), nature must be law-governed, somehow or other. Later (in the Analytic of Principles), Kant will deliver on the promise of synthetic a priori *knowledge*, by extracting from the categories principles that dictate the *forms* of such laws.

• Note first that *synthesis of apprehension* is [by definition] that composition of the manifold in an empirical intuition whereby perception (that is, empirical consciousness of the intuition as appearance) is possible.

— The synthesis of apprehension must, of course, always conform to the a priori forms of outer and inner sensible intuition, space and time.

— But space and time are not only a priori *forms* of intuition—they are also themselves *intuited* a priori [that is, they are the *objects* of intuitive representations].

— These a priori [formal] intuitions contain manifolds, and therefore are represented with the determination of the *unity* of these manifolds (see the Aesthetic).

• This unity, given a priori, is a *unity of the synthesis* of the [whole] manifold of space and time.

— Consequently, it is also a combination to which everything that is to be represented as determined *in* space and time must conform. [I think, both *each* thing and *all together.*]

— As such it is the condition of the synthesis of all *apprehension* [that is, *not even* mere apprehension—mere perceptible appearances—escape *this* condition].

— This unity is given not *in* but *with* the [formal] intuitions [of space and time].

> The unity is not the *content* of the intuitions (that would be conceptual—see §15), but rather a unity that they themselves *have* and so bring *with* them.

• This unity, however, can be no other than the unity, which (as we saw in the first half) is necessary for any combination of the manifold of a given *intuition in general*, simply by virtue of being subject to the OSUA—but applied, in this case, to our *sensible intuition*.

> In other words, the formal intuitions (which we have a priori) are instances of "intuitions in general," hence fall under the argument of the first half.

— And that unity is a unity in accord with the categories.

— All synthesis, therefore, even that which makes perception [mere apprehension of appearances] possible is subject to the categories.

— And, since experience is knowledge via connected perceptions, the categories are conditions of the possibility of experience.

— Hence, they are valid a priori for all objects of experience.

Notes

1. Marginal note: "For rep[resentations], x is obj[ectively] valid if x it can figure in an obj[ectively] valid judgment (i.e., if it is a rep[resentation] _of_ something). Not always used in this precise sense elsewhere in the Critique (i.e., this is the meaning in the context of the TD)."

2. Marginal note: "Object considered under some level of abstraction. [First half: from our particular forms of sensibility, but not of sensibility as such (disregarding character of _our_ manifold).]

3. Marginal note: "Full Blooded object of intuition (e.g., known object)."

4. Marginal note: "Second half: lifting abstraction."

5. Marginal note: "AKA Categorial Unity."

6. Marginal note: "Two phases of this unity: (a) unity of consciousness, (b) unity of the thought. ←These are one and the same unity.

7. Marginal note: "To be an object (Geg[enstand]) for us, it must be an Objekt (i.e., fall under unity prescribed by cat[egories]."

Contributors

Zed Adams, Department of Philosophy, New School for Social Research

William Blattner, Department of Philosophy, Georgetown University

Jacob Browning, Department of Philosophy, New School for Social Research

Steven Crowell, Department of Philosophy, Rice University

John Haugeland(†), Department of Philosophy, University of Chicago

Bennett W. Helm, Department of Philosophy, Franklin and Marshall College

Rebecca Kukla, Department of Philosophy, Georgetown University

John Kulvicki, Department of Philosophy, Dartmouth College

Mark Lance, Department of Philosophy, Georgetown University

Danielle Macbeth, Department of Philosophy, Haverford College

Chauncey Maher, Department of Philosophy, Dickinson College

John McDowell, Department of Philosophy, University of Pittsburgh

Joseph Rouse, Department of Philosophy, Wesleyan University

† Deceased June 23, 2010.

Index